Physical Theory in Biology

Foundations and Explorations

STUDIES OF NONLINEAR PHENOMENA IN LIFE SCIENCE

Editor-in-Charge: Bruce J. West

Vol. 1: Fractal Physiology and Chaos in Medicine
by B J West

Vol. 2: Patterns, Information and Chaos in Neuronal Systems
edited by B J West

Vol. 3: The Lure of Modern Science — Fractal Thinking
by B J West & B Deering

Vol. 4: Physical Theory in Biology — Foundations and Explorations
edited by C J Lumsden, W A Brandts & L E H Trainor

Vol. 5: Nonlinear Dynamics in Human Behaviour
edited by W Sulis & A Combs

Studies of Nonlinear Phenomena in Life Science – Vol. 4

Physical Theory in Biology

Foundations and Explorations

Charles J Lumsden
Institute of Medical Science, University of Toronto

Wendy A Brandts
Department of Physics, University of Ottawa

Lynn EH Trainor
Department of Physics, University of Toronto

Editors

World Scientific
Singapore • New Jersey • London • Hong Kong

Published by

World Scientific Publishing Co. Pte. Ltd.

P O Box 128, Farrer Road, Singapore 912805

USA office: Suite 1B, 1060 Main Street, River Edge, NJ 07661

UK office: 57 Shelton Street, Covent Garden, London WC2H 9HE

Library of Congress Cataloging-in-Publication Data

Physical theory in biology ; foundations and explorations / Charles J. Lumsden, Wendy A. Brandts, Lynn E. H. Trainor, editors.
p. cm. -- (Studies of nonlinear phenomena in life science ; vol. 4)
Includes index.
ISBN 9810230826 -- ISBN 9810231210 (pbk.)
1. Biophysics. I. Lumsden, Charles J., 1949– . II. Brandts, Wendy A. III. Trainor, Lynn E. H., 1921– . IV. Series.
QH505.P462 1997
571.4--dc21 97-7676
CIP

British Library Cataloguing-in-Publication Data

A catalogue record for this book is available from the British Library.

Printed in Singapore by Uto-Print

Preface

The essays in this book illustrate and celebrate the vitality of modern biology's dialogue with theoretical physics. The dialogue is part of a great scientific adventure, in which biological understanding is enriched and deepened by physical theory without losing its own inherent traditions and perspectives. This essay collection makes no attempt to be encyclopedic. Any didactic virtue these essays may have is secondary to their principal intent of exciting the interest and promoting the involvement of physics majors in thinking about biology and the nature of living matter. More seasoned investigators in both the physical and life sciences will, we hope, also find challenging ideas and relevant examples in the essays presented in this volume.

This book is not a text, nor is it in any way a substitute for the, by now, many excellent introductions to biophysics and mathematical biology so effectively used in undergraduate instruction today. Yet, despite the quality of the available literature important gaps remain, and one of these — in our minds a primary one — is a source in which may be found a selection of case studies in what can happen when theoretical physics and biology connect. Our contributors, physicists and biologists alike, are all actively engaged in building this connection. The reader new to the subject of physical theory in biology can, in addition, much profit from a reading of Schrödinger's classic *What is Life?* (Cambridge: Cambridge University Press, 1944) and the provocative modern sequels: Mae-Wan Ho, *The Rainbow and the Worm: the Physics of Organisms* (Singapore: World Scientific, 1993) and Michael P. Murphy and Luke A.J. O'Neill, editors, *What is Life? The Next Fifty Years* (Cambridge: Cambridge University Press, 1995). These three earlier works beautifully exemplify the spirit of intellectual adventure that one meets at the frontiers between the physical and life sciences.

Since the mathematical biosciences and theoretical biology have been established largely by the efforts of scholars first trained as physical scientists, engineers and applied mathematicians, there is strong overlap between the mathematics picked up by today's physics and engineering undergraduates and the mathematical apparatus actually used to understand living systems: Analytical dynamics and field theory are powerful tools for probing the mysteries of animate as well as inanimate matter. Thus the formal content of the reports gathered here should be readily accessible to students of the physical sciences who are in their senior undergraduate or graduate years, as well as to similar readers specializing in one of the branches of mathematical biology.

There are exceptions to this "overlap rule," of course: Mathematical game theory, so important to modern evolutionary thinking about animal conflict and cooperation in populations and ecosystems, is more often encountered in the ethology or political science class than in the relativity seminar. Similarly, the formal theory of computation and abstract machines, which is turning out to be a linchpin in the "new science of complexity" and its connections to physics, may seem strange unless one has been lucky enough to get off on the wrong floor of the math building and stays to "get hooked." And although exposure to nonlinear dynamics, iterated systems and the calculus of bifurcations is increasingly common in the physics curriculum, packed teaching agendas can leave scant time for tracing the deep, beautiful links between bifurcation theory and the mathematics of groups and symmetries, itself central to the structure and priorities of contemporary theoretical physics. We have therefore included material introducing mathematical game theory, computation theory and bifurcation symmetries in a manner that hopefully will make apparent their cutting-edge importance for biology while setting the stage for further reading in advanced texts and the research literature. Anticipating that the essays taken as a whole may be of interest to readers from a range of backgrounds in physics, biology, mathematics, philosophy and computer science, we have attempted wherever possible to give more detailed citations of the literature sources in the bibliographies than would be natural if we saw the treatment as suited to one narrow discipline. Those trips to the library of the "other department" may therefore prove not quite so mystifying!

Our sincere thanks and appreciation are due the institutions and agencies, their directors and personnel, whose support of our work over the years has made the creation of this book possible: the Departments of Physics at the University of Toronto (CJL, LEHT) and the University of Ottawa (WB), the Institute of Medical Science (CJL, LEHT) and the Department of Medicine (CJL) at the University of Toronto, the Natural Sciences and Engineering Research Council of Canada (CJL, WB, LEHT), and the Medical Research Council of Canada (CJL). The staff of World Scientific has been helpful and supportive at every turn, making the production of this volume a pleasant undertaking. To Kim Tan of WSP especially, our warm gratitude is due. We are honored by Bruce West's decision to select this book as the fourth volume in his Series on modern frontiers in physics and biology; his guidance has been most helpful in giving the work its final form.

It is fair to say that the book would not have come to press as you see it here without the dedicated efforts of two special people: Judy Fitzgerald of the Department of Medicine, and Jennifer Tam of the Physics Department at the University of Toronto. Ms. Fitzgerald worked closely with the editors and authors to prepare most of the book's illustrations for publishing and helped lay them out in the text. We are the beneficiaries of her skill and taste. Ms. Tam, our swift-fingered LaTeXpert, endured the canonical cycle of revisions^N with patience and good humor, coaxing raw words into composed type.

In organizing the book's intellectual content we have been privileged to draw on a unique group, scholars from many parts of the world who have been associated in one way or another with the theoretical physics/biology program established at Toronto by one of us (LEHT) in the mid-1970's. We are especially indebted to these busy, creative people for responding to our idea for a book about physical theory and biology with the essays that follow.

As teachers we have been blessed, in addition, by the company of the many bright, curious physics majors (and, from time to time, highly motivated biology and medicine majors) who over the past twenty years have participated in our courses and seminars and actively contributed to the excitement of doing "physical theory in biology," almost to the invention of a new discipline. We will be happy indeed if we have captured some of their boundless confidence in the unity of knowledge and made it come to life in these pages. It is our hope that like-minded students in the years to come will be stimulated and especially benefited by these essays, and by the challenges and perspectives generated by them.

CJL, LEHT
Toronto, Canada

WB
Ottawa, Canada

Autumn, 1996

Table of Contents

The Contributors

Wendy A.M. Brandts
Department of Physics
University of Ottawa
150 Louis Pasteur
Ottawa Ontario
Canada K1N 6N5 — *wendy@physics.uottawa.ca*

Allan E. Busch
Ontario Science Centre
770 Don Mills Road
Toronto Ontario
Canada M3C 1T3 — *allanb@osc.on.ca*

Fred W. Cummings
2365 Virginia Street #4
Berkeley CA
USA 94709-1338 — *cummings@nature.berkeley.edu*

Paul A. Dufort
Institute of Medical Science
University of Toronto
Toronto Ontario
Canada M5S 1A8 — *paul@isengard.med.utoronto.ca*

Joseph Frankel
Department of Biological Sciences
University of Iowa
Iowa City IA
USA 52242 — *joseph-frankel@uiowa.edu*

Brian Goodwin
Biology Department
The Open University
Walton Hall
Milton Keynes
U.K., MK7 6AA — *B.C.Goodwin@open.ac.uk*

Julian Haffegee
Bioelectrodynamics Laboratory
The Open University
Walton Hall
Milton Keynes
U.K., MK7 6AA

Trevor N. Hart
Divison of Medical Microbiology
and Immunology
1-41 Medical Sciences Building
University of Alberta
Edmonton AL
Canada T6G 2B7 *hart@maigret.mmid.ualberta.ca*

Mae-Wan Ho
Bioelectrodynamics Laboratory
The Open University
Walton Hall
Milton Keynes
U.K., MK7 6AA *m.w.ho@open.ac.uk*

Francis P. Jones
42 Heath Street West
Toronto Ontario
Canada M4V 1T3 *fjones@empress.com*

Ellen W. Larsen
Department of Zoology
Ramsay Wright Zoological
Laboratories
University of Toronto
Toronto Ontario
Canada, M5S 3G5 *ellenw@zoo.utoronto.ca*

Charles J. Lumsden
Department of Medicine and
Institute of Medical Science
University of Toronto
Toronto Ontario
Canada M5S 1A8 *cjl@vortex.med.utoronto.ca*

Liaofu Luo
Physics Department
Inner Mongolia University
Huhehote 010021
Inner Mongolia
China *lfluo@nmg2.imu.edu.cn*

G.J. Nelson
Computer Laboratory
New Museums Site
University of Cambridge
Pembroke Street
Cambridge
U.K.

G.W. Rowe
Department of Mathematics
and Computer Science
University of Dundee
Dundee
Scotland, DD1 4HN *growe@mic.dundee.ac.uk*

Jeffrey P. Sutton
MGH Neural Systems Group
Harvard Medical School
Building 149, 13th Street
Boston, MA
USA 02129 *sutton@nmr.mgh.harvard.edu*

Patrick Tevlin
Ontario Science Centre
770 Don Mills Road
Toronto Ontario
Canada M3C 1T3 *patrickt@osc.on.ca*

John Totafurno
30 Denvale Road
Toronto Ontario
Canada M4B 3B5

José-Leonel Torres
Instituto de Fisica y Matematicas
Universidad Michoacana
58060 Morelia, Michoacana
Mexico *jltorres@ifm1.ifm.umich.mx*

Lynn E.H. Trainor
Department of Physics
University of Toronto
Toronto Ontario
Canada *ltrainor@physics.utoronto.ca*

Yu-Ming Zhou
Bioelectrodynamics Laboratory
The Open University
Walton Hall
Milton Keynes MK7 6AA
U.K.

Physical Theory in Biology

Foundations and Explorations

Introduction

The Editors

Searches for understanding traditionally have followed very different paths in physics and biology. Might they have more in common than once thought? Can physical knowledge and biological knowledge merge into a common whole?

Success in physical theory is reckoned in terms of the discovery and elucidation of physical laws, mathematical in form and universal in application. The primary characteristic and advantage of mathematical laws is that they are both synthetic and predictive: not only do they combine many observations into one framework, but this framework then enables new behaviors to be forecast. When tested experimentally, the predictions may support the theory or cause it to be modified. Theoretical physics thus provides a mathematical road map of nature, charting the principles that shape matter and energy everywhere in the cosmos. Essentially, physical theory bridges between phenomenology and philosophy.

The remarkable effectiveness of this search for universal laws can be attributed in part to the simplicity (compared, say, to living organisms) of the systems traditionally dealt with by physicists. They typically consist in either very few parts, or large numbers of identical components interacting in rather similar ways, so that one can reduce the behavior of the many-part complexity through the use of space and time averages, and through symmetry principles with respect to spacc and time transformations.

To the biologist either of these conceptual templates — the system of but few parts or the one of many identical components — may seem inadequate, even as a beginning approximation to the kinds of material organization found in living things. Living organisms are strikingly complex in structure, and in the incredible variety of meaningful processes their structures support.

In the biological world, this diversity occurs not only between species or between individuals of one species, but also within each individual's cells, tissues, and organs. An orderliness of form is nonetheless evident, stretching from the DNA molecule to the whole biosphere. Patterns and symmetries abound in space and time. Biological structures are dynamic, changing on millisecond scales (as does the nerve impulse), life cycle intervals, and so on up to evolutionary epochs.

It comes as no surprise that biological thought has grounded itself in the need for classification and rational taxonomy, a search for commonalties and systematic differences. While physics has pursued matter down to smaller and smaller divisions for which simplicity and uniformity appear, biology has sought understanding of variety and organization at more global levels in seemingly irreducible systems. Biologists struggle with phenomena from which it may not be possible (or useful) to distil out general laws of either structure or behavior. Accordingly, biological theories have traditionally been seen as suggestive rather than predictive, explanatory and inductive rather than rigorously deductive.

But all of this is changing rapidly. A new partnership between physical theory and the biological sciences — between the simple and lawful, and the complex and diverse — is being forged by the information revolution. The advent of high speed powerful computers and globally accessible networks offers new ways of handling — processing, storing and transmitting — information. Thus theorists in all fields are able to tackle more complex problems: problems with more data, more variables, more structure, more sensitivity — particularly the nonlinearities so endemic in biology. And as a result of exploring more complex behaviors, the traditionally precise sciences like physics and chemistry are discovering the limits to their dominant view (epitomized by the simple deterministic systems of classical mechanics). Hence they are now expanding to include, for example, nonlinear dynamics and nonequilibrium physics. Consequently, the boundaries separating the scientific disciplines are shifting and in some cases will probably disappear. For example, pattern-forming mechanisms in dynamic meteorology require the solutions to systems of nonlinear differential equations that bear a family resemblance to the chemical reaction and diffusion equations that may underlie pattern formation in biological organisms. Today one finds researchers in virtually all academic fields, including the humanities and social sciences, using ideas about nonlinearity, self-organization, fractal behavior, strange attractors, symmetry breaking bifurcations, chaos, and the like. The possibility of discovering laws of structure and form in biology is not now as distant as it formerly seemed.

At the same time, biological research has undergone a kind of conceptual condensation, the plurality of its originating subjects self-organizing into two great, synthesizing disciplines. In the first, physiology, biochemistry, molecular genetics, pathology, anatomy, and histology have merged to create cellular and organismic biology. This discipline focuses on *how* questions: how cells divide, how genes encode information, how physical and chemical mechanisms are enjoined to provide structure, process and coherence in biological organization. In the second, the joining of ecology, taxonomy, systematics, palaeontology, comparative anatomy, genetics, population biology, and sociobiology creates evolutionary science. This discipline focuses on *why* questions: why genes encode information in one way rather than another, why parents help their offspring. The question "Why?" can be answered primarily by the study of history; and the history of biological process is by definition evolution.

In cellular and organismic biology the hegemony of physics is almost complete: the modern biology laboratory is crammed with instrumentation meant to dissolve organisms into molecules and reveal their stunningly intricate pattern of chemical reactions. With a full molecular inventory and a detailed map of the reactions, complete living cells, or major parts thereof, might one day crawl out of a test tube or a supercomputer simulation. That, anyway, is part of a vision which sees modern biology as the chemical physics of living matter, today pursued with a vigor unparalleled in the history of science. Pushed to an extreme, an effort like this could have the consequence of equating all interesting and important biological questions to those solvable strictly in terms of notions from the physical sciences — admitting as biology only those phenomena described by chemistry and physics and ignoring all the integrative, high-order phenomena (like self-reproduction) that help define biology as a science in its own right.

That this is unlikely to happen can be attributed in part to the blatantly integrative, in-your-face nature of biological complexity alluded to above. It refuses to be swept under the intellectual rug. We must acknowledge as well the irreducibly historical nature of living matter: Explanations of organisms or species are not found merely by turning to the universal laws of the physical sciences. The luxuriantly idiosyncratic whirl of events that, over time, shapes individuals (development) and populations (evolution) must also be taken into account. Perhaps the greatest contribution of molecular biology to the history of ideas is the merger it is forcing between evolutionary science and cellular and organismic biology. Faced with massive compilations of gene sequences and protein folding diagrams, meticulously gathered from many different kinds of living things, molecular biologists have turned not away from evolution but toward it, in an effort to make some sense of the data.

Although at first blush the goals of evolutionary science, rooted as they are in the unravelling of historical detail, seem to have little in common with the needs and interests of the physical sciences, a glance at the mathematical models used in evolutionary studies to explain history reveals once again a physicist's conception of nature. Admittedly, the variates in equations for evolutionary change have odd-sounding names like meme frequency, ethological transition rate, and fecundity, and familiar concepts like angular momentum and the density matrix are notably absent. But the reigning metaphor of nature as a vast mathematical machine is little disputed. The limits of this metaphor's validity in the life sciences, as well as the social sciences and humanities, are sure be tested relentlessly in the years ahead.

The map of the chemistry of life lies unfinished, and its completion must count among the great intellectual enterprises of our times. Yet, unlike the situation three decades ago, at the time of the birth of molecular biology, experimentalists and theoreticians alike can already marshal an abundance of facts about the molecules and chemical reactions associated with the living state. Even with a complete map, however, who will put all the bits and pieces back together in a way that explains (rather than merely mimics or simulates) life itself? The essential intellectual task of synthesis must at this point join reductionism, and as we noted above it is with the task of synthesis, the production of coherent patterns of explanation, that theoretical work is essentially engaged.

Can reductionism itself suffice? Likely not. Life is a collective property of certain complex systems and loses its meaning in a strictly reductionistic approach in which one studies the parts and their interactions more or less independently of the whole. The pursuit of the complete chemical map of any organism, or indeed its genetic map, however worthy and useful these efforts, does not in itself subvert the demand for the integrative principles at work in intact cells, tissues, and organisms, including their development and their coherent interaction with one another over time.

How are these collective, organized properties of living systems to be set down, and rationally related to the physicochemical map? Physicists, in their efforts to understand non-living matter, have discovered as least two key concepts for the description of collective properties: that of the *field*, by which structures and functions distributed throughout the fabric of a complex system are related one to another in a coherent manner, and that of the *ensemble* of statistical mechanics, by which redundant details of a many-part system like the physicochemical map might be subsumed, averaged, and recombined to express a dynamics of integrated behavior.

Two field concepts in particular have been influential in the history of modern biology: that of the developmental field, and that of the evolutionary field. The developmental field, vividly portrayed in mathematical form by Turing (1952), is hypothesized to carry regularities in signal transduction between populations of cells, sustaining an orderly progression of events in cell division, movement, differentiation and death from fertilized egg to pristine newborn. Long regarded as a quantitative abstraction by cell biologists, the developmental field is now being rooted to the physicochemical map by the discovery of molecules that distribute through developing tissues and exert organizing effects on cell populations (e.g. Nüsslein-Volhard 1996).

The evolutionary field (or "fitness landscape" or "payoff matrix" as it is more commonly called) was introduced into population biology by Sewall Wright (1932; 1969 for review). It expresses the evolutionary forces (natural selection especially, but also randomizing influences like genetic mutation and recombination) that buffet a population and create patterns of phylogenetic change. Unlike the developmental field, which is grounded in the spatial structure of the intact tissue or organ, the evolutionary field stands rooted in an abstract information space, whose dimensions are defined by the possible combinatorial variations of an individual's genome as constrained within the laws of physics and chemistry. Also once viewed as a facilitating abstraction, the fitness landscape too is in a period of rapid revitalization as complexity science ties it more directly to the organization of the cell's physicochemical map and ecological field work makes it possible to correlate quantitative measures of fitness with specific genetic varieties.

It is our view that key integrative concepts like the field and the ensemble, originating from physical theory, have been sufficiently successful in their applications to the life sciences to warrant serious examination as part of an emerging language for integrative biology. The essays comprising this book consider this possibility from the vantage points afforded by biology's two great organizing disciplines: cellular and organismic biology, and evolutionary science. Our attention therefore turns first to matters of theory itself — to the challenge of making notions like complexity, integration, and emergence useful ideas in theoretical physics — and then in subsequent sections to their role in elucidating organic development, and the physicochemical mechanisms of integration in cell and tissue, as well as their descriptive value in problems relating to evolution and the origins of life. Despite their topical range, the essays share the viewpoint expressed here, namely that the theoretical physicist's idea of the universe as a place filled with order at once strange, yet open to the mathematical mind, is valuable in the life sciences. Conversely, the theoretical physicist finds continual surprise thinking about matter that is alive (and at times itself also thinking and feeling!), where complex, nonlinear mechanisms cannot be ignored.

We laid out the book very much aware that until recently subjects like complexity and emergence were more often found amidst the debates of philosophers and other professional "science analysts" than in scientific reports per se. We are motivated to take these matters on at least provisionally not because we fancy ourselves metaphysicians, but rather because it seems to us that scientific progress itself has made them too important for scientists now to leave entirely to others. Arguably for the first time in the history of physics and biology, the way we use terms like complexity, emergence and holism is beginning really to matter to our civilization's burgeoning ability to understand and ultimately synthesize living things. This change in the practical relevance of once-obtuse notions is a striking sign of the impact between nonlinear mathematics and all areas of human thinking. Tractable mathematical models, of practical import yet rich in behaviors to which terms like "complex," "integrative," "emergent" seem legitimately to apply, are at last widely available, and while not yet a communal art form are already attracting intense scrutiny from scientists and philosophers alike. Even when the media hype is fractioned off the claims and arguments swirling around these new efforts, there remains evidence of a nascent "science of complexity," described more fully in the chapters to follow, that may profoundly effect how we understand ourselves and our universe. We therefore offer the essays in Part I because we are advocates of a process in which science seeks to reclaim once- powerful synthetic ideas from natural philosophy and with them reshapes itself, even at the expense of crossing alien ground in the effort to reach new frontiers.

Biology, then, presents the theorist with profound challenges, physics the traditions and experiences to help meet them. Is it possible to put in a nutshell what "doing" physical theory in biology is like? Difficult, but we are drawn to the great evolutionary biologist George Gaylord Simpson (1964), who in writing of the differences between scientific discovery in physics and biology noted, with typical pungency, the irreducible diversity of living matter as realized not only by the uncountable number of species, but as well by the richness of phenomena found in every biological thing.

Although Simpson ultimately was skeptical of the role physics might play in the life sciences and, rightly, urged biologists to liberate themselves from "physics envy," we find his insight about diversity on target: In physical theory in biology, the study of the principles that apply to all things must join the study of things to which all principles apply. The sharp glitter of this seeming paradox, and the energy created by those seeking its intellectual extremes are, as of this writing, much evident in our young discipline.

References

Nüsslein-Volhard, C. (1996, August) Gradients that organize embryo development. *Scientific American* **275**, 54–61.

Simpson, G.G. (1964) *This View of Life.* New York: Harcourt, Brace & World. p. 107.

Turing, A.M. (1952) The chemical basis of morphogenesis. *Philosophical Transactions of the Royal Society of London, Series B* **237**, 37–72.

Wright, S. (1932) The roles of mutation, inbreeding, crossbreeding and selection in evolution. In D.F. Jones, ed., *Proceedings of the Sixth International Congress of Genetics, Ithaca, N.Y., 1930.* Volume 1. Brooklyn, NY: Brooklyn Botanic Gardens, pp. 356–366.

Wright, S. (1969) *Evolution and the Genetics of Populations.* Volume 2. *The Theory of Gene Frequencies.* Chicago: University of Chicago Press.

I

Foundations

Emergence in Physics and Biology

L.E.H. Trainor

1 Nature of Emergence in Physics and Biology

A central question running through much of theoretical biology is the role of emergence in explanation. By emergence, we shall mean here the appearance of unexpected structures and/or the occurrence of surprising behaviors in large systems composed from microscopic parts, whether physical or biological. By unexpected and surprising we mean structures and behaviors which are not intuitive and are not simply predictable. From a reductionist point of view, one seeks understanding of emergent phenomena through analysis of subsystems and their interactions, whose properties are either known or can be taken for granted. For example, in the physics of water one can understand the occurrence of low temperature crystallization, given the nature and properties of individual water molecules.

Emergence in biology is neither as well defined nor as well understood as emergence in the physical context. Nonetheless, there is a tendency in modeling complex biological systems to account for the behavior of the composite system as emergent from the known properties of subunits. Several such examples, appear in this collection. In the essay by Trainor, Brandts and Torres, for example, a model is reviewed (Gordon, Goodwin and Trainor 1992) in which the behavior of an ant colony is regarded as emergent from an assumed interaction between individual ant pairs. This model is an extension and outgrowth of the approach to associative memory in brain modeling initiated by Hopfield (1982) in terms of a "neural network" of interacting neuron pairs. A second example, perhaps less obvious, is the essay by Ho and her associates in this volume, in which the behavior of developing embryos is attributed to some underlying, coherent morphogenetic field.

Perhaps, the most fascinating and challenging of biological problems is the question of what is mind and what is the nature of consciousness, and how are these related to the brain and neural organization. The prevalent point of view (for example, Crick 1994 and Penrose 1989, 1994) is to regard the mind as an emergent property of brain operation, in contradistinction to the dualist point of view espoused by Descartes which postulated mind stuff as well as brain stuff. The dualist point of view is generally treated with scorn in the AI (artificial intelligence) community (see for example, Dennett 1991). Presumably any AI claim that computers can, in principle, have consciousness given a sufficiently large and complex system, then rests on a belief that consciousness is an emergent phenomenon.

2 Emergence Classes

The purpose of the present essay is to point out to the community of theoretical biologists that also in physics, emergence is a subtle and controversial concept. In 1985 the present author (Trainor 1985) proposed two broad and distinctive classes of emergence, which he termed *reducible* and *a priori irreducible*, respectively.

Reducible emergence arises when one can deduce the properties of large composite systems, perhaps seemingly holistic, from the known and accepted properties of subunits. The crystallization of water at low temperatures, an example given above, falls into this class. Another familiar example would be the condensation of water vapor into water in its liquid form.

Individual water molecules in isolation have no liquid properties of their own, and even in small clusters as vapor the same holds true. But large numbers of water molecules in interaction with each other in appropriate temperature and pressure ranges, do display liquid properties. In this sense liquidity is an emergent property associated with large numbers of water molecules in interaction. Nonetheless, liquidity would be difficult to derive from the known interactions between water molecules. Indeed, many properties of water are not yet well understood, e.g. the possibility that self-maintained structures can be imprinted in water by large organic molecules remains a contentious issue to this day. But these difficulties have more to do with the complexity of calculations on the structure of water rather than with any disagreement among researchers on the reducibility of these emergent properties.

By contrast, the atoms of helium, which are also well understood and simpler in principle than water molecules, nonetheless display superfluid properties in bulk below four degrees Kelvin, which are not emergent from a consideration of the dynamical interactions between helium atoms. The properties of superfluid helium, which are remarkable and, by any reasonable criterion, surprising, cannot be derived from the known properties of individual helium atoms, however precise and careful the calculations.

Individual helium atoms do not have superfluid properties, so superfluidity is clearly an emergent phenomenon, yet it cannot be deduced from the dynamical interactions alone between helium atoms. The emergence is *a priori irreducible.* The reason is that there is a holistic principle which applies to large numbers of helium atoms, which is not reducible to the properties of individual helium atoms or their dynamical interactions. This principle is the requirement that helium atoms shall obey the so-called Bose-Einstein statistics, namely that the wave function describing a collection of helium atoms, however large, must be symmetrized in the coordinates of all atoms in the collection.

3 Tangled Wave Functions and Separability in Quantum Mechanics

In classical mechanics, two particles separated beyond the range of their mutual interaction forces behave as independent particles. For example, one can make measurements on one particle without in any way affecting the state of motion of the second particle. In quantum mechanics this is not necessarily so. The state of a system is represented by a wave function (in effect, a probability amplitude), which for the simple case of two particles might be designated by a function $\psi(x_1, x_2)$ whose value depends on the coordinates x_1 of particle one (for example position and spin coordinates) and x_2 of particle two. In the quantum theory, if the two particles are independent, it can be shown that ψ can be expressed as a product function

$$\psi(x_1, x_2) = \phi_1(x_1)\phi_2(x_2) \tag{1}$$

where $\phi_1(x_1)$ is the wave function for particle one and $\phi_2(x_2)$ is the wave function for particle two. If one cannot express the wave function as a product in this manner, the state is referred to as a tangled state, since the coordinates of either particle cannot be separated out. This bothered Einstein among others. In the famous Einstein, Rosen, Podolsky (EPR) paper (1935) it was argued that quantum mechanics could not be a complete theory, so by implication there must be hidden (undetected) variables which should be taken into account (but which are not) if one is to get a theory that has sensible notions of reality in its formulation. One implication seemed to be that quantum mechanics does not satisfy a simple separability criterion, that widely separated particles should behave independently. In a theorem of John Bell (known as Bell's theorem) (Bell 1964) it was shown that quantum measurements do not satisfy a simple inequality which in some sense is a consequence of the separability requirement, whether there are hidden variables or not. Applied to actual experimental systems, satisfaction of Bell's inequality could be taken as a proof of the invalidity of quantum mechanics. Subsequent experiments in many laboratories have born out the validity of quantum mechanics, particularly experiments by the French group (Aspect, Grangier and Roger 1981).

But the point we wish to emphasize here is that tangled states are already inherent in the application of quantum statistics, as apparently had already bothered Einstein (Howard 1985). Quantum statistics requires that the wave function for identical fermions (particles of half odd integer spins) must be antisymmetrical in the coordinates of all particles in the system, while for identical bosons (particles of integral spin) the wave function must be symmetrical in the coordinates of all particles in the system. This is the deep connection between spin and statistics in quantum mechanics first clarified by Pauli (1940). Unless one obeys these rules in quantum calculations, one does not obtain results in accord with experiments. (For a simple example in scattering theory see Trainor and Wu 1953.)

For two He_4 particles (bosons), for example, if the Schrödinger equation yields a solution $\psi(x_1, x_2)$, this wave function must be replaced by the symmetrized function

$$\psi_s(x_1, x_2) = \frac{1}{\sqrt{2}}(\psi(x_1, x_2) + \psi(x_2, x_1)) \tag{2}$$

since this function has the property of remaining invariant when the coordinates of particles one and two are interchanged. The factor $1/\sqrt{2}$ is just a normalizing factor which guarantees conservation of overall probability.

One might well ask is this symmetrization procedure important or is it rather trivial? Indeed, it is utterly important, as shown in countless experiments involving identical bosons. Perhaps, most notably the remarkable superfluid behavior of liquid helium below four degrees Kelvin would not be explained without this requirement. In this case where one has billions upon billions upon billions of helium atoms in a liter or so of the liquid, the quantum entanglement is impressive and profound.

But the quantum statistical effects on entanglement can already be seen in our humble two particle system represented in equation (2). Except in relatively trivial cases the symmetrized wave function ψ_s cannot be expressed as a product of two functions, one depending only on the coordinates of particle one, and the other depending only on the coordinates of particle two.

Clearly, superfluidity in liquid helium provides an impressive example of what we have called an *a priori irreducible* emergence, since the particle wave functions are irreducibly entangled from the outset by the quantum edict of symmetrization, which itself is required for theoretical and experimental agreement in a multitude of experimental situations.

4 Does Irreducible Emergence Occur in Biology?

Superfluidity is a dramatic example of the problem of entangled wave functions in quantum mechanics. This bothered Einstein and indirectly lead to the so-called EPR controversy (1935). This controversy relates to Bell's theorem and the issue of non-locality in quantum phenomena (Bell 1964). The essential point here is that quantum statistics is a very holistic and powerful consideration in physics, and leads irreducibly to emergent properties of macroscopic systems.

If physics has irreducible emergent properties, is it possible that biology also has examples of irreducible emergent properties? One tradition in biology has been to seek laws of growth and form (Antonelli 1985) which complement the modernized version of Darwin's theory of selection by mutation and survival, in which developing systems represent stabilities in the landscape of possible forms consistent with the laws of chemistry and physics. A question raised by the physical examples of reducible and irreducible emergences is, are there irreducible laws of growth and form in biology? Indeed, the experimental work of Ho on developing embryos (essay in this volume) suggests the possible presence of a coherent field throughout the embryo that somehow directs or constrains the embryonic development. Can one regard such a coherent field as a reducible emergence phenomenon, or is this possibly an example of an irreducible biological emergence?

Not enough is yet known about such systems, or indeed about the systems discussed in the essays by Brandts and Totafurno and by Totafurno and Trainor in this volume, to be certain whether these field phenomena can be understood as reducible emergence phenomena, or whether there is an element of irreducibility in them as well — more than meets the eye as it were. If the development of quantum mechanics is any guide, perhaps theoretical biologists should not be too limiting in their approach to large scale and strange biological phenomena. Indeed, in his approach to the problem of consciousness, Roger Penrose (1989, 1994) raises the question of whether or not one has not only to modify thinking in biology, but perhaps as well in modern theoretical physics. But, this is not the first time that studies of biological phenomena have lead to the development of new ideas in both physics and mathematics (see for example Turing 1952, Thom 1975, Lewin 1992).

References

Antonelli, P.L., ed. (1985) *Mathematical Essays on Growth and the Emergence of Form.* Edmonton: University of Alberta Press.

Aspect, A., Grangier, P. and Roger, G. (1981) Experimental test of realistic local theories via Bell's theorem. *Physical Review Letters* **47**, 460–467.

Bell, J.S. (1964) On the Einstein Podolsky, Rosen paradox. *Physics* **1**, 195–200.

Crick, F. (1994) *The Astonishing Hypothesis.* New York: Charles Scribner and Sons.

Dennett, D.C. (1991) *Consciousness Explained.* London: Allen Lane, The Penguin Press.

Einstein, A., Podolsky, B. and Rosen, N. (1935) Can quantum-mechanical description of physical reality be considered complete? *Physical Review* **47**, 777–780.

Gordon, D.M., Goodwin, B.C. and Trainor, L.E.H. (1992) A parallel distributed model of the behavior of ant colonies. *Journal of Theoretical Biology* **156**, 293–307.

Hopfield, J.J. (1982) Neural networks and physical systems with emergent collective computational abilities. *Proceedings of the U.S. National Academy of Sciences* **79**, 2554–2556.

Howard, D. (1985) Einstein on locality and separability. *Studies in History and Philosophy of Science* **16**, 171–201.

Lewin, R. (1992) *Complexity: Life at the Edge of Chaos.* New York: Macmillan.

Pauli, W. (1940) The connection between spin and statistics. *Physical Review* **58**, 716–722.

Penrose, R. (1989) *The Emperor's New Mind.* New York: Oxford University Press.

Penrose, R. (1994) *Shadows of the Mind.* New York: Oxford University Press.

Thom, R. (1975) *Structural Stability and Morphogenesis.* Reading, MA: W.A. Benjamin.

Trainor, L.E.H. (1985) Remarks on emergence in physics and biology. In P.L. Antonelli, ed., *Mathematical Essays on Growth and the Emergence of Form.* Edmonton: University of Alberta Press, pp. 201–206.

Trainor, L.E.H. and Wu, T.Y. (1953) Symmetry requirements in electron scattering by an atom. *Physical Review* **89**, 273–274.

Turing, A.M. (1952) The chemical basis of morphogenesis. *Philosophical Transactions of the Royal Society of London, Series B* **237**, 37–72.

Holism and Reduction

Charles J. Lumsden

1 The Chapels of San Carlo

In 1643, Francesco Borromini (1599–1667) began work on *San Carlo alle Quattro Fontane*, the Church and Monastery of the Spanish Reformed Displaced Trinitarians in Rome (Figure 1). On foundations laid between two roadways, the Baroque master constructed one of Rome's "tiniest architectural wonders" (Steinberg 1977, iv). A wonder it remains to this day. To the great good fortune of scholars everywhere, Borromini never divulged the secrets of *San Carlo's* aesthetic unity. So began a debate among art historians, which continues still, about how the little church should be conceived in order to understand that unity. Covering the centuries between Borromini and us are no less than a dozen proposals, each decoding *San Carlo* as a design of simple interlocked shapes — stretched ovals, joined rings, quatrefoils, snipped ellipses and, of course, the cross (Figure 2).

Even physicists, accustomed as they are to the merits of analysis through simplification, might argue that such scholarly efforts are misguided because they reduce an intact whole, an emergent product of human creativity, to the geometrically trivial. In experiencing *San Carlo*, we are surely not engaged just by the bits and pieces, however pretty or agreeable each may be to our senses. We respond as well to the meaningful juxtaposition of the structural elements, all working together on our minds to regenerate Borromini's original vision of *San Carlo*, the church as a whole.

This small tale of the Baroque makes a simple but essential point: Physics and the other natural sciences are not the only human practices seeking to understand complex things by teasing them into parts. The strategy is universal. Carried too far it risks accusations of *reductionism*, of equating the whole to its parts. As the essays by Trainor and by Brandts (this volume) point out, we are rather used to the reductionism charge being leveled at the natural sciences, but no field of inquiry seems immune, and looking beyond the confines of any single discipline one sees the reductionism debate going on everywhere (Simon 1981, Argyros 1991, Lumsden 1996). The counter-accusation is *holism*, of refusing to see anything except the whole itself.

Holism is an idea that has haunted philosophy and the sciences — especially biology and the human sciences (psychology, anthropology, sociology, literary theory) — for well over a century without coming into clear focus.

Figure 1: Holism in action: Francesco Borromini's *San Carlo alle Quattro Fontane*, Rome, 1643. Pencil drawing by John Ruskin (1819–1900). Reproduced by kind permission of the estate of Mr. Louis G. Davidson, courtesy of Mrs. Louis B. Davidson.

Holism implies emergence, the existence of forms and processes at one level of organization that can be explained only with difficulty, if at all, by descriptions of processes occurring among their parts. *San Carlo* makes no sense, we feel compelled to say, when experienced only in terms of its component geometric forms. Similarly, growth and form in biological tissues cannot be understood on the basis of what we know about the component cells alone, just as an ecosystem cannot be predicted entirely from what we know about the individual animals in it. A residue of phenomena remains after each such effort at reduction to the contiguous (or even deeper) level. It would seem that nature is made to hold back on us, that new phenomena will spring up just beyond the limits of our latest grasp of cause and effect. While these limits hold, we can only describe the residual events, perhaps capturing them in a scientific

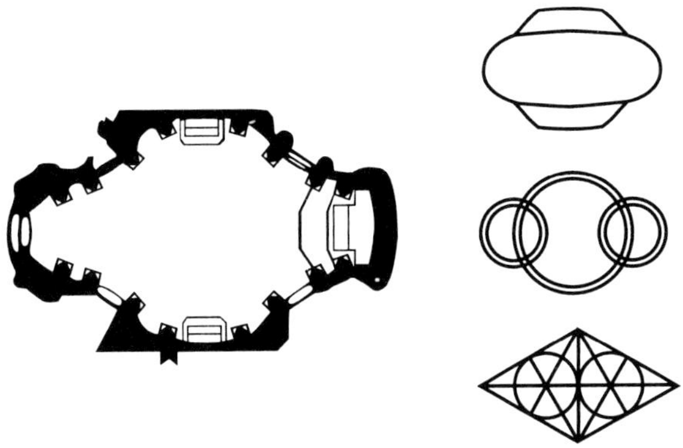

Figure 2: *San Carlo's* ground plan, and some of the geometric themes that have, over the centuries, been championed as the secret of *San Carlo's* aesthetic unity. After Steinberg (1977).

theory of their own.

2 The Romance of Reduction

The intractability and haziness surrounding the notion of holism tend to fade to the extent that parts become known and their consequences made computable. Indeed, in a famous definition Nagel (1961) characterized reduction as the procedure whereby the content of a so-called "secondary" discipline is somehow shown to be the logical consequence of base axioms in a "primary" discipline. Thus chemistry from physics, biology from chemistry, civics from sociobiology, and so on. Were it in truth so simple.

"Primary" and "secondary" all too readily become fighting words when scholars use them to describe each other's disciplines, so I must be clear that the terms will be used here strictly as descriptive assertions, not as preemptive claims about intellectual merit. To emphasize this I will keep the terms surrounded by quotes for the next few lines, as a reminder that we are not taking their valuational capacities seriously — just their descriptive ones. In this sense, the "primary" discipline is the one first encountered on moving up though levels of organization from smaller to larger and more complex systems and interactions. During this progression, a discipline can find itself "secondary" to a "primary" one when somebody notices that the phenomena dealt with in the "primary" discipline might be taken as meaningful building blocks of the phenomena dealt with by the "secondary" discipline.

The operative word here is "meaningful," and there are at least two dimensions to the term's significance for physical theory in biology. The first serves to contrast description, taken as a human activity, with valuation. Assigning a term like "meaningful" *is* undeniably an evaluative act, performed in the midst of debates about how the arts and the sciences work. But to ascribe meaning does not, in itself, reserve priority solely for a primary discipline. All contributing disciplines are potentially of the utmost significance, and it is easy to see why. On one level, for example, *King Lear* certainly is a set of words. But just listing the words we find in some edition of *Lear* does not exhaust what we can, with significance, say about the work. Viewed as a whole, *King Lear* is also a Shakespearean tragedy, and this descriptive term belongs to the secondary discipline (literary criticism, say), not to the primary (such as psycholinguistics). Yet, unless we are equipped with the notion of tragedy and Shakespearean tragedy, we are, quite simply, lost in the details should we attempt the life-long task of appreciating and comprehending *Lear*.

The second dimension of significance to the term meaningful concerns some pivotal distinctions between deduction and interpretation. Deduction, as human activity, makes new things by applying rules of inference to starting assumptions. Interpretation does it by assigning new meaning to whatever our human imaginations find engaging, including the consequences of deduction. Connecting a primary and a secondary discipline does, as Nagel foresaw, undeniably involve a certain amount of logical consequence. However, deduction in itself will never reach out of the primary discipline to join with the secondary one — no matter how long a deductive chain we forge. An imaginative leap sooner or later is needed to make the jump and link the disciplines. The reason can be seen in the example above: Secondary disciplines are usually organized around concepts that, at first encounter, have no meaning at all in the originating vocabulary and axioms of the primary discipline. Thus literary criticism uses notions like tragedy, while psycholinguistics deals with terms like morpheme, phoneme, grammar and syntax — apparently a world away from the realm of drama and theater!

Scientific studies of creative insight and its role in human understanding (e.g. Sternberg and Davidson 1995) hint that the imaginative leap away from pure deduction involves an act of interpretation, which assigns new significance to clusters of primary-discipline symbols and concepts. At some point in the logical chain, a deduced pattern of relations among symbols in the primary discipline is grasped in a new way: It is observed to resemble a pattern in the core principles of the secondary discipline. Principles analogous in meaning are then obtained from the primary discipline *if* we interpret the appropriate combinations of its symbols, suitably transformed, in the same way we do corresponding symbols in the secondary discipline. Thus reduction in fact synthesizes new meanings, which align the ideas of complementary subjects. It does not dissolve one discipline using the semantic apparatus of another. The fictive physicist trying to stop the stampede of highly intelligent physics majors into biology and other areas of condensed matter studies with the cry "It's all just quarks!" really does merit our sympathy for having come up so short: Minimally, it's all just quarks plus interpretations. Indeed, the well-known relativity theorist Roger Penrose has boldly argued that a proper understanding of how the human mind can grasp creative new truths will require fresh theories in *physics*, as well as in biology, as novel and as fundamental as any we have seen thus far (Penrose 1994, and see Cairns-Smith 1996).

Of course, to take just one instance, what we know of developmental biology conforms to the theoretical tenets of molecular physiology. The received view in the life sciences is that reduction of this sort is feasible in principle across every level of organization from ecosystems to molecules. But, as with *Lear* and the list of words, the comfortable security afforded by such knowledge usually provides little fresh insight. That is because the opposite thrust, from primary discipline to secondary discipline (for example from molecular physiology to developmental biology) is far more difficult. The phenomena studied in the secondary discipline are often grounded in dauntingly complex patterns of interaction among elements on the primary level. At least currently, these patterns of interaction are too often mathematically intractable, as well as disturbingly opaque in the logic of their design.

Biology is also a historical discipline. Events shaping the history of organic growth or evolution may be too numerous in kind, or too poorly understood, to encompass within the language of either primary or secondary discipline. Thus cells shape themselves in response not just to local molecular events but to the particular geometric configurations in which they happen to reside. For historical reasons as well it is natural to speak of a residue of descriptive generalizations in a secondary discipline, even after the theoretical principles of the primary discipline have been thoroughly and rigorously applied.

Holism, as an approach to the complexity we observe in the universe, involves this recognition that understanding a higher level of organization requires not only the theoretical assumptions of primary and secondary disciplines but also pivotal, interpretive hypotheses, along with peculiarities of history that may be meaningful for the secondary discipline alone. Holism is nontrivial to the extent that the underlying interactions are strong, nonlinear, and, in the pattern of their organization, complex. To take a table top example from insect sociobiology, the interactions of the foraging and nurse workers in colonies of fire ants follow rather precise probabilistic rules, but these in turn depend on the recent nutritional history *of the colony*, as well as on the spatiotemporal patterning of food supply outside the nest. But once those circumstances are known, the emergent pattern of the colony's behavior as a whole — the ethology of the superorganism — can be predicted with some precision (e.g. Hölldobler and Wilson 1994, and Trainor, Brandts and Torres, this volume). This feat is possible because entomologists have already given us a rich description of the superorganism as well as or the details of individual ant behavior. In other words, a secondary discipline is in place alongside the primary.

Although many such tractable cases exist, interpretive holists like myself readily concede that other interactions (like those resonating through the textual field of a Shakespearean tragedy or within the architectural spaces of *San Carlo*) lie far beyond the grasp of present theory and method. They are the grand challenges of complexity theory.

If the required deductive acts, taken one after another, form an impressively lengthy sequence, and if the interpretive link can, with any fairness, be defended as notably bold or novel, the contrasting alternations of deductions and interpretations bear some resemblance to the pattern of change posited by theories of so-called "punctuated equilibrium" (Gould and Eldredge 1977, Eldredge 1989, Prothero 1992). Could punctuated equilibrium lie at the heart of understanding how reduction works in science?

Systems dominated by punctuated equilibrium can evolve new forms of organization and action, but do so in two related modes. The first mode of change takes place smoothly; changes are small, new forms and functions are seemingly reasonable consequences of what has come before them. Since the steps are small, significant change can take a long time (maybe forever) to accumulate.

The second mode is exactly the opposite in both tempo and mode. Big changes happen fast. They *may* be of a kind that merely extrapolates the scale of the smaller changes to more grandiose proportions (such as a huge earthquake following a long period of minor tremors), or, more interestingly, they may produce something quite new and unexpected as judged by the preceding small changes (such as the emergence of a remarkable new taxonomic category of life forms from a community of previously interbreeding populations).

Nothing in principle decrees that the two domains, small gradualistic steps and novel fast leaps, have to be entirely different and unrelated, and indeed there have been exciting suggestions that in general the two will be found to be linked by an entire spectrum of scaled versions of themselves, much as the thermodynamic state of a fluid poised at a critical point may experience density fluctuations on all spatial scales. In other words, there is a critical state, or something very much like one, which acts as an attractor for the dynamics governing form and function within the system. Such "self-organized criticality" (Bak and Chen 1991, Bak and Paczuski 1995) can sustain vast diversity in how complex systems change over time, from relatively frequent small shifts to rare big ones.

It is much too early to judge whether theory reduction can be validly understood as a nonlinear process in which gradualistic, deductive steps are interrupted and cross-linked by interpretive leaps that weave previously isolated disciplines into new, more unified intellectual formations. The hard, incisive data on how real scientists do science, including theory reduction and synthesis, are just starting to flow as science-doing begins shedding its reputation as a philosopher's favored plaything and attracts hard-nosed experimental, historiographic and theoretical attention (such as we see, for instance, in the work of scholars like Langley, Simon, Bradshaw and Zytkow 1987; Schaffner 1993; Sternberg and Davidson 1995; Buchwald 1995; and Koslowski 1996). For the present, these parallels with punctuated equilibrium offer a starting point for further analysis, an analogy that even now may help in understanding how a discipline can be reduced in the sense we've discussed, but emerge from the process not only intact but greater than it was.

3 Layered Webs

I have noted that disciplinary reduction and synthesis, by which holistic phenomena may be approached across scientific disciplines, depend fundamentally on acts of interpretation. These interpretive acts are not arbitrary or whimsical; they create new patterns of association among elements of the primary discipline while preparing it for linkage to the secondary discipline. At this point it will prove useful to implement some straightforward notation, in order to see better how the interpretive process might work when its subject is a system with complex, hierarchical organization. As in all such formal schematics pointed at the mind's best work, the simplifications are (at least currently) extreme, but even a premature forcing of rigor where none may ultimately be justified will help to establish some important observations about holism and synthesis.

Let us use the familiar symbol $\mathbf{x}$ to refer *en masse* to the elements $x_1, \ldots, x_N$ of the primary discipline's theoretical vocabulary, so that $\mathbf{x} = (x_1, \ldots, x_N)$. When the primary discipline is one of the branches of physics, the x_k are usually coordinate and momentum variables describing the motions of the particles comprising the system, such as the atoms in a crystal lattice or the protons and neutrons in an atomic nucleus. In the case of a primary discipline from biology, however, the vocabulary is potentially much larger. The $\mathbf{x}$ could be chemical concentrations, population densities, rates of growth, gene frequencies, shapes, and so on, depending on the level of organization embodied in the primary discipline. (They need not necessarily be numeric, or possessed of sharp logical boundaries.) The secondary discipline also needs some notation, and I will use the symbols $\mathcal{A}_1, \ldots, \mathcal{A}_M$ to label the relevant elements of its theoretical vocabulary, with $\vec{\mathcal{A}} \equiv (\mathcal{A}_1, \ldots, \mathcal{A}_M)$. Very often the number of primary discipline elements x_k greatly exceeds the number of secondary discipline elements $\mathcal{A}_m$, so that $M \ll N$, but this need not be the case.

Suppose that $\mathcal{T}_P$ and $\mathcal{T}_S$ are, respectively, the deductive theories created by the primary and secondary disciplines to explain their subject matter. A reasonable starting point for appreciating what reduction and synthesis involve in practice is then to ask how we might [re]generate $\mathcal{T}_S$ from $\mathcal{T}_P$. Is there some type of linkage operation, which I will denote by $\mathbb{L}$, that takes us from the world as described by the primary discipline to the one described by the secondary, in other words, for which $\mathbb{L} : \mathcal{T}_P \rightarrow \mathcal{T}_S$?

The answer, so far as it currently goes, is a guarded "Yes." The guard is needed because interpretive holism is a young field and because not all parts of a discipline's theory seem equally amenable to reduction and synthesis. When viewed from afar, deductive theories like $\mathcal{T}_P$ and $\mathcal{T}_S$ can be seen to contain at least three sorts of text. First there is text describing the elements, such as $\mathbf{x}$, which ultimately denote structures and processes in nature according to that discipline. Then there is text that describes patterned regularities of association among the elements. Natural scientists conventionally refer to such patterned regularities as laws that govern the behavior of the elements. Regretably, the law metaphor has received very bad press in the human sciences, where it is argued to encourage the insensitive application of physics models where they do not belong, degrading human uniqueness and imprisoning human free will — such as it is. The natural science use of the law metaphor of course has no such necessary consequent, but this deeply misguided objection remains influential, and a significant impediment to the proper study of holism. An introduction to its history can be found in the final chapter of Lumsden and Wilson (1983). For the purposes of this essay I will use the symbols $\mathcal{L}_P$ and $\mathcal{L}_S$ to denote the relevant parts of the theories $\mathcal{T}_P$ and $\mathcal{T}_S$, and refer to them as laws with no further discussion.

A third part of the text in $\mathcal{T}$ comprises what might be called a discipline's natural history, $\mathcal{H}$. The natural history contains the unique and particular information needed to produce meaningful predictions and explanations from the laws. The natural history includes, but may go far beyond, temporal successions of initial conditions and boundary events. In practice the natural histories $\mathcal{H}$ are probably best understood as irreducible emergents in the sense of Trainor (this volume). They are things we must simply know (such as the alleged terrestrial impact of a large meteor initiating the late Cretaceous mass extinctions) if a discipline is to be worked deductively. As I noted earlier, linkage of primary and secondary disciplines does tend eventually to shrink the amount of information in the natural history, as more and more is subsumed by the laws $\mathcal{L}$ and the formal procedures for reduction and synthesis. But there is always a residuum.

To summarize: Each discipline, which may find itself primary to some other disciplines and secondary to yet others, we shall envisage as described by its theoretical structure $\mathcal{T}$; and $\mathcal{T}$, in the crudest possible terms, has a three-part textual anatomy. (Die-harder formalists take this *much* further; two notorious classics in the genre are Woodger 1937 and Bunge 1967; but see Schaffner 1993, esp. Chapter 9.) For a primary discipline this is $\mathcal{T}_P = < \mathbf{x}, \mathcal{L}_P, \mathcal{H}_P >$, while for a secondary, $\mathcal{T}_S = < \vec{\mathcal{A}}, \mathcal{L}_S, \mathcal{H}_S >$. The action of the operator $\mathbb{L}$ that links the primary to the secondary discipline can be expressed by the diagram

$$\mathbb{L} : \mathcal{T}_P \longrightarrow \mathcal{T}_S, \quad \text{also written as} \quad \mathcal{T}_P \xrightarrow{\mathbb{L}} \mathcal{T}_S \qquad (D1)$$

If our system is even more complex, with hierarchically organized levels of structure and function $K = 1, 2, 3, \ldots$ with each described by a corresponding theory $\mathcal{T}_1$, $\mathcal{T}_2$, $\mathcal{T}_3$, ... in some chain of disciplines, then

$$\mathcal{T}_1 \xrightarrow{\mathbb{L}} \mathcal{T}_2 \xrightarrow{\mathbb{L}} \mathcal{T}_3 \xrightarrow{\mathbb{L}} \mathcal{T}_4 \xrightarrow{\mathbb{L}} \cdots \qquad (D2)$$

where the generalization in notation is obvious.

We can now taste some of the sweet irony that makes holism a scholar's delight. To analyze how disciplines might be linked one to another, we must take $\mathbb{L}$ which, like *San Carlo*, embodies a human act of interpretive, synthetic insight, and break it down — take *it* apart in order to understand its action by rebuilding it. This may be impossible (Findlay and Lumsden 1988, Penrose 1994). We know, however, that any such reconstituted procedure, if achieved, will accomplish two things: First, it will make explicit the act of interpretation by which the contents of the primary discipline are reworked and prepared for linkage with the second discipline. Second, it actually will *leave* the primary discipline and enter the phenomenal world described by the secondary discipline, synthesizing the elements of its vocabulary and its laws.

One of the truly remarkable discoveries of modern theoretical physics is that principal levels of emergent structure and function, even in complex nonlinear physical systems far from thermodynamic equilibrium, can be reached by a linkage operation IL that has a very simple form (but even here there are still many key unsolved problems and fundamental questions; a fine survey is Sklar 1995). IL need have just two parts! The first I will denote $\mathcal{O}$ ("$\mathcal{O}$" as in "o"rganizing). This portion encodes the process of interpreting the primary discipline in terms of the secondary discipline. The operator $\mathcal{O}$ accepts as input the elements $\mathbf{x}$ of the primary discipline's theoretical vocabulary. Its output is a set of functions $A_1(\mathbf{x})$, ..., $A_M(\mathbf{x})$. In the simplest cases, if the theoretical vocabulary of the secondary discipline contains M elements $\mathcal{A}_m$, there are M functions $A_m(\mathbf{x})$, which represent the secondary discipline elements interpreted in terms of the primary discipline elements. I will refer to the $A_m(\mathbf{x})$ as *interpretors*. Although a neologism, the term seems helpful in the present setting, and naturally motivated.

The second part of IL takes the interpretors as input and maps them into the secondary discipline. In addition to the interpretor functions $A_m(\mathbf{x})$, the other essential input for this operator is the law set $\mathcal{L}_P$ of the primary discipline, without which a description of system behavior is impossible. A symbol often chosen in physical theory to denote this operator is ρ, which generally carries an explicit time dependence, so that for a primary and a secondary discipline

$$\rho(t) : \{A_m(\mathbf{x}), \mathcal{L}_P\} \rightarrow \langle A_m \rangle(t) \in \mathcal{T}_S$$

In physical thermodynamics, ρ is known in more familiar terms as the *statistical ensemble*, since physicists take the practical view that one gets from a primary to a secondary discipline by recognizing most events on the primary level to be, from the standpoint of the secondary, irrelevant details: The exact position of every particle and the exact momentum exchange of every collision are no longer important. They are gotten rid of by averaging them out. ρ quantifies the amount of averaging needed, while the interpretors $A_m(\mathbf{x})$ are the objects averaged. The secondary discipline construct $\langle A_m \rangle(t)$ produced by the action of ρ on $A_m(\mathbf{x})$ is called the *expectation value* of $A_m(\mathbf{x})$ given the statistical ensemble $\rho(t)$. The set $\{\langle A_m \rangle(t)\}$ is the predicted content of the secondary discipline. Although there is no reason to anticipate that the operator ρ will take the form of a probability distribution in all cases germane to biology and the human sciences, no practical alternate terminology is currently in use. I will when necessary refer to ρ as the ensemble and $\langle A_m \rangle(t)$ as expectation values, while fully recognizing the potentially "non-statistical" nature of these objects in novel applications.

Under $\mathcal{O}$ the elements $\mathbf{x}$ of the primary discipline are carried into the interpretors $A_m(\mathbf{x})$. The process $A_m(t) = A_m(\mathbf{x}(t))$ is often called the *symbolic dynamics* of the system, since the level hypersurfaces of the $A_m(\mathbf{x})$ define the walls of "cells" in the coordinate space of $\mathbf{x}$. The system, carried along the trajectory $\mathbf{x}(t)$, moves among these cells, to each of which is attached a "label" or "symbol" effectively denoting that the system is somewhere between $\mathbf{A}$ and $\mathbf{A} + \delta\mathbf{A}$. Much as a worker bee scuttling over the face of honeycomb visits many mutli-sided cells as she makes her rounds, the symbolic dynamics affords a complete record of the system's wanderings, at a level of resolution defined by the magnitude δ and written in terms of the values of the interpretors $A_m(\mathbf{x})$. A system's symbolic dynamics turns out to be fundamental to understanding the relationship between its inner mechanism and the complexity of its potential behavior, as well as its potential for computational activity and information processing. See Dufort and Lumsden, this volume, for further discussion and a detailed example.

Under ρ the interpretors go over into the expectation values $\langle A_m \rangle(t)$:

$$\mathcal{T}_P \ni \mathbf{x} \xrightarrow{\mathcal{O}} \mathbf{A}(\mathbf{x}) \xrightarrow{\rho} \langle \mathbf{A} \rangle(t) \in \mathcal{T}_S \qquad (D3)$$

If we define the linkage operator to be $\mathbb{L} \equiv \rho \bullet \mathcal{O}$ and rewrite diagram $(D3)$ with a kink in it,

$$\begin{array}{ccc} \mathbf{x} & \xrightarrow{\mathcal{O}} & \mathbf{A}(\mathbf{x}) \\ & \overset{\mathbb{L}}{\searrow} & \Big\downarrow {\scriptstyle \rho :: \mathcal{L}_P} \\ & & \langle \mathbf{A} \rangle(t) \end{array} \qquad (D4)$$

This commutative diagram states that the operator formed by the sequential composition of $\mathcal{O}$ and ρ links the primary to the secondary discipline. I have used the notation $\rho{::}\mathcal{L}_P$ in $(D4)$ as a reminder that the ensemble minimally has two input slots — one for intrepretors and one for the laws $\mathcal{L}_P$ of the primary discipline. The laws $\mathcal{L}_P$ determine the dynamics of both the interpretors $A_m(\mathbf{x})$ and ensemble ρ itself. A specific example will be considered below.

The arguments I have used in arriving at diagram $(D4)$ are quite general, so there is no reason to stop with just one step of inter-discipline linkage. Iterating gives a pattern something like

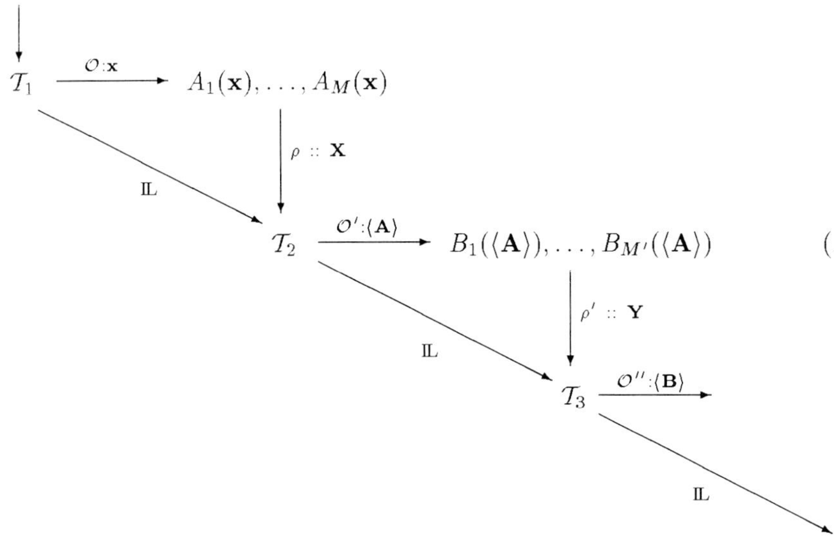

$(D5)$

In this diagram a web of explanatory links between disciplines goes off in both directions: at the top, toward the dynamical variables of ever more "primary" disciplines; at the bottom, toward ever more "secondary" ones. From the level corresponding to our original secondary discipline, the steps of reduction and synthesis can be repeated to seek descriptions in which the $\langle A_m \rangle$ play the role of primary-level variables, now sustaining a further level of emergence, to be contacted through the expectation values $\langle B_k \rangle$, and so on. Drawing as it does on the terminology of physical statistical mechanics and on ideas about hierarchies of increasingly macroscopic domains of order, the approach modeled by diagram $(D4)$ might be called *layered statistical mechanics* (LSM). The **Strong LSM Conjecture** is that there exist effective (and therefore computable) interpretation procedures $\mathcal{O}, \mathcal{O}', \mathcal{O}'', \ldots$ and linkage operations ρ such that diagram $(D5)$ commutes. In other words, $\mathbb{L}$ exists, $\mathcal{A}_m(t) \equiv \langle A_m \rangle(t)$ for all levels, and LSM works.

Happily, the Strong Conjecture in the naive form just stated may be dead where it stands. Robert Rosen (1985, 1991) has presented a very important argument that may apply to systems of relationship like $(D5)$ in the case where complexity is very high. "Very high" would, for our purposes here, mean that the number of meaningfully distinct levels captured in diagram $(D5)$ — the number of times N that $\mathbb{L}$ is applied — is very large, so large that it can be treated as effectively infinite. Then, the kind of system we get in the *limit* as N approaches infinity may bear little or no resemblance to the familiar, mechanism-like entities we see joined by any single application of IL. The limit could, in principle, be something new, and stranger than any system describable by any set of mechanistic equations, however large. If I understand him correctly, Rosen is suggesting that it is at the outer reaches of such limits, rather than in the depths of dynamical equations, that will be found the mathematical essence of the living state. The thorough exploration of such exciting proposals can be expected to play a leading role in formal theories of holism and emergence in the years to come.

Although my strategy will soon change, I have been careful in setting down the above to use the more obtuse notion of *element*, rather than a more familiar and precise mathematical notion like *variable* in my references to $\mathbf{x}$ and $\vec{\mathcal{A}}$. This is because in biology and the human sciences the $\mathbf{x}$ and $\vec{\mathcal{A}}$ are not always elements whose specific meaning is best expressed by numbers or sets of numbers. They may be objects, such as sonatas or churches or passions, whose rigorous description best takes place in a symbolic but non-numerical language of inference and expression. I will have more to say about these incredibly interesting and challenging cases below. For now we can stay with the far better understood case in which all the $\mathbf{x}$ and $\vec{\mathcal{A}}$ have numerical ranges.

4 Vector Field Hierarchies and the Dirac Manifold

LSM can be illustrated by one of the simplest cases relevant to the study of emergence and holism in biology: chains of primary and secondary disciplines that each use ordinary differential equations (ODEs) as their laws $\mathcal{L}$ (Lumsden and Trainor 1985). The purpose of this section is to further develop the notion of LSM for such systems by highlighting its links to mathematical ideas crucial in physical statistical mechanics. A prior acquaintance with analytical mechanics is presumed (for the remainder of the section only), but this section can be skipped without losing the main argument.

On the primary level, the laws $\mathcal{L}_P$ for $\mathbf{x}$ are then the ODEs

$$\frac{dx_k}{dt} \equiv \dot{x}_k = X_k(\mathbf{x}, t), \qquad k = 1, \ldots, N \tag{1}$$

where the X_k are continuous and suitably smooth functions of the primary-discipline variables $\mathbf{x} = (x_1, \ldots, x_N)$ and possibly the time t. The secondary and higher-order disciplines also use ODEs like equation (1) to describe the relations amongst *their* variables. The character of Rosen limits (see above) is not presently understood even for emergence systems this simple, so I will stay away from infinite iterations of $\mathbb{IL}$ and discuss the better understood case of primary-secondary system linkage, i.e. $N = 1$ in $\mathbb{IL}^N$.

Now biology steps in, albeit in an unexpected way: In diverse instances — the transfer of information between cells, the chemical reactions of metabolism, the flow of blood, the search for food by predators, the territorial strategies of insect societies are a few examples — the biological process takes place in an efficient manner described by an optimization principle. This "fit design" for life in the wild is made possible by natural selection acting, over time, on the endless, heritable variations intrinsic to biological populations to facilitate replication efficiency at multiple levels of structure and organization (Lumsden and Wilson 1983; and see the chapter by Rowe, this volume).

Suppose that vector fields such as $\mathbf{X}(\mathbf{x})$ express this principle of fit design. In the simplest case, which for clarity I use here, the design is specified by a single function $L(\mathbf{x}, \dot{\mathbf{x}}, t)$ (the objective function, usually called the Lagrangian by physicists), which is maximized over the process in question, i.e. for which there is a variational principle

$$\delta \int_{t_1}^{t_2} L \, dt = 0. \tag{2}$$

Equation (2) will serve as a covering law in the theory $\mathcal{T}_P$ of the primary discipline. The structure of the vector field $\mathbf{X}(\mathbf{x})$, and all temporal change in $\mathbf{x}$, must be consistent with it. If the Euler-Lagrange equations corresponding to the variational principle for L are to reproduce $\mathbf{X}(\mathbf{x})$, then $\partial L/\partial \dot{x}_k$ must in general be independent of $\dot{x}_k$ for all k. The Lagrangian itself will therefore be at most linear in the rates,

$$L = U_k(\mathbf{x}, t)\dot{x}_k - U_0(\mathbf{x}, t), \tag{3}$$

such that

$$\dot{x}_k = -\Gamma^{-1}_{kn} \frac{\partial U_0}{\partial x_n} \tag{4}$$

on the primary level, where

$$\Gamma_{kn} = -\Gamma_{nk} \equiv \frac{\partial U_k}{\partial x_n} - \frac{\partial U_n}{\partial x_k} \tag{5}$$

and

$$\frac{\partial \Gamma_{kn}}{\partial t} = \frac{\partial}{\partial x_k}\Gamma_{ni}X_i - \frac{\partial}{\partial x_n}\Gamma_{ki}X_i \tag{6}$$

with

$$\Lambda_k \equiv -\frac{\partial U_k}{\partial t} - \frac{\partial U_0}{\partial x_k} \tag{7}$$

and all the indices running between 1 and N. Here and elsewhere repeated indices imply the summation convention. The antisymmetry of the Γ_{kn} guarantees that the function $U_0(\mathbf{x})$ associated with the vector field $\mathbf{X}(\mathbf{x})$ of the primary discipline is a constant of the motion:

$$\frac{dU_0}{dt} = \frac{\partial U_0}{\partial x_k}\dot{x}_k = \frac{\partial U_0}{\partial x_k}(-\Gamma^{-1}_{kn})\frac{\partial U_0}{\partial x_n} \equiv 0. \tag{8}$$

From the physicist's perspective, it is striking that a blind force like natural selection should create biological populations whose members are usefully described by the familiar language of variational principles, Lagrangians, and Euler-Lagrange equations. But such is the case. It is of course unusual for physical statistical mechanics, and the study of emergent behavior in complex physical systems like crystals and fluids, to be carried out in Lagrangian form; the Hamiltonian description is generally much preferred. The Hamiltonian description, when valid, asserts a tight unity among the generators of a system's temporal evolution, the fundamental symmetries of that dynamics, and the quantities conserved during the system's time development.

However, a problem in bringing the impressive unity of this picture to $\mathbf{X}(\mathbf{x})$ and the other vector fields on our model hierarchy is at once evident. The usual construction,

$$H = p_k\dot{x}_k - L, \quad p_k \equiv \partial L/\partial \dot{x}_k, \tag{9}$$

leads to the identification $H = U_0(\mathbf{x})$. But then the Hamiltonian equations imply that

$$\dot{x}_k = \partial U_0/\partial p_k \equiv 0, \tag{10}$$

which is clearly a contradiction (compare equation (1)!). The source of difficulty is the linearity of the objective function L in the rate variables $\dot{x}_k$. This linearity makes the transformation from the Euler-Lagrange to Hamiltonian pictures completely singular: The so-called Dirac matrix for $\mathbf{X}(\mathbf{x})$, whose elements are

$$D_{kn} \equiv \partial^2 L/\partial \dot{x}_k \partial \dot{x}_n, \tag{11}$$

has all D_{kn} identically zero on the ODEs of the primary discipline. For physical systems, the Lagrangians L are (typically) comfortably quadratic in the $\dot{x}_k$, and singularity is not an issue. But there is no reason why natural selection (which in biomathematical scenarios seems to have a penchant for rates rather than accelerations) should ape molecular physics in designing cells, organisms, or ecosystems — even "optimal" ones — and it has not done so.

Fortunately, singular maps from Lagrangian to Hamiltonian have also caused trouble for physics, particularly in the origins of quantum field theory. Dirac (1950, 1964; a modern introduction is Govaerts 1991) discovered a generalization of the usual Hamiltonian procedure suited to the needs of quantizing fields. It also turns out to work beautifully on the fit vector fields of equation (3).

A detailed description of the Dirac approach would take us well beyond the scope of the essay. In essence, Dirac's method produces a Hamiltonian picture by an embedding procedure, which places the original vector field in a phase space with excess dimensions, and then locks the phase flow down onto a constraining surface or manifold, on which the flow is fully consistent with the requirements of equations (1) and (2). The generalized Hamiltonian theory thus begins in a normal $2N$-dimensional phase space of independent canonical coordinates x_k and p_k. The actual motion, associated with the vector field $\mathbf{X}(\mathbf{x})$, is confined to a hypersurface $\mathcal{M}$ of lower dimensionality, defined by the so-called *primary constraints* $\phi_\nu(\mathbf{x},\mathbf{p}) = 0$, $\nu = 1,\ldots,N$ and embedded in the $2N$-dimensional phase space.

The momentum variables for $\mathbf{X}(\mathbf{x})$ in this phase space are then given by the rate-independent expressions

$$p_\nu = \partial L/\partial \dot{x}_\nu = U_\nu(\mathbf{x}). \tag{12}$$

For each p_ν there is a primary constraint

$$\phi_\nu = p_\nu - U_\nu(\mathbf{x}), \qquad \nu = 1,\ldots,N. \tag{13}$$

The Legendre transform for the construction of the Hamiltonian form of the optimization principle (2) is

$$H_0 = p_k \dot{x}_k - L = U_0(\mathbf{x}) \tag{14}$$

with the generalized Hamiltonian equations

$$\dot{x}_k = \frac{\partial U_0}{\partial p_k} - \dot{x}_\nu \frac{\partial U_\nu}{\partial p_k} \tag{15a}$$

and

$$\dot{p}_k = -\frac{\partial U_0}{\partial x_k} + \dot{x}_\nu \frac{\partial U_\nu}{\partial x_k} = \frac{\partial U_k}{\partial x_n}\dot{x}_n \tag{15b}$$

for $k = 1,\ldots,N$. The summation convention for repeated indices is assumed throughout. Equation (15b) specifies the relation between $\mathbf{X}(\mathbf{x})$ and the generalized canonical elements Γ_{nk} and $U_0(\mathbf{x})$:

$$\dot{x}_k = -\Gamma^{-1}_{kn}\frac{\partial U_0}{\partial x_n}, \tag{16}$$

recovering equation (4), while (15a) give the identities $\dot{x}_\nu = \dot{x}_\nu$, $\nu = 1, \ldots, N$.

The final Hamiltonian of the primary-level dynamics is obtained with the aid of the primary constraints:

$$H = H_0 + \dot{x}_\nu \phi_\nu \;=\; U_0(\mathbf{x}) + (p_\nu - U_\nu(\mathbf{x}))\dot{x}_\nu, \tag{17}$$

with

$$\dot{H} = 0. \tag{18}$$

In terms of the Poisson bracket $[\cdot,\cdot]$,

$$\dot{p}_k = [p_k,\; H] \quad k = 1, \ldots, N \tag{19}$$

for the vector field $\mathbf{X}(\mathbf{x})$ and

$$\dot{A}_m(\mathbf{x}) = [A(\mathbf{x}), H] \quad m = 1, \ldots, M \tag{20}$$

for the dynamics of the interpretors $A_m(\mathbf{x})$.

For deterministic vector field theories like equation (1), the minimal capability consistent with the Strong LSM Conjecture would be that sufficient to answer the questions like the following: Suppose that at time $t = t_0$ our system is described by secondary-discipline properties $\mathcal{A}_1, \ldots, \mathcal{A}_M$ having initial values $\mathcal{A}_1(t_0), \ldots, \mathcal{A}_M(t_0)$, and by primary-discipline values $x_1(t_0), \ldots, x_N(t_0)$, such that the interpretor dynamics $\dot{\mathbf{A}}(\mathbf{x})$ is as in equation (20) and $\vec{\mathcal{A}}(t_0) = \langle\mathbf{A}\rangle(t_0)$. What then are the values of the $\vec{\mathcal{A}}(t)$ for $t > t_0$? Does the difference $|\vec{\mathcal{A}} - \langle\mathbf{A}\rangle|(t)$ between the observed and predicted behavior of the secondary-discipline properties remain zero for all $t > t_0$? Or do the observed values diverge from the predicted values of the interpretors? Similarly, if the flow $d\vec{\mathcal{A}}/dt$ follows the vector field $\mathbf{Y}$ as in ($D5$), then how closely does $\langle\mathbf{A}\rangle(t)$ follow $\mathbf{Y}$; i.e., is it true that $d\langle\ \mathbf{A}\ \rangle\ /dt = \mathbf{Y}(\langle\ \mathbf{A}\ \rangle)$, at least to some very close approximation?

In cases like this, LSM must in effect explain the patterns of temporal change in the system on both the primary and secondary levels, given the dynamics (1) and (20) along with the initial state of the secondary-discipline variables. The appropriate ensemble ρ for the projection $\rho(t){:}A_m(\mathbf{x}) \rightarrow \langle A_m\rangle(t)$ is the normalized probability density $\rho(\mathbf{x}, t)$ that the system is in primary-variable state $\mathbf{x}$ at time t, given that it was in secondary variable state $(\mathcal{A}_1(t_0), \ldots, \mathcal{A}_M(t_0))$ at time t_0:

$$\rho(t) : A_m(\mathbf{x}) \rightarrow\; \langle A_m\rangle(t) = \int d\mathbf{x}\; A_m(\mathbf{x})\rho(\mathbf{x}, t), \quad t \geq t_0. \tag{21}$$

Conservation of probability guarantees that $\mathbf{X}(\mathbf{x})$ induces the ensemble dynamics

$$\partial\rho/\partial t \; = \; -[\rho, H] + \rho D(\mathbf{x}) \tag{22}$$

where $-D(\mathbf{x})$ is the compressibility of the vector field $\mathbf{X}(\mathbf{x})$. If $\mathbf{X}$ is incompressible then $D \equiv 0$ and $\partial\rho/\partial t$ reduces to $-[\rho, H]$, the familiar Liouville equation. By imposing the Dirac constraints $\phi_\rho = 0$ it is possible to show that on the Dirac manifold $\mathcal{M}$ the flow driving ρ has a generalized canonical structure characterized by the generator $U_0(\mathbf{x})$ and the skew-symmetric tensor Γ^{jk}, which defines a symplectic metric on $\mathcal{M}$. This fully links the temporal development of ρ to the symmetries and generators of the vector field. For many classes of nonlinearity in $\mathbf{X}(\mathbf{x})$, $\mathcal{M}$ is not a static surface defining a fixed shape in phase space. It is time-dependent with a morphogenetic pattern all its own (still little understood), its shape changes playing out the effects of the primary constraints, which now have their own dynamics.

5 The Signs of Synthesis

For most of biology and the human sciences the structure of the interpretors $A_m(\mathbf{x})$ is anything but obvious. Indeed, perhaps the biggest tease in diagram $(D5)$ is the interpretation operator $\mathcal{O}$, which stands for a procedure that builds interpretors *ab initio*, in other words, which actually reduces to a formalized symbolic order the human act of creative interpretation. No such procedure yet exists. If, to continue our example from section 4, the vector field of the primary discipline is highly dissipative, the pattern of attractor basins and repellor regions associated with its bifurcation diagram does contain important information about the system's capacity for emergent long-range order (e.g. Haken 1977). If the field is conservative, with zero dissipation everywhere, then analogies to the familiar Hamiltonian statistical mechanics of physical systems can be used to identify potentially useful interpretors. But these are just hints. A strong theory of $\mathcal{O}$ is, in a central sense, the Grail of Hopes here, since the interpretors are the keys to aligning primary with secondary disciplines. For now, metaphor, intuition, natural history, and a lot of symmetry breaking (Anderson 1972) are our guides to the interpretors for each discipline.

The second big tease in $(D5)$ is the streamlined look it gives to each layer or level of organization. The notation, $\mathcal{L}_P$, suggests that the full story about the dynamics of the primary variables $\mathbf{x}$ can be read off in terms of the $\mathbf{x}$-variables themselves. One does not have to "leave the level" to understand what is happening in the primary discipline. This certainly is so in the pristine "textbook" hierarchies of classical conception: Like a set of nested Chinese puzzle boxes, each level of organization, with its assigned theoretical description, supports those above it and can, if desired, sustain the interpretive acts that produce the $A_m(\mathbf{x})$ via $\mathcal{O}$ and thus provide for the reduction of secondary to primary discipline.

For many systems of the kind studied in biology and the human sciences, the levels of organization do not come apart quite as easily as the textbook view would have us believe. In these cases, models posed in the primary discipline are found to contain concepts and propositions of the secondary discipline. The tidy pattern of supervening levels dissolves into a formalist's haze as the levels cross and twist together. In the simple vector field theories of the previous section, for example, the primary discipline is no longer

$$\dot{\mathbf{x}} = \mathbf{X}(\mathbf{x}),$$

but something more like

$$\dot{\mathbf{x}} = \mathbf{X}(\mathbf{x},\ \langle \mathbf{A} \rangle).$$

Events in the secondary discipline feed back to influence the primary. Cause-effect relations among disciplines now run both up and down the hierarchy. The commutative diagram ($D5$) would have to be rewritten as something like

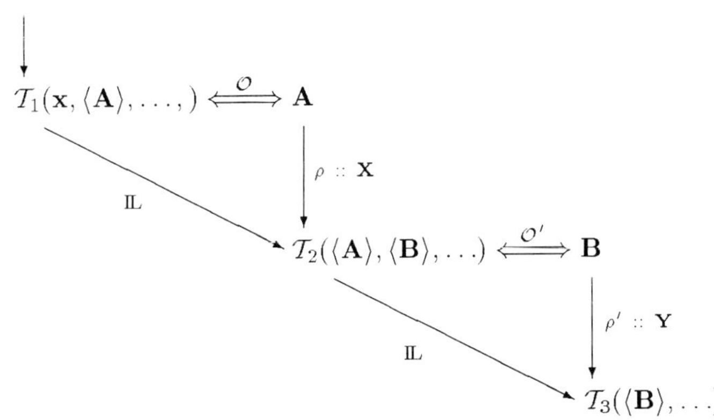

($D6$)

in order to express the change, where the symbol $\overset{\mathcal{O}}{\Longleftrightarrow}$ is meant to imply that the the relation between two levels is causally and interpretively bidirectional. The Chinese boxes form a Klein bottle! Warren McCulloch (1965), thinking about how the brain's neural circuitry might work, called such tangled-up hierarchies *heterarchies*. Douglas Hofstadter (1979) has treated in depth their potential importance for understanding radically emergent phenomena like human creativity and consciousness. I consider them basic to understanding how

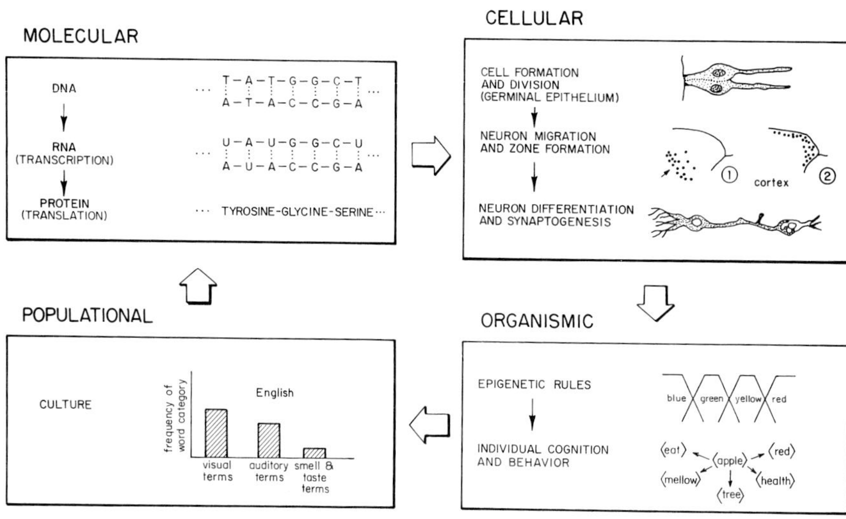

Figure 3: A heterarchical model of gene-culture coevolution. Genome expression in the brain makes some outcomes of culture learning and cognitive development more likely than others, creating a characteristic gound plan for the human mind, whose created culture is the principal environment of human mental evolution. From *Genes, Mind and Culture: The Coevolutionary Process* by Charles J. Lumsden and Edward O. Wilson. Copyright © 1981 by the President and Fellows of Harvard College. Reprinted by permission of Harvard University Press.

human mind and culture might have evolved from our primate beginnings (Figure 3).

A moment's reflection shows that a system described by assertions like "$\mathbf{X}(\mathbf{x}, \langle \mathbf{A} \rangle)$" can be said to contain, or encode, a sort of compressed model of its world. This model is carried by the $\langle \mathbf{A} \rangle$. At first it may seem odd to say that a physical system can be driven by an encoded meaningful representation, $\langle \mathbf{A} \rangle$, of a world. But such systems actually are very familiar; we usually call them *computational* (Bennett 1982, Pylyshyn 1984, Dufort and Lumsden, this volume). Naturally occuring instances of the utmost importance to biology are diverse: intracellular and transmembrane signaling systems, neural networks for cognition and learning, the dynamics of individual behavior, the relation between person and culture.

Diagram ($D6$) implies a complexity yardstick that is quite natural to measuring emergence and reduction in heterarchies, although it is quite different from the measures tendered to date. One might call it "complexity by depth of representational embedding." In describing a heterarchical tangle there is in principle no reason to stop with

$$\mathbf{X}(\mathbf{x}), \qquad \mathbf{X}(\mathbf{x},\ \langle\mathbf{A}\rangle).$$

In general it will be more like

$$\mathbf{X}(\mathbf{x}),\ \ \mathbf{X}(\mathbf{x},\langle\mathbf{A}\rangle),\ \ \mathbf{X}(\mathbf{x},\langle\mathbf{A}\langle\mathbf{B}\rangle\rangle,\langle\mathbf{B}\rangle),\ \ \mathbf{X}(\mathbf{x},\langle\mathbf{A}\langle\mathbf{B}\langle\mathbf{C}\rangle,\mathbf{C}\rangle\rangle,\langle\mathbf{B},\langle\mathbf{C}\rangle\rangle,\langle\mathbf{C}\rangle),$$

and so on, depending on which supra-primary level is registered in the encoded representation $\langle\bullet\rangle$ at the primary level. Intuitively, the greater the number of such levels represented, or

pushed down into the dynamics of the primary discipline, the greater the complexity of the instantiated world model and the resulting behavior. To take a concrete example, conscious thought and choice use representations embodying beliefs and values about the world as organized on successive levels beyond that of the person: the "family," the "workplace," the "government," and so on. Each noun placed in quotes denotes a collective, emergent entity, socially defined in this case (Lumsden and Wilson 1981, Searle 1995), that are everyday components of our notions for interpreting (and creating!) human action. Human minds and their products are heterarchy-complex in the sense that they can rise across such multiple levels. Counting the $\langle\bullet\rangle$ in the rightward slots of $\mathbf{X}$ would be a first, modest step in measuring heterarchical complexity. Going further is an exciting task for the future.

6 The Limits of Expression

Theories of holism that frame the world as hierarchies of interacting parts may at first appear apt, but a world of discrete parts is full of the holes left between the parts. What could be slipping through those semantic gaps? It has been argued that an effective theory of holism will use the notion of continuity, and build its hypotheses around the pivotal metaphor of the *field* (e.g. Hayles 1984, 1991; see also the Introduction to this volume).

Now, fields and continuity can be all to the good, but getting rid of parts and their interactions *altogether* strikes me as too extreme a move. Holism and emergence do engender interactions in which parts form wholes, at least in living systems. Physics, biology, and sociobiology are all deeply concerned with this "vertical" approach to emergence in complex systems, and we have seen it represented, albeit crudely, in diagrams ($D5$) and ($D6$) and contrasted with the notion of "irreducible emergence" by Trainor in his essay on emergent

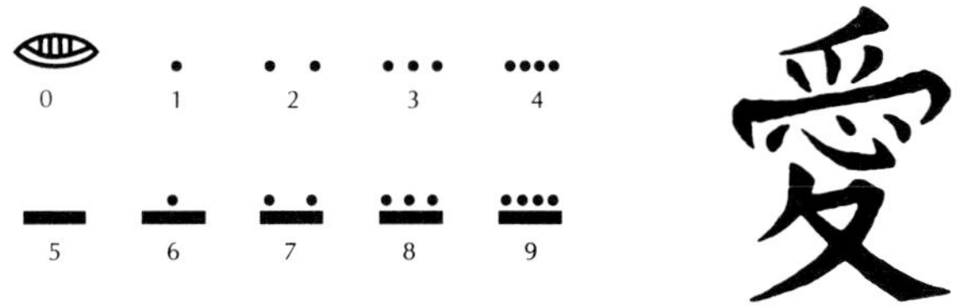

Figure 4: Discrete versus dense symbols. See text.

phenomena. On the other hand, the widening discussion of the field metaphor does remind us that emergence has a "horizontal" aspect as well — one that determines the organization of meaningful relations *on* any level, whether it be cellular or societal.

The use of the field metaphor makes a radical break with the conventional, divide-and-conquer strategy of naive reductionism. Already foiled by the heterarchical tangle, divide-and-conquer meets the assertion that at, some level of organization, meaningful organization arises from variables interacting through continuous extension, and this regardless of the presence (possibly) of discrete bits and pieces on other organizational levels.

While field metaphors are just beginning to influence the humanities, they are already powerful tools in biology and the social sciences. One thinks of the astonishing productivity of field models of biological growth and development, so lavishly illustrated by the essays in this volume. Similarly, social interactions may have emergent properties that are field-like. Personal mobility within the conventionally recognized boundaries of a society, for example, may determine an effectively continuous variation of commodity production, gene-exchange, innovation flow, and the like. Horizontally integrated things are also common results of the human mind's activity. They include writing systems, paintings, poems, concertos, buildings — the stuff of culture. Surprisingly, these entities are all of a kind, elements in what Nelson Goodman (1976; also Elgin 1983) calls dense symbol systems.

I have tried to illustrate some of the intrigue of symbolic density in Figures 4 and 5. Put briefly, dense symbol systems are a breeding ground for holism and emergence because they support continuously shaded gradients of meaning. The case of symbolic *discreteness* is nicely illustrated by the Mayan number glyphs (on the left in Figure 4). Thanks to the use of discrete lines and dots, each glyph belongs to a distinct category of marks for which membership critera are all-or-none. We know what number the scribe has in mind. But, in the hands of master calligraphers, Chinese orthography (Figure 4, right) is a dense symbol system: Every nuance of shape, shading, and relationship in and among the component strokes matters, and so the potential meaning of the mark varies continuously over many dimensions of signification, emotion, and value. Interpretation in such dense symbols systems must deal with the infinite ambiguity implicit in nuance, even in matters of signification.

When symbols are dense, as they are throughout our cultural information systems, opportunities for heterarchical emergence abound. That is because the relations among parts provides a context that can affect the part and its interactions, even on the level of a primary discipline. The orthography supporting our own familiar alphabetic writing system provides a vivid example. On the left side of Figure 5 you see a progression of marks in which "a" shades into "d." The meaning of the mark on each end seems easy to resolve, but what about the two in the middle? On the right, the middle mark is in a context where the long-range order (induced by the word-meaning we cannot help assigning to each triad of marks) influences our experience of the individual mark. Compare the horizontal with the vertical arrangements. English-speakers I've tried this out on usually "see" the middle mark as "d" in the horizontal arrangement, and as "a" in the vertical one. The emergence of the word affects the attributions we assign to its parts, but these attributions emerge in part from our attention to likely meaningful word contexts for the constituent marks.

7 Conclusions

Emergence and holism count among the great intellectual challenges of our time, their expression uniting the sciences with the humanities with the arts. In this essay I have suggested that the study of holism needs pushing beyond traditional issues like part-versus-whole. The following beckon:

- **Vertical emergence**, the pull between the thing-in-itself and its decomposition into parts.
- **Interpretation**, by which are created the ideas that ultimately align primary and secondary disciplines.
- **Horizontal emergence**, the tightening of (nonlinear?) cohesion within a level of organization, so that mechanisms such as bifurcation lead to a dynamics of long-range order in a level worth summarizing by a secondary level.
- **Tangling**, the bifurcation of a lower-level dynamics into computation, and with that the necessary dependence of the dynamics on representational models of the world.

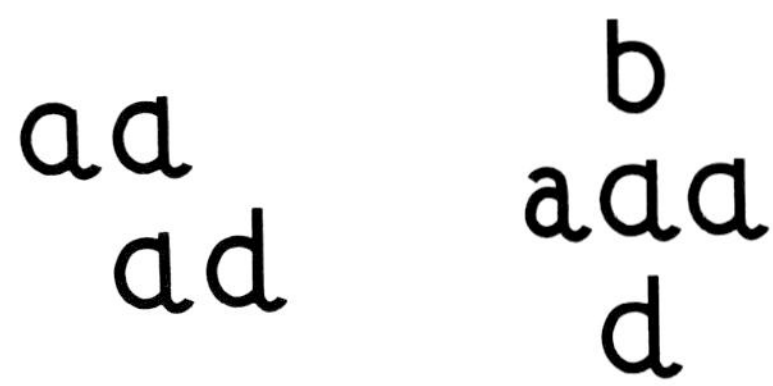

Figure 5: Heterarchical effects among dense symbols. *Left.* Which marks are "a" and which "d"? *Right.* Look at the middle letter in the display. It is usually "seen" as "d" in the horizontal reading and "a" in the vertical reading. The whole determines the part which shapes the whole. From Nelson Goodman, *Languages of Art*, Hackett Publishing Company, Inc., 1976, with permission.

- **Meaning and mechanism**, the irreducible ingredients of emergence between the sciences, humanities, and arts.

In the face of evolved complexity, we cannot hope to exhaust the meaning we find in nature, persons, and culture. This means, surely, that the study of holism and emergence, while partially tractable, is no finite task. With each hint of interpretation, we add to the world's diversity while affirming its basic unity. This is not to say, however, that the happy looting of physics for cool models will stop in the face of taboos against physics envy. Indeed quite the opposite: chaos and its many edges (Langton 1991; Peitgen, Jürgens and Saupe 1992; Kauffman 1993) have arrived (Figure 6). *San Carlo* will never look the same again.

Acknowledgments

My thanks to Lynn Trainor, Mel Silverman, and Ed Wilson for discussions over the years. The Harry Frank Guggenheim Foundation, the Natural Science and Engineering Research Council of Canada, the Medical Research Council of Canada and the Bickell Foundation have generously supported aspects of the work reported.

References

Anderson, P.W. (1972) More is different: broken symmetry and the nature of the hierarchical structure of science. *Science* **177**, 393–396.

Argyros, A.J. (1991) *A Blessed Rage for Order: Deconstruction, Evolution, and Chaos.* Ann Arbor, MI: University of Michigan

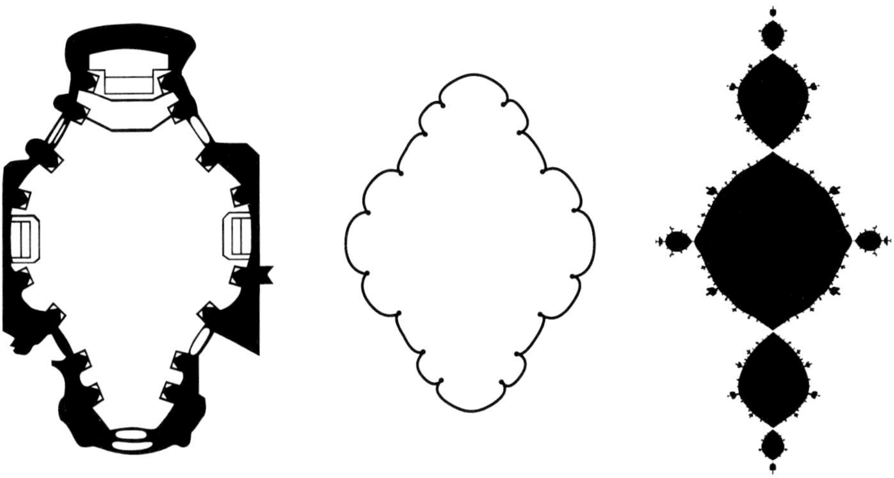

Figure 6: The *San Carlo* debate made modern. *Left, middle, right*: the church's ground plan, an outline of the gently rounded chapels and bays making up its periphery, and a fractal Julia set for $z = z^2 + c$. The study of holism, emergence, and complexity will continue to strengthen the ties among all areas of human enquiry.

Press.

Bak, P. and Chen, K. (1991, January) Self-organized criticality. *Scientific American* **264**, 46–53.

Bak, P. and Paczuski, M. (1995) Complexity, contingency, and criticality. *Proceedings of the U.S. National Academy of Sciences* **92**, 6689–6696.

Bennett, C.H. (1982) The thermodynamics of computation — a review. *International Journal of Theoretical Physics* **21**, 905–921.

Buchwald, J.Z., ed. (1995) *Scientific Practice: Theories and Stories of Doing Physics.* Chicago: University of Chicago Press.

Bunge, M. (1967) *Foundations of Physics.* New York: Springer-Verlag.

Cairns-Smith, A.G. (1996) *Evolving the Mind: On the Nature of Matter and the Origin of Consciousness.* New York: Cambridge University Press.

Dirac, P.A.M. (1950) Generalized Hamiltonian dynamics. *Canadian Journal of Mathematics* **2**, 129–148.

Dirac, P.A.M. (1964) *Lectures on Quantum Mechanics.* Belfer Graduate School of Science Monograph No. 2. New York: Yeshiva University.

Eldredge, N. (1989) *Time Frames: The Evolution of Punctuated Equilibria.* Princeton, NJ: Princeton University Press.

Elgin, C.Z. (1983) *With Reference to Reference.* Indianapolis, IL: Hackett Publishing Company.

Findlay, C.S. and Lumsden, C.J. (1988) *The Creative Mind.* London: Academic Press.

Goodman, N. (1976) *Languages of Art. An Approach to a Theory of Symbols.* Indianapolis, IL: Hackett Publishing Company.

Gould, S.J. and Eldredge, N. (1977) Punctuated equilibria: tempo and mode in evolution reconsidered. *Paleobiology* **3**, 115–151.

Govaerts, J. (1991) *Hamiltonian Quantization and Constrained Dynamics.* Leuven Notes in Mathematical and Theoretical Physics. Volume 4, Series B: Theoretical Particle Physics. Leuven, Belgium: Leuven University Press.

Haken, H. (1977) *Synergetics: An Introduction. Nonequilibrium Phase Transitions and Self-Organization in Physics, Chemistry, and Biology.* New York: Springer-Verlag.

Hayles, N.K. (1984) *The Cosmic Web: Scientific Field Theories and Literary Strategies in the Twentieth Century.* Ithaca, NY: Cornell University Press.

Hayles, N.K., ed. (1991) *Chaos and Order: Complex Dynamics in Literature and Science.* Chicago: University of Chicago Press.

Hofstadter, D.R. (1979) *Gödel, Escher, Bach: an Eternal Golden Braid.* New York: Basic Books.

Hölldobler, B. and Wilson, E.O. (1994) *Journey to the Ants.* Cambridge, MA: Belknap Press of Harvard University Press.

Kauffman, S.A. (1993) *The Origins of Order: Self-Organization and Selection in Evolution.* New York: Oxford University Press.

Koslowski, B. (1996) *Theory and Evidence: The Development of Scientific Reasoning.* Cambridge, MA: MIT Press.

Langley, P., Simon, H.A., Bradshaw, G.L. and Zytkow, J.M. (1987) *Scientific Discovery: Computational Explorations of the Creative Process.* Cambridge, MA: MIT Press.

Langton, C. (1991) *Computation at the Edge of Chaos: Phase-Transitions and Emergent Computation.* PhD Thesis, Department of Computer and Communication Sciences, The University of Michigan, Ann Arbor, MI.

Lumsden, C.J. (1996) Sociobiological futures for an on-line species. *Telepolis*, http://cellmath.med.utoronto.ca/TELEPOLIS

Lumsden, C.J. and Trainor, L.E.H. (1985) Hamiltonian statistical mechanics and biological order: problems and progress. In P.L. Antonelli, ed., *Mathematical Essays on Growth and the Emergence of Form.* Edmonton: University of Alberta Press, pp. 3–42.

Lumsden, C.J. and Wilson, E.O. (1981) *Genes, Mind, and Culture: The Coevolutionary Process.* Cambridge, MA: Harvard University Press.

Lumsden, C.J. and Wilson, E.O. (1983) *Promethean Fire: Reflections on the Origin of Mind.* Cambridge, MA: Harvard Universtiy Press.

McCulloch, W.S. (1965) *Embodiments of Mind.* Cambridge, MA: MIT Press.

Nagel, E. (1961) *The Structure of Science: Problems in the Logic of Scientific Explanation.* New York: Harcourt, Brace and World.

Peitgen, H.-O., Jürgens, H. and Saupe, D. (1992) *Chaos and Fractals: New Frontiers of Science.* New York: Springer-Verlag.

Penrose, R. (1994) *Shadows of the Mind. A Search for the Missing Science of Consciousness.* New York: Oxford University Press.

Prothero, D.R. (1992) Punctuated equilibrium at twenty: a paleontological perspective. *Skeptic* **1**, 38–47. See also http://www.skeptic.com/01.3.prothero-punc-eq.html.

Pylyshyn, Z.W. (1984) *Computation and Cognition. Toward a Foundation for Cognitive Science.* Cambridge, MA: MIT Press

Rosen, R. (1985) *Theoretical Biology and Complexity: Three Essays on the Natural Philosophy of Complex Systems.* New York: Academic Press.

Rosen, R. (1991) *Life Itself: A Comprehensive Inquiry Into the Nature, Origin, and Fabrication of Life.* New York: Columbia University Press.

Schaffner, K.F. (1993) *Discovery and Explanation in Biology and Medicine.* Chicago: University of Chicago Press.

Searle, J.R. (1995) *The Construction of Social Reality.* New York: The Free Press.

Simon, H.A. (1981) *The Sciences of the Artificial.* Cambridge, MA: MIT Press.

Sklar, L. (1995) *Physics and Chance: Philosophical Issues in the Foundations of Statistical Mechanics.* New York: Cambridge University Press.

Steinberg, L. (1977) *Borromini's San Carlo alle Quattro Fontane: A Study in Multiple Form and Architectural Symbolism.* New York: Garland Publishing.

Sternberg, R.J. and Davidson, J.E., eds. (1995) *The Nature of Insight.* Cambridge, MA: MIT Press.

Woodger, J.H. (1937) *The Axiomatic Method in Biology.* Cambridge: Cambridge University Press.

Complexity: A Pluralistic Perspective

Wendy A.M. Brandts

1 Introduction

Complexity is manifested in the living world as diversity and organization, particularly self-organization. No two biological organisms are exactly alike — vast diversity is displayed between members of a population as well as between the parts of a single organism. This is true for both the structures and the behaviors of organisms. The organization characteristic of complexity in biological organisms can exist to an amazing degree on many levels or scales of biological form. Patterns and symmetries abound in space and in time, changing on scales ranging from milliseconds (e.g. nerve impulses) to the millenia of evolutionary times. Conversely, the very complexity of biological organization demands that different levels of explanation be employed.

Complexity is so pervasive in the biological world that it has come to be considered as a characteristic of biological systems. Biological systems in turn are regarded as ultimate realizations of complex systems. However, while concepts of complexity have been developed in fields such as mathematics and computer science, and more recently physics, the usefulness of complexity as a formalized concept to describe biological systems has not yet been thoroughly demonstrated. To forge such a connection between fields, such as the application of notions of complexity in physics to the biological realm, presents a great challenge. In addition, this bridging foreshadows a larger meeting of the physical and biological points of view and the dissolution of the boundaries between physics and biology.

Recent enthusiasm in the study of complex systems (Horgan 1995) has generated hopes of attaining a unified theory of complex systems and thus a characterization of complexity itself. The aim is to secure general principles which, for example, transcend the boundaries between physics, biology, and even economics, and thus apply to complex systems as diverse as alloys, embryos and economies. For example, Bak and his associates (1988) have developed a theory of self-organized criticality which is suggestive of both extinction of species and stock market crashes. Its central metaphor is a sandpile. When grains are continuously added onto the top, the sand self-organizes itself into a critical state which can be characterized quantitatively via a power law describing size and frequency of avalanches. In this way, systems are supposed to drive themselves to the edge of a critical state where complexity and adaptability is maximal. Another approach (Nicolis and Prigogine 1989) arises from the assumption that complexity in the natural world is rooted in nonequilibrium nonlinear behavior. In this case, a unified formulation of complexity aims to reconcile deterministic with probabilistic laws, and conservative with dissipative systems, by embedding dynamics in a broader formalism that includes stochasticity and irreversibility.

But just the task of *defining* complexity has proved to be extremely difficult. The dynamical view (Langton 1990, Kauffman 1993, and others) defines complex systems as residing at the "edge of chaos," i.e. between highly ordered stable phenomena, and randomness. Although measures of "order" and "randomness" abound, this crucial intermediate-zone of complexity appears to defy pinning down. Many definitions of complexity that involve concepts of information, order, entropy, and randomness have been proposed; but all are beset by difficulties.

Thus, in spite of the focus and enthusiasm over the last decades in the field of complexity studies, many basic problems remain unsolved. In this essay we will address some issues pertaining to complexity as a formalized concept and as a descriptor of biological systems. We will argue for a pluralistic approach where no single level of analysis or viewpoint has priority.

With the aim of employing complexity as a formal notion or physical descriptor, we outline (section 2) the need for a measure or measures of complexity. To expand on the nature of complexity in biological systems, we discuss it's origins, in particular, the processes of evolution and of dynamical behavior (section 3) as well as some often encountered characteristics of complexity (section 4). In section 5 the vast task of reducing complexity for modeling purposes is addressed. We present two methods: a bottom-up approach of averaging, and a top-down approach of employing fields and symmetries.

Drawing on this background, in section 6 the need for a pluralistic approach to complexity is identified. Our argument proceeds via several examples of the many different viewpoints that can be taken of a biological system, which in turn leads to a search for natural descriptions which would endow some "preferred" description or evaluation of complexity. However, we then emphasize that in even the "simplest" and most clearly defined systems, formal systems, fundamental problems are encountered when attempting to devise "unique" or "unified" complexity measures. Finally, our pluralistic approach touches on the issue of reductionism, which we highlight via the problem of differentiating dynamics from computation.

2 Measures of Complexity

Complexity may be a real quality or characteristic of biological systems. The term has long been employed descriptively, in accord with its colloquial sense, where it denotes, for example, more genes or a greater number of connections between cells. But in order to incorporate the concept of complexity in quantitative models of physical and biological systems we seek a more precise definition of complexity. Ultimately, we want to devise a measure, or measures, of complexity.

Some measures have already been used in the context of particular modeling situations. We give several examples here. In a biological context, May has used the number of species or connections in the interaction matrix (May, 1973) of an ecosystem model, and a dynamic quantity (May and Oster, 1976) as measures of complexity; Banerjee and co-workers (1990) applied a formal measure of information to seed development and branch growth data. In a more general dynamical context, Wolfram (1984a,b) developed an algorithmic complexity measure in terms of the minimal graph of the regular language defined by a cellular automata; Kaspar and Schuster (1987) have used an algorithmic complexity measure, a measure that counts the number of production steps to create a string of digits under a restricted set of operations, to characterize spatiotemporal patterns in nonlinear systems. And in a more formal context, Huberman and Hogg (1986) have based a physical measure of complexity on diversity. It applies to discrete hierarchical structures and reaches a maximum for systems which are intermediate between perfect order and complete disorder.

An apparently insurmountable shortcoming of the measures of complexity based on information and order is that they fail to take into account the difference between syntax and semantics. For example, algorithmic complexity (Chaitin 1982) measures the computer program required to generate a function, but it cannot reflect the "meaning" of the function. Thus application of this measure to say, strings of DNA, cannot take into account coding and non-coding sequences. Moreover, algorithmic complexity is problematic even in formal regimes, suffering from deficiencies such as the "Halting Problem" which implies that even if a program of the least number of lines were found, one would not know whether it would ever stop running on the computer to produce an output (Brandts 1986, and the following essay by Dufort and Lumsden, this volume).

If finding even one adequate complexity measure is difficult, the issue of a universal measure is open. A universal, or at least widely applicable, measure of complexity is desirable as an aid in comparing different systems or the properties of a given system at different times. Thus aspects of growth, self-organization, development, learning, genetic evolution and so on, processes which seem to exhibit an increase in complexity, could be potentially related. However, a universal measure of complexity may not be attainable; there seem to be many kinds of complexity, and unless a law analogous to "conservation of energy" is discovered, complexity may not be transformable among them.

3 Origins of Complexity

Prebiotic and early biological systems were simpler than the organisms we know now. Complexity has apparently increased over those eons of time. What processes have lead to this increase?

Two biological processes through which progressively higher levels of organization of matter are attained can been identified: development and evolution. These processes are interconnected because the possible forms that arise in development — the patterns or configurations that form readily or stably under constraint of physical law — will limit the path of evolution. Conversely, developmental sequences are restricted by the history of evolution, by the actual road traveled and "remembered" in the genetics. The relative importance of such contingent factors and selection, as compared against the forms determined by structural laws of physics and chemistry, is a matter of continuing debate among the "neodarwinists" and the "structuralists" (see Frankel, this volume).

From the point of view of complex systems, it is of great interest to determine whether there is some common source or dynamic which directs both these processes of increasing complexity. Evolution and development may be part of a universal dynamic which leads to successively more complex states. There might, for instance, be a law that indicates an increase in complexity, or as in the case of self-organization where attractors limit the system to a small part of its phase space, theorems that limit motion in some high-dimensional space. On the other hand, there may instead be many conceptually and practically distinct "routes to complexity," as for example has been found with the routes to chaos (transition to a strange attractor, period doubling, intermittency, and so on).

One recent formulation which aims to treat evolution within a larger dynamical framework introduces the idea of a "fitness landscape" for evolution (Kauffman 1993). A new form of statistical mechanics is hypothesized, wherein the units are systems rather than particles, and the phase space they evolve or flow through is a fitness landscape. Mutations in a model organism change its fitness in an erratic fashion in accordance with the "ruggedness" of its landscape. Selection leads biological systems into the complex regime poised between order and chaos because complex systems are adaptive, optimizing their evolvability and the complexity of tasks they can perform.

In a network realization of the model, the number of elements or number of connections is used as a measure of complexity (of structure). It is found that as complexity is increased, fitness peaks become poorer. Thus the model suggests that as systems become more complex, evolution via natural selection becomes less important and the intrinsic dynamical features of a system become more so. That is, the complexity in physical systems may not differ fundamentally from complexity in biological systems.

Exploring the dynamics of complex systems, their formation or evolution and their resultant abilities or behaviors, may lead to the discovery of wide-ranging complexity laws that unify systems in a new way, such as by transcending the boundaries between physics and biology.

Thus the structure or architecture of complexity itself can ultimately be examined, and the influence of additional constraints (such as stability criteria, adaptability, evolvability) addressed.

4 Characteristics of Complexity

Complexity can reside in the structure of a system (e.g. having many diverse parts with varying interactions or an intricate architecture) or in its dynamics (e.g. alternating between many modes of behavior such as attaining stable states, oscillating in limit cycles or wandering through chaotic regimes). Often complexity in structure and in dynamics go together. For example, a large highly connected network tends to exhibit richer behavior than a small sparse network. But this is not always the case; as discussed below, complex behavior can arise from the dynamics of structurally simple systems. Complexity and complex systems can thus be characterized by structural and/or dynamical aspects. For example, Poon and Grebogi (1995) list traits of complex systems that contain both: (1) many parts interconnected in a complicated manner, (2) ordered and random behaviors, (3) a hierarchy of structures.

Dynamical complexity is typically associated with (1) nonlinearity, (2) instability, and (3) lack of smoothness. Nonlinearity makes possible the rich variety of behaviors characteristic of dynamical systems. It also entails increased difficulty in analysis and prediction. As biological systems are highly nonlinear — a feature which has made them largely inaccessible to quantitative analysis prior to the advent of high power computing — they usually exhibit dynamical complexity.

In addition to nonlinearity, stability plays a crucial role in dynamical systems and their apparent complexity since stability, and the consequent predictability, intuitively connote "simple" behavior. Stability is generally defined by the effect that a perturbation has on the evolution of a system. If a system is stable, an infinitesimal change in its initial state will alter only infinitesimally its state at some future time. On the other hand, when an infinitesimal variation in the initial state results in a finite (possibly large) difference in the future state, the system is said to be unstable. It is this great sensitivity to initial conditions that makes the actual evolution of unstable systems difficult to predict. Thus while there may be a simple causal law governing the behavior of some system, unless there is stability associated with it (e.g. some of the solutions or orbits are stable), there will be no deterministic predictability and the evolution of the system will appear complex and even erratic. For instance, the elastic collisions in a pinball machine are chaotic although they are presumably ruled by Newton's laws.

On the other hand, in contrast to this association of stability with "simple" in the dynamical systems point of view, from the point of view of biological evolution, stability (adaptability, evolvability) may be associated with complexity (e.g. Kauffman's view (1993)). The issue is whether the states that are manifested are those which have persisted because of their great stability. Then, since evolved (biological) states tend to be complex ones, complexity would now appear to contribute in some way to the potential stability of a state, thus joining complexity with stability (rather than instability). Similar questions concerning the relation between complexity and stability have also been raised for ecosystems, though these are often described by linear models (Pimm 1984). The one thing that emerges clearly from these questions is that multiple interpretations of complexity need to be entertained.

Finally, we discuss lack of smoothness as another characteristic or source of complexity in dynamical systems. Non-smoothness as a source of complexity may have to do with the kind of mathematics used for dynamical systems and thus be somewhat an artifact of description (a point we will return to again — complexity as a description-dependent phenomenon). In following the Newtonian paradigm by using the language of calculus to formulate physical laws as equations, smoothness has often become subtly embedded in physical laws (Rosen 1978). As application of calculus requires functions to depend smoothly on their variables, only phenomena which exhibit smoothness can be modeled easily and naturally — and hence appear "simple" — within this formalism.

There are, however, a great number of phenomena which have a discontinuous or a multilevel structure, so that examination on successively finer scales reveals more and more detailed structure, not a smoothing out. Hence, they do not lend themselves so obviously to description via the calculus. Examples of this phenomenon are: (1) density fluctuations which exist at all scales in a fluid at the critical point, due to molecular clusters of all sizes, (2) geographical topography, (3) hierarchical organisation in biological systems.

Some advances have been made in developing formalisms not underlain by a "smoothness prejudice." Mandelbrot's fractals, self-similar or scale invariant structures having fractional dimensions, are the fundamental structures in his geometry (Mandelbrot 1982). Another development is Thom's catastrophe theory (Thom 1975) which is essentially a classification of singularities (important in dynamical systems because singularities can give rise to instabilities).

In discussing dynamical systems above we have neglected to consider that in reality, dynamics or evolution happens in a noisy world. Noise alters phase space exploration, especially around bifurcation points. Thus a more adequate treatment of complexity from the dynamical point of view would also incorporate fundamental issues of probability and randomness.

5 Reducing Complexity

The task of dealing with the complexity of biological systems is daunting. The real complexity present in biological world must somehow be reduced in the modeling process. While many approaches may be *a priori* valid, each will also be limited; thus we anticipate that a pluralistic and multidisciplinary treatment of complexity will be necessary. In this plurality of approaches, we would not expect to reduce all theories to one universal formulation.

In attempting to account for the (emergent) properties of biosystems by physical principles, we are immediately faced with the difficulty of quantitative description. How can we precisely describe such complex systems? Any system large enough to be identified as biological has too many physical/chemical variables for us to be able to construct a physical model in a straightforward manner (e.g. attempt a Newtonian point particle approximation or a quantum mechanical wavefunction superposition for $\sim 10^{23}$ elements!). In addition, the nonlinear relations between these variables make analysis prohibitive. Even if it were feasible, it seems unlikely that such a description could be meaningfully interpreted.

The reductionist approach would be to divide biological systems into smaller subsystems (such as chemical networks) and study them piecewise, and then reconnect all the pieces. When this method is foiled by inherent connectivities, the "internal structures" might be approximated and simplified, and more of the connections (the "whole") retained in modeling.

Means of reducing the number of variables can also be tried. This can potentially be achieved by several different approaches. Our terminology here derives from a hierarchical view of systems, where "level" refers to the resolution or degree of coarse-graining in a description, such as spatial resolution:

1. "bottom-up" or averaging techniques,
2. "top-down" or macroscopic theory, in which only certain characteristic gross features, such as symmetries, are included from the outset.

In section 6 we will also suggest that a pluralistic approach to modeling, via a multiplicity of descriptions in which each description naturally emphasises its own (limited) number of variables, should be considered as a legitimate method of modeling, without requiring descriptions and their language to translate into each other in a simple reductionistic fashion.

5.1 Averaging

A bottom-up approach eliminates variables by combining them via some sort of averaging procedure. In this way, many lower level variables may form the parameters or variables of a higher level description. For example, the velocity of gas particles might be averaged to assign a single temperature measure to the whole gas. In statistical mechanical systems such as this gas, one is dealing with a large number of similar or even identical subsystems. However, this approach is difficult to implement in biological modeling because (among other things) we do not know what are or how to make the relevant averages for biological systems with nonsimilar components.

This averaging technique has proved fruitful in cases where the complexity of the system is disorganized, for example, in formulating a statistical mechanical foundation of thermodynamics where the (disorganized) macrostate does not depend on the details of the microstates, such as which of the many possible microstates the system might have passed through in a small time interval. For any macroscopic measurement, the physical system will have sampled an enormous number of the microstates so that what are empirically accessible and of interest are average properties. In contrast, in cases where the characteristic complexity of the system is organized, such as in biological systems, the system (micro)variables cannot be meaningfully averaged over because the macroscopic state does depend on the details of the organization at all levels. For example, if we create a metabolic system of "average proteins" by using the average locations and electrical properties of the active sites (the biologically important microscopic system variables) of a large collection of real proteins, we do not get an equivalent macrostate, a sort of "averaged-out metabolism." We get no metabolism at all because the biological functioning of metabolic processes (the macrostate) relies on highly stereospecific recognition in enzymes and their space and time coordinated activity (the details of the microstates).

In spite of the difficulties of averaging techniques applied to systems of non-identical components and organized complexity, some schemes have been proposed (e.g. statistical mechanics of complex systems (Kauffman 1993), where each "particle" is a system; see also Lumsden, this volume).

5.2 Fields and Symmetries

An alternative route, the top-down approach, is to develop higher level descriptions or models not explicitly bound to the concepts and observables of the lower levels. These higher level descriptions might be *a posteriori* related to their lower level (hardware, physical, chemical, ...) embodiments (the reductionist's ambition). Upper level variables may turn out to be the previously sought appropriate averages of lower level variables, or the lower levels might act as constraints on the upper ones via boundary conditions.

Several examples of this approach are illustrated in this volume:

1. *Developmental fields in biology* — systematizes much experimental data without relying on any detailed molecular mechanisms (Brandts and Totafurno).

2. *Information Theory and DNA* — statistical analysis of nucleotide sequences without requiring intractable quantum chemistry. (Luo; and Trainor, Rowe and Nelson).

3. *Artificial Intelligence and computer architecture analogies* — high level properties of computational systems without specifying the lower level hardware. (Busch and Trainor, Sutton).

6 A Pluralistic Approach to Complexity

In these final sections we will draw together our support for a pluralistic approach to complexity and complex systems, an approach in which a multiplicity of descriptions must be considered, often at different levels or on different space and time scales, and some kind of synthesis achieved. By reviewing some of the many different approaches taken in the study of complexity — each starting from its own "viewpoint," employing its own tools and techniques of analysis, and leading to relevant, though necessarily limited, insights — we will demonstrate that no single level of analysis or viewpoint can be given priority and thus that a universal definition of complexity, based on some universal analysis or model, cannot be supported.

First we review several classifications of complexity. The existence of many descriptions of complexity prompts us to ask whether any single description can be selected or preferred, a description which is somehow endowed with unique significance. To this end we examine natural descriptions and formal systems. However, we then discover that even for these simplest and most precisely defined systems (formal systems), a robust definition of complexity has not been attained. Finally, we discuss an example in which two viewpoints of a system, the dynamics and the computation interpretations, are so integral and intertwined that we are forced to abandon a reductionist approach and accept a pluralistic viewpoint.

6.1 Viewpoints, Descriptions, and the Observer

The profusion of viewpoints (ways of looking at a system) from which a complex system can generally be examined leads to a multiplicity of possible descriptions:

> complex systems ... admit many descriptions, each of which is only partially true ... each way of looking at a complex system requires its own description, its own mode of analysis, its own breaking down of the system into parts.
>
> — Rosen (1979; ellipses added)

In addition to the multitude of viewpoints on which a description or model of a complex system can be based, any given model can have numerous different representations (Figure 1). The choice of appropriate viewpoint depends on which features of a system are to be explained (an observer-dependent issue), such as the energy flows, the cybernetic connections, or the biochemical constitution. Which representation to employ is largely dictated by pragmatic concerns, like mathematical manipulability or relation to the expression of the empirical results.

Various classifications of viewpoints have been proposed. We mention the *thermodynamic* and *kinetic* interpretations of living organisms (Peacocke 1983) as our first example. Thermodynamics (TD), in the formulations of Network TD, is concerned with energy flow in living systems and with the generation and evolution of complexity in open systems far from equilibrium (see also Nicolis and Prigogine (1989)). Kinetic interpretations, i.e. systems of reacting and diffusing molecules and oscillating networks of biochemical reactions, have indicated possible mechanisms for symmetry breaking — which

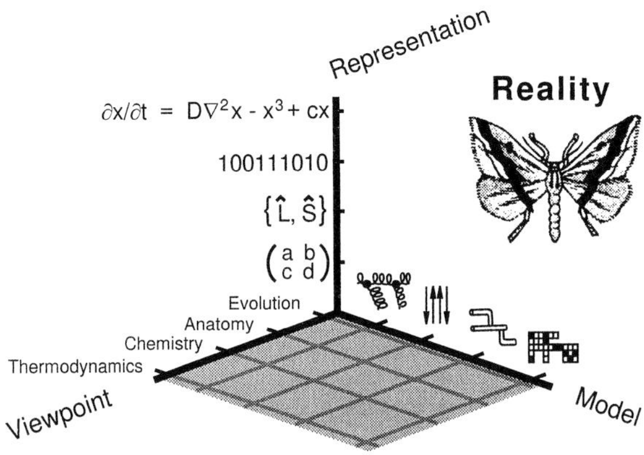

Figure 1: A whimsical picture depicting the richness of describing complex systems. A complex system can be seen from a multitude of *viewpoints*, each viewpoint forming the basis for numerous potential *models*, and any given model having various different *representations* or mathematical expressions.

leads to pattern formation — and have shown how reaction processes interlock to support behavior characteristic of the living state (self-reproduction, energy utilization, etc.).

These two viewpoints, thermodynamic and kinetic, are seen as complementary. Since the kinetic details of many biochemical processes are not known, much of theoretical biology is based on general mathematical (kinetic) models, which are designed to be compatible with the laws of macroscopic physics by imposing TD constraints. TD does not prescribe the mechanisms; kinetics does not reflect uniquely the physics. Both approaches are needed to attain a full physico-chemical description of biological systems.

In a more abstract approach, a system's possible decompositions are used to classify it according to two types of complexity: *descriptive* and *interactional* (Wimsatt 1976). In this scheme, if various decompositions of a system (e.g. spatial decomposition of a rock according to (1) chemical composition, (2) thermal conductivity, (3) electrical conductivity, (4) density) have coincident boundaries, then the system is said to be "descriptively simple." Alternatively, various mapping relations can hold between decompositions with non-coincident boundaries and such systems are termed "descriptively complex." The second type of complexity in this classification, interactional complexity, involves the causal interactions of the subsystems of a particular decomposition. If the causal interactions within a subsystem are stronger than those between different subsystems then the system is "interactionally simple."

Several formal works have also distinguished between two types of complexity. In an information theory treatment of complexity, (1) *static* versus (2) *dynamic* measures of complexity are proposed (De Luca 1975). In an analysis of the complexity of sequences of symbols (bit strings) two approaches are considered (Chaitin 1975). We label them the (1) *product* and the (2) *process* viewpoints. (Note that this corresponds to a noun/thing — verb/action duality in language; see below.) The product viewpoint only looks at the syntactic arrangement of symbols. The process viewpoint considers how the sequence was generated. This duality of approaches corresponds to a perspective of biological science summarized by two groups of questions posed by living organisms (Peacocke 1983, p. 272):

1. Product viewpoint: What is it now? How can we most accurately describe it as we see it now?

2. Process viewpoint: How did it come to be the way it is? How did it evolve?

While these product and process viewpoints may be treated as independent in a formal analysis of bit strings, they are intimately linked in biological situations.

In a similar way, two types of descriptions, "state" and "process," of complex systems can be identified:

1. *State* descriptions characterize the world as sensed (compare the product viewpoint, above); they provide the criteria for identifying objects, often by modelling the objects themselves (e.g. blueprints, chemical structural formulae).

2. *Process* descriptions characterize the world as acted upon (compare the process viewpoint above); they provide the means for producing or generating objects having the desired characteristics (e.g. recipes).

A biologically significant aspect of those conceptual distinctions involves the survival of adaptive organisms, which is based on their ability to correlate a process description with a state description, i.e. find a means to a desired end. "Thus, problem solving requires continual translation between the state and process descriptions of the same complex reality" (Simon 1962, p. 479). This analysis can also be applied to biological theory itself: human theorists are biological organisms and biological theory represents one of the problems we seek to solve!

The above discussion has outlined several dual viewpoints that can be taken of complex systems and complexity. The divisions are somewhat arbitrary, i.e. descriptions which inextricably combine two viewpoints, opposed in some classification, could be formulated. The prevalence of *dualities* in these scheme deserves comment. These dualities may be more a reflection of the nature of the human mind and the functioning of the brain than of any intrinsic property of the systems examined; thus the role or nature of the observer enters. Significantly, a tendency for dualities can be seen right in the fundamental structure of human language: the basic classification of phenomena into "things" (nouns) and "actions" (verbs). The possibility that language and mind are inextricably connected with brain structure and that the science of the mind and of life may ultimately play a fundamental role in our understanding of the world has been recognized by many. For example:

> The classical view of science was to regard the world as an "object", to try to describe the physical world as if it were being seen from the outside as an object of analysis to which we do not belong ... To progress further, we must have a better understanding of our position, the point of view from which we start our description of the physical universe ... we must relate knowing to characteristic features of life.
>
> — Prigogine (1980; ellipses added)

> ... biology will develop a point of view which may radically alter the interpretation of physics ... the interpretation of quantum mechanics is very strange ... living systems produce this interpretation. Only by understanding how they do it will this interpretation become clear.
>
> — Pattee (1979; ellipses added)

On one hand, it may seem amazing that constructs of the mind, such as mathematics, should be so applicable to the physical/real world. Yet on the other hand, the brain deals with and has evolved under the constraints and selective pressure of these physical influences. So we might expect some fundamental correspondences between mental constructs and physical laws. We might also expect our understanding itself to follow a sort of feedback procedure where, for example, dualities will be used to make sense of living systems and the new insight gained thereby will in turn explain why the dualities were identified in the first place.

An extension of these observations suggests that the internal descriptions which organisms presumably have (e.g. genetic information and cognitive images of their environment) can be expected to be composed of various types of descriptions. For example, Pattee (1961) discusses process descriptions governing molecular structure, and Simon (1962) suggests how RNA can be seen as a template/state descriptor for protein, while protein is a process description, via enzyme action, for metabolism.

6.2 Natural Descriptions

In view of the dependence that system descriptors such as complexity have on the (arbitrarily chosen) viewpoints, models and representations of a system, we are induced to seek preferred fundamental or *natural* descriptions and variables. A natural description for biosystems might be indicated by the terms of their own internal description, which may differ greatly from our arbitrary external ones. The idea of a unique preferred "natural description" is illustrated in the following two examples.

6.2.1 A biological system: Drosophila development

Traditionally, the description of living organisms has been based on morphology (often anthropomorphically) rather than embryology. (This is an example of a product viewpoint instead of a process viewpoint.) This approach leads to a very subjective picture, i.e. an arbitrary description, of an organism:

> But when the morphologist compares one animal with another, point by point or character by character, these are too often the mere outcome of artificial dissection and analysis.
>
> — Thompson (1942)

One aim of developmental genetic studies of the fruit fly *Drosophila* is to objectify this picture by understanding the difference between a fly's internal (self-)description and our external description (Martinez-Arias and Lawrence 1985). Such studies have indicated that *Drosophila* is constructed from compartments which become sequentially restricted through successive boundary definitions during embryonic development. These compartments leave no trace of their borders in the body pattern and hence cannot be identified in a morphological study.

As a particular example, consider the eye. In our external or morphological description, the eye would be regarded as a distinct organ with a unique function. However, throughout development it is merely a part of the anterior compartment of the antennal segment of the head, and a group of cells becomes committed to give rise to the eye only during the very last stage. Thus the fly's internal description probably does not even have a word corresponding to "eye."

This example illustrates the arbitrariness of our (external) descriptions, inasmuch as they do not reflect the same emphasis, or division of a system into parts, that the system itself recognizes.

6.2.2 A formal-logic system: Primitives and the grue–bleen problem

In Goodman's (1972a) calculus for measuring the simplicity of extralogical bases (an "extralogical basis" is all the primitive, undefined terms — aside from the logical terms — which are used to define the other terms in a system), he presents the grue/bleen problem. It is based on these definitions:

1. A thing is *grue* if and only if: a) before time t, it is green; or b) it is blue.
2. A thing is *bleen* if and only if: a) before time t, it is blue; or b) it is green.
3. A thing is *green* if and only if: a) before time t, it is grue; or b) it is bleen.
4. A thing is *blue* if and only if: a) before time t, it is bleen; or b) it is grue.

Now suppose that we have two descriptions (formal systems) that are identical except that in the first one "blue" and "green" are the primitive terms, "grue" and "bleen" being derived terms, and in the second one "grue" and "bleen" are the primitives and "blue" and "green" the derived. The symmetry between the definitions (1),(2) and (3),(4) makes it impossible to distinguish between these two systems on the basis of syntactic complexity, i.e. there is no formal justification for preferring one description over the other. (At the core of this curious problem is the fact that the difference between these two systems is *semantic*, not *syntactic*.) Nevertheless, the terms "blue" and "green" might arguably be preferred as descriptors since they seem more natural. Or, in Goodman's terms, "blue" and "green" are more *entrenched* in the English language than "grue" and "bleen" are. Accordingly, he developed a theory of entrenchment which was intended to justify choosing the more natural (entrenched) terms as the primitives. However, it turned out that entrenchment was essentially a simplicity criteria (Goodman 1972b) and so measuring it seems to be at least as difficult as measuring complexity.

As illustrated by these examples, there may be ways of distinguishing "natural" or "preferred" descriptions for real systems but it is not apparent whether that intuition can be made formally rigorous. The interpretation that biological systems have their own descriptions suggests that one should study biological systems to uncover the format of their internal descriptions. It has been stated that descriptions must have a physical embodiment, such as DNA in the cell, and that, for example, the 3-body problem of dynamics is not self-description (Pattee 1979). But even this intuitively reasonable proposal is difficult to assess rigorously. For instance, if genes are considered to pick between different possible solutions of physical (morphogenetic) equations, then genes are a description of the organism within a given physical context or environment. Similarly, for the 3-body problem, the "gene" could as well be the mass, shape, charge, etc., of the bodies, which determines their behavior (picks a possible trajectory) in the environment of gravity, wind currents, friction, and so on. Perhaps internal or self description is just a different way of describing a dynamical system (cf. the problem of dynamics versus computation discussed below).

6.3 Formal Systems

Given the problems of tackling concepts of complexity arising in natural systems, we examine whether issues of complexity can be studied more rigorously from within a well defined, limited artifice. Thus we turn to formal systems and examine their connection to descriptions and complexity.

A formal system is a restricted language (including its rules) in which mathematical proofs can be evaluated algorithmically, i.e. by a recursive procedure involving only logical and arithmetical manipulations. Specifically, a formal system includes: (1) a finite alphabet of symbols, (2) an unambiguous grammar (syntax) specifying how a meaningful statement can be formed, (3) a finite list of axioms, or initial assumptions, and (4) a finite and computable list of rules of inference for deducing theorems from the axioms or from other theorems (Chaitin 1975). Mathematical systems (algebras, geometries, logics) are all self-consistent formal systems. Mathematical models, such as those associated with physical theories, are formal systems. A formal system can be exhibited in many different *representations*, which may be transformed into each other via the formal rules if they are *equivalent.*

A formal system may serve as a model for a real system over some range of the real behavior. Many different formal systems may be required to cover the whole range of activity exhibited by the real system. Even for a narrow range of real system conditions, a real system may map to numerous formal systems, each one capturing a different aspect of the real behavior. This is usually the case for a complex system.

In addition to its interpretation as a model, a formal system may have properties of its own which are independent of any real system behavior. This means that paradoxes which are present in the formalism could be completely unconnected with its role as a model of a real system. Due to such ambiguities in the relation between a formal system and the real system it models, conclusions about a formal system can not always be extrapolated to the real system. Thus a description or model must be distinguished from the actual process which it describes. Making this distinction greatly influences one's evaluation of complexity and, as we shall see, this point is also central in addressing reductionism.

By demanding that a distinction be made between a description or model and the actual process that it refers to, we divide the consideration of complexity into two aspects:

1. Quantifying the complexity in our description, i.e. assigning a complexity to a formal system.

2. Determining whether that description and its complexity capture some important feature of the real system.

We note that this somewhat unsatisfying division is not limited to the property of complexity. Other physical variables seem also to be only properties of the formal systems that are used as models of reality. For example, the position and momentum of a classical object, although measurable objectively with instruments, are seemingly only elements of a formal system, the Newtonian formal system of mechanics as it refers to the idealized point particle. This view is forced upon us because when one goes into the quantum mechanical regime — a more exact model — position and momentum are no longer simultaneous properties of the formal system nor, presumably, of the real system.

The first aspect of the complexity problem, measuring the complexity of a formal system, has been the goal of many philosophy of science and computation theory studies (see Brandts 1986 for a review). The second aspect, relating the complexity exhibited by a formal system to complexity in the real world, is tackled by physical and biological modeling. Although these two aspects of the complexity problem are conceivably independent problems, examining real systems might even suggest a useful approach to formal systems.

Given the numerous ways of describing a real system, due to the many viewpoints, models, and representations that a real system may map to, we could decide that only one of such possible descriptions reflects the "true" complexity of the real system. A natural or fundamental formal model on which to base a complexity measure might be discovered. Alternatively, we propose that many equally valid and relevant complexity measures can be defined.

6.4 Reductionism and the Dynamics Versus Computation Problem

The question of reductionism bears on the complexity problem because complexity measures can be based on any of many different system descriptions and it is not clear how these measures are to be related. If all such multiple descriptions could be reduced to one fundamental description, then presumably so could also the various measures of complexity be reduced. In practice, the various descriptions are not always clearly distinguished and proposed measures for the complexity of a system will include, for example, some dynamical factors as well as some computational factors. If descriptions cannot be reduced, then there is unavoidable arbitrariness in assigning one complexity measure to a real system.

Reductionism as "reduction of scientific theories" involves relating concepts from higher level descriptions to those of lower level ones, and deriving laws at the higher level from those of the lower; explaining emergent biological behavior in terms of physical processes is an example. The controversy revolving around reductionism, which has long been a concern of philosophers of science (e.g. Nagel 1961), can often be traced to a failure to distinguish between a real system and a description or theory: the autonomy of a *theory* (i.e. the fact that the theory is not fully determined by the laws of lower level theories) does not necessarily imply an autonomy of the *processes* occurring in the system.

For example, a biological system may be composed of purely physical parts (atoms, molecules, energy, etc.) and interactions (established physical forces) yet a theory which explains the biological behavior may not be reducible to physical terms. This is *irreducibility of theories*, and it is indeed generally agreed that there are distinctive ideas in biology which cannot just be translated into physical or chemical concepts (Medawar 1974, Jacob 1974). On the other hand, there is the view that the *processes themselves are irreducible* because *new* interactions arise at higher levels. For example, the exchange symmetry of fermions, which influences the structure of the periodic table, is a rule which applies to multi–fermion systems but cannot be derived from the behavior of a single fermion. If the processes are irreducible, then they will certainly not be explicable by a fully reducible theory.

We take thinking as an example of a real behavior that can be (partially) described by two apparently irreducible theories, dynamics and computation. The two complementary languages of dynamics and computation represent two complementary viewpoints which are often employed in biological descriptions. We label behavior *dynamics* when the system is simply following dynamical laws; alternatively, we may recognize that *computation* takes place in the system (i.e. the system itself is calculating — we are not talking here about how *we* simulate the system). As a particular case consider cellular automata which have been described both as dynamical systems and as computers (Wolfram 1984b). Reducibility of theories in this instance would imply that they are equivalent descriptions differing only at some level of interpretation; however, no explicit demonstration of an equivalence has yet been given.

While there are extremes for which the appropriateness of a dynamical interpretation over a computational one, or *vice versa*, is clear, this is certainly not always the case, especially when dealing with systems that show cognitive features, such as the human brain which can calculate or execute algorithms.

The question of dynamics versus computation or "thinking" is further addressed in Artificial Intelligence (AI). The traditional aim of AI research has been to simulate human thinking with computer programs. Emphasizing symbol manipulation (rather than neuronal excitations), the approach taken is to represent knowledge of the world by formal symbols and then use algorithmic manipulation in order to make inferences. The "hard line" interpretation of this approach is the supposition that if intelligent behavior could be simulated in this way, the algorithmic nature of intelligence would be demonstrated. (However, this approach to AI remains an area of unresolved conflict, challenged by arguments that there are non-algorithmic dimensions to human thought (e.g. Penrose 1989 and Kelso 1995).

The idea that an accurate simulation of a system's behavior by a computer ought to be regarded as equivalent to the real system behavior has been expressed for systems in general (i.e. not solely for "cognitive systems"): So far as it is possible to simulate the behavior of natural systems with a digital computer, we can say that the natural system computes. Dynamics and computation can be viewed as equivalent (Conrad and Liberman 1982).

Nevertheless an algorithmic emulation of a system behavior does not necessarily correspond to what the system actually *does*. A distinction must be made between "simulation" and "duplication": It doesn't matter how good the technology is, or how rapid the calculations made by the computer are. If it really is a computer, its operations have to be defined syntactically, whereas consciousness, thoughts, and all the rest of it involve more than a syntax. Those features, by definition, the computer is unable to *duplicate* however powerful may be its ability to *simulate*. After all we can do computer simulations of any process whatever that can be given a formal description, and nobody supposes that the computer simulation is actually the real thing (Searle 1984, p. 37).

Acknowledgement of the limitations of a purely formal approach in AI-related fields (computer science, neurophysiology, biophysics, etc.) has lead to the "connectionist" approach. The connectionists have challenged at least two of the assumptions underlying abstract symbol processing AI. One is that the problem may be treated solely on the formal level, i.e. that computation or thinking done by living organisms (humans) can be represented by, or executed as well by, a digital computer and that the biological peculiarities only enter in a determination of a particular .physical embodiment or hardware. For example, research on color perception shows that people categorize color not simply on the basis of wavelength, but that idiosyncracies of neurophysiology *are* an important determinant. The other suspect assumption is that all symbols meaningful to the mind have counterparts in objective reality. Psychological studies of human categorization on the basis not only of shared properties (the abstract symbols) but also personal experience (including that of the body) and imagination (e.g. metaphors and other vague associations) make doubtful the existence of a direct connection between such classes of symbols.

In the connectionist paradigm, thinking is not a manipulation of formal symbols (computation). On the contrary, it is an emergent feature of the process whereby neural connections are established and strengthened. The models, densely-linked, massively parallel architectures, differ from traditional AI schemes in that the networks attain the final state (output corresponding to a "solution") by settling down into a stable state rather than by being prescribed in detail by a deterministic, stepwise, algorithm (although connectionist models are nevertheless executed on a digital computer).

For example, the "Boltzmann machine" "learns" appropriate associations between inputs and outputs by adjusting the strengths of its internal connections until a global minimum energy state is attained (Hinton and Anderson 1989). This type of "computation" probably emulates more closely the information processing of living systems because it operates on a global, dynamical, energy-process basis rather than the single level mode of symbol manipulation in digital computers.

Thus, not only are we incapable of clearly distinguishing between dynamics and computation — i.e. when we should regard a system as "calculating" or, on the other hand, when it is simply "following dynamics" — but that our idea of *what, in general, constitutes a computation* is not well defined. It may even turn out that obeying dynamical laws is most usefully treated as a form of cognition.

In summary, unless we adopt a pluralistic approach to complexity in which multiple definitions of complexity are accepted, each connected to a particular viewpoint or model, a comprehensive treatment of complexity in physics and biology will depend on resolving reducibility of theories issues, such as the dynamics-computation problem.

7 Conclusion

In view of the complexity present in the real world and in our perceptual processes, we have argued that complexity can not be treated as a single concept, a uniquely definable or measurable property which will submit to one unified treatment. On the contrary, in this essay we have advocated a pluralistic approach to complex systems. Laws and models are simplifications of the world. They are bought at the price of exclusion and can only capture part of the whole story. As a result, measures of complexity will also only be useful in a restricted sense, applying to one aspect of behavior (e.g. information content in a digital signal or stability of a chaotic attractor in N-dimensions). Thus an evaluation of the complexity of a system will ultimately depend upon what one wants to understand about the system, i.e. on what models one chooses to apply, because complexity cannot be separated from its models and their vocabulary.

Nevertheless, models of and approaches to complexity *can* successfully be borrowed from different fields of science — science in the broadest sense: including perhaps mathematics, physical and life sciences, economics, semiotics, linguistics, etc. — and principles *will* be found to apply across different levels or in different systems of organization. That is, boundaries can be crossed. And they need to be crossed, for it does seem that "over and over again nature uses the same principles of self-organization to produce dynamic patterns on all scales" (Kelso 1995). Thus we would anticipate that, for example, while the algorithmic complexity measures germane to computer science and formal systems might be inappropriate to characterize cell nuclei from a thermodynamic point of view, such measures might meaningfully shed light on genetic coding sequences.

In conclusion, the wide-reaching applicability of concepts of complexity, capable of unifying aspects of behavior in systems viewed as disparate in the usual divisions of the world (such as Newtonian particles and the stock market, or changes in weather patterns and the mind), does not however infuse a "universal" status into any one concept of complexity, nor imply its relevance to *all* systems. We must ultimately expect many theories of complexity.

Acknowledgements

Thanks to Peter Turney for introducing me to the formal studies of complexity, to Andre Longtin for suggesting references and to CJL for recommending that I express my personal viewpoint. Financial support from the Natural Sciences and Engineering Research Council of Canada is gratefully acknowledged.

References

Bak, P., Tang, C. and Wiesenfeld, K. (1988) Self-organized criticality. *Physical Review A* **38**, 364–374.

Banerjee, S., Sibbald, P.R. and Maze, J. (1990) Quantifying the dynamics of order and organization in biological systems. *Journal of Theoretical Biology* **143**, 91–111.

Brandts, W.A.M. (1986) *On Models of Complexity in Biophysical Systems.* MSc Thesis, Department of Physics, University of Toronto, Canada.

Chaitin, G.J. (1975, May) Randomness and mathematical proof. *Scientific American* **232**, 47–52.

Chaitin, G.J. (1982) Gödel's theorem and information. *International Journal of Theoretical Physics* **21**, 941–954.

Conrad, M. and Liberman, E.A. (1982) Molecular computing as a link between biological and physical theory. *Journal of Theoretical Biology* **98**, 239–252.

De Luca, A. (1975) Complexity and information theory. In G. Longo, ed., *Coding and Complexity.* New York: Springer-Verlag, pp. 207–269.

Goodman, N. (1972a) *Problems and Projects.* Indianapolis: Bobbs-Merrill, pp. 371–388: The new riddle of induction.

Goodman, N. (1972b) *Problems and Projects.* Indianapolis: Bobbs-Merrill, pp. 334–336: Safety, strength, simplicity.

Hinton, G.E. and Anderson, J.A., eds. (1989) *Parallel Models of Associative Memory.* Hillsdale, NJ: Erlbaum.

Horgan, J. (1995, June) From complexity to perplexity. *Scientific American* **272**, 104–109.

Huberman, B.A. and Hogg, T. (1986) Complexity and adaptation. *Physica D* **22**, 376–384.

Jacob, F. (1974) *The Logic of Living Systems: A History of Heredity.* London: Allen Lane.

Kaspar, F. and Schuster, H.G. (1987) Easily calculable measure for the complexity of spatiotemporal patterns. *Physical Review A* **36**, 842–848.

Kauffman, S. (1993) *The Origins of Order: Self-Organization and Selection in Evolution.* New York: Oxford University Press.

Kelso, J.A.S. (1995) *Dynamic Patterns: The Self-Organization of Brain and Behavior.* Cambridge, MA: MIT Press.

Langton, C.G. (1990) Computation at the edge of chaos: Phase transitions and emergent computation. *Physica D* **42**, 12–37

Mandelbrot, B.B. (1982) *The Fractal Geometry of Nature.* New York: W.H. Freeman.

Martinez-Arias, A. and Lawrence, P. (1985) Parasegments and compartments in the *Drosophila* embryo. *Nature* **313**, 639–642.

May R.M. (1973) *Stability and Complexity in Model Ecosystems.* Princeton, NJ: Princeton University Press.

May, R.M. and Oster, G.F. (1976) Bifurcations and dynamic complexity in simple ecological models. *American Naturalist* **110**, 573–599.

Medawar, P. (1974) A geometric model of reduction and emergence. In F.J. Ayala and T. Dobzhansky, eds., *Studies in the Philosophy of Biology: Reduction and Related Problems.* Berkeley, CA: University of California Press, pp. 57–63.

Nagel, E. (1961) *The Structure of Science: Problems in the Logic of Scientific Explanation.* New York: Harcourt, Brace & World.

Nicolis, G. and Prigogine, I. (1989) *Exploring Complexity.* New York: W.H. Freeman.

Pattee, H.H. (1979) Interview. Compiled together with interviews of R. Rosen and R.L. Somorjai under the title "A symposium in theoretical biology." In P. Buckley and F.D. Peat, eds., *A Question of Physics: Conversations in Physics and Biology.* Toronto: University of Toronto Press, pp. 84–123.

Pattee, H.H. (1961) On the origin of macromolecular sequences. *Biophysical Journal* **1**, 683–710.

Peacocke, A.R. (1983) *The Physical Chemistry of Biological Organization.* Oxford: Clarendon Press.

Penrose, R. (1989) *The Emperor's New Mind.* New York: Oxford University Press.

Pimm, S. (1984) The complexity and stability of ecosystems. *Nature* **307**, 321–326.

Poon, L. and Grebogi, C. (1995) Controlling complexity. *Physical Review Letters* **75**, 4023–4026.

Prigogine, I. (1980) *From Being to Becoming: Time and Complexity in the Physical Sciences.* New York: W.H. Freeman.

Rosen, R. (1978) *Fundamentals of Measurement and Representation of Natural Systems.* New York: Elsevier North-Holland.

Rosen, R. (1979) Interview. Compiled together with interviews of H.H. Pattee and R.L. Somorjai under the title "A symposium in theoretical biology." In P. Buckley and F.D. Peat, eds., *A Question of Physics: Conversations in Physics and Biology.* Toronto: University of Toronto Press.

Searle, J.R. (1984) *Minds, Brains, and Science.* Cambridge, MA: Harvard University Press.

Simon, H.A. (1962) The architecture of complexity. *Proceedings of the American Philosophical Society* **106**, 467–482.

Thom, R. (1975) *Structural Stability and Morphogenesis.* Reading, MA: W.A. Benjamin.

Thompson, D'A.W. (1942) *On Growth and Form.* New York: Cambridge University Press.

Wimsatt, W.C. (1976) Complexity and organisation. In M. Grene and E. Mendelsohn, eds., *Topics in the Philosophy of Biology.* Boston Studies in the Philosophy of Science, Volume XXVII. Boston: D. Reidel, pp. 174–193.

Wolfram, S. (1984a) Computation theory of cellular automata. *Communications in Mathematical Physics* **96**, 15–57.

Wolfram, S. (1984b) Cellular automata as models of complexity. *Nature* **311**, 419–424.

Dynamics, Complexity and Computation

Paul A. Dufort and Charles J. Lumsden

1 Introduction

The young physical theorist spends a great deal of time learning about linear ordinary and partial differential equations and how to solve them analytically. There are many examples in the sciences where the power and beauty of exact solutions have been used fruitfully to understand and predict the behavior of physical systems. Why is it, then, that we cannot simply model a computer with a set of ODEs or PDEs, solve them and predict the computer's behavior in the same way we do for a simple harmonic oscillator? If this were possible we would never again have to run a complex computer model for days or weeks or months in order to obtain a result. Given the initial conditions we could simply look up the future state of the computer in our equations to find the answer.

Of course, this is not possible, but the reason why is far deeper than the simple fact that modeling a computer with differential equations would be a nontrivial task. (This has already been accomplished with only three ODEs (Moore 1990, 1991; Branicky 1994, see also Blum, Shub and Smale 1989)!) The discovery of the reason why we cannot beat the computer at its own game is fundamental to understanding why theoretical computer science is becoming so important to physics and biology. It dates back to the 1930's, when mathematician Alan Turing proved a famous result called the Halting Problem. One of the implications of this result is that there is no faster way to determine the output of an arbitrary computer program than to run the program. We cannot just "skip to the end" to find the result — instead we must go to the trouble of executing every instruction until the program finishes. We will examine his result in more detail below.

Perhaps the quintessential realization in the study of physics and computation is that if a computer can be considered a dynamical system, then perhaps a dynamical system can also be thought of as a computer (see also the essays by Lumsden and by Brandts, this volume). Computation and dynamics may, as Brandts points out, be dual representations of the same underlying phenomenon, somewhat like waves and particles in quantum mechanics. Of course, this does not mean that every simple pendulum is a supercomputer. Rather, it turns out that there is an entire spectrum of computational capacities, and that dynamical systems for which there exist analytical solutions occupy the bottom levels. Conversely, extremely complicated dynamical systems which come close to matching the computational capacity of a modern computer obey the Halting Problem, in the sense that there is no faster way in general to predict their dynamical behavior than to simply let it unfold, either naturally or piggybacked onto a computer simulation. There are no shortcuts.

It is in the study of biological systems that the potential duality between symbolic, digital computation versus real-numbered, analog dynamical systems becomes the most pronounced. There is no question that naturally occurring analog systems with continuously varying state variables like the brain and the cytoplasm do in fact process information, or "compute," at some level: They receive information from the outside world, combine it with stored information and manipulate it according to a set of rules, using the result as further input for additional processing and to guide their future actions. This, in general terms, is exactly what computers do. In the case of the human brain, a simple demonstration of the level of its computational abilities is the fact that any human can easily simulate a universal computer, and is therefore by definition capable of universal computation. So biological systems do compute, and at least some of them are universal computers. The potential duality between dynamical and computational systems motivates careful discussion of computation as studied in theoretical computer science and nonlinear dynamics as studied in theoretical physics and biomathematics. When approached in this way, computer science may provide a powerful new set of models and metaphors for understanding the patterns of behavior that we see in complex systems, including those that are indisputably alive.

2 Machines and Grammars

Computability theory asks for which functions there exists an algorithm or *effective procedure*, consisting in a list of instructions to be carried out step by step according to a set of well-defined rules, which will upon completion yield the value of the function for a specified input value.

One might at first think that the limitations on what a computer can do depend only on the programmer's ingenuity and the level of technology available. However, much as no level of ingenuity or technology will allow an inventor to build a perpetual motion machine, there are many well-posed mathematical problems which simply cannot be solved by any algorithm running on any computer. The most famous of these so-called noncomputable problems is the Halting Problem, mentioned briefly above. We will discuss below the causes of noncomputability, but first let us examine the conceptual machine that was created for the express purpose of addressing the question of what can be computed.

2.1 Turing Machines and the Halting Problem

Searching for a way to formalize the essence of an effective procedure, Turing invented a deceptively simple device now called the *Turing machine* (TM) (Hopcroft and Ullman 1979, Barwise and Etchemendy 1993). The TM consists in three pieces: (1) a *tape*, extending to infinity and partitioned into equal-sized cells, each of which contains one of b symbols a from a finite alphabet S; (2) a *read/write* head that scans the tape, capable of moving left or right one cell at a time and reading or writing on the tape a symbol a from S; and (3) *a finite control*, connected to the read/write head and capable of being in one of c states q from a finite set of states Q. At each discrete time step, the TM reads the symbol a currently under the read/write head and determines the state q that the finite control is in. It then takes the pair (a, q) and maps it to an output triple (a', q', d') via the *transition function* $\delta{:}S \times Q \rightarrow S \times Q \times \{L, R\}$. The triple tells the TM which new symbol a' from S to write in the current cell, which new state q' from Q the finite control should enter next, and which direction d' to move the read/write head ($d' = L$ for left or R for right). At the beginning of operation, the finite control is in a special start state $q_s \in Q$ and the read/write head is positioned at the leftmost end of the tape. A finite number of symbols comprising the input are written on the leftmost cells of the tape, and all cells after the input are blank. The TM then begins its operation and continues until it enters another special state marking the end of the computation, called the *halt state*.

Figure 1 is a schematic depiction of a TM that computes the integer multiplication function $f(x, y) = xy$, for $\{x, y | x, y \in Z \text{ and } x, y \geq 0\}$. This TM has 12 states and 4 symbols comprising the alphabet $S = \{0, 1, 2, B\}$, where B is the symbol for a blank cell. The input to the tape consists in a set of x zeros and a set of y zeros separated by a 1. On output, this is replaced by xy zeros. (The integers in this TM are thus coded in unary, a number system in which the integer n is represented by n zeros. This TM and the details of its operation can be found in Hopcroft and Ullman 1979, p. 158).

For those familiar with programming but unfamiliar with theoretical computer science, the TM probably seems an unusual device, unlike any real computer presently in existence. Indeed, the vast majority of modern computers evolved from a different model of computation called the *von Neumann machine* (after its inventor, John von Neumann) which is essentially the familiar random access machine. The TM, it is generally agreed, would make an extremely inefficient computer for real calculations.

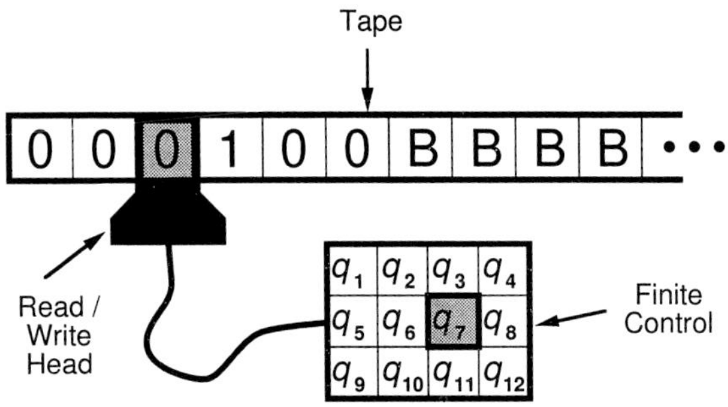

Transition Function δ

Symbol

State	0	1	2	B
1	(7, B, R)	—	—	—
2	(3, 2, R)	(5, 1, L)	—	—
3	(3, 0, R)	(3, 1, R)	—	(4, 0, L)
4	(4, 0, L)	(4, 1, L)	(2, 2, R)	—
5	—	(6, B, R)	(5, 0, L)	—
6	(8, 0, L)	—	—	—
7	(7, 0, R)	(2, 1, R)	—	—
8	—	(9, 1, L)	—	—
9	(10, 0, L)	—	—	(11, B, R)
10	(10, 0, L)	—	—	(1, B, R)
11	—	(12, B, R)	—	—
12	(12, B, R)	(Halt, B, R)	—	—

Figure 1: A Turing machine that computes the product of two integers. *Top*. Basic components, common to all Turing machines. *Bottom*. The transition function δ, shown here in matrix form, enabling integer multiplication. See text for details.

Nevertheless, the TM was the first formal model of computation, and was the device in terms of which many of the most important theorems and proofs concerning computation were originally stated. Furthermore, the remarkable fact is that all subsequent models of computation have been shown to be no more powerful in their computational capability than the Turing machine. It must be emphasized that the issue here is not speed, for it is undeniable that some computer architectures are faster than others. The issue is rather what can or cannot be computed regardless of how long it takes, and it is a fact that there is absolutely no computation that a thousand supercomputers can do that is beyond the capabilities of a single TM.

The equivalence in capability of the TM with all real computing machines means that, for any algorithm that can be written in any programming language, a corresponding TM can be constructed that will carry out the same computation. This assertion has been proved by showing that all models of computation define the same set of computable functions, the so-called *partial recursive functions*. They are called "recursive" because they are allowed to appear in their own definition (for example, $f(0) = 0, f(n) = 2f(n-1)+n, n \in Z$ defines a recursive function), while they are called "partial" because they may not be defined on all inputs (for example, most real computer programs rely on their input data obeying certain restrictions or being in a particular format, or else the computation may never terminate). If a function is defined on all possible inputs then it is called *total recursive*. The set of total recursive functions is a proper subset of the set of partial recursive functions. Some computable functions are total recursive, but the class of all computable functions is partial recursive.

Among the set of all computable functions and their corresponding TMs there are certain special machines, called *universal Turing machines* (UTMs), whose sole function is to be able to simulate any other Turing machine. These universal machines take as their input two arguments, the first being the transition function δ of the machine M to be simulated, and the second being the input to be processed by M. Although there are many TMs that are universal, the smallest known universal Turing machine has only seven states and four symbols (Minsky 1967).

There is an even stronger assertion about the power of TMs than the statement of their equivalence with other formal models of computation. This statement, called the *Church-Turing thesis*, claims that the TM also captures exactly our intuitive notion of an algorithm or list of instructions. It is a thesis rather than a rigorous mathematical result because of our inability to precisely quantify the intuitive sense of what constitutes an algorithm.

We turn now to the issue of why there exist functions that cannot be computed by any algorithm. There are essentially two different kinds of infinite sets, the *countably* infinite sets and the *uncountably* infinite sets. The countably infinite sets are so called because their elements can be put into a one-to-one correspondence with the non-negative integers, allowing you to "count" them, while the uncountably infinite sets must be matched up with the real number line.

The set of all algorithms can be put into a one-to-one correspondence with the integers, so that this set is countably infinite. To see this, imagine writing out a list of all integers $n \in Z$ in increasing order. Beside each number write its representation in binary, the base 2 number system. Since there is a one-to-one correspondence between the integers and the set of all possible sequences of zeros and ones that represent them in binary, every possible sequence of zeros and ones appears exactly once in the list. Now, take each number whose binary representation contains a number of binary digits (or *bits*) that is a multiple of 8, and separate these sequences of zeros and ones into blocks of 8 bits each. Each of these blocks of 8 bits can be identified with a character in ASCII, the coding system that is used by many computers to represent text. Thus, every possible text file that can be created using the ASCII character set appears exactly once in our list beside a unique integer, which is called its *index*. However, every possible computer program that can be written in C, FORTRAN, LISP or any other programming language, has a representation as a plain text file made of ASCII characters (Kfoury, Moll and Arbib 1982). The result is that every possible computer program appears at some point in our list of integers. We have thus *enumerated* the set of all possible computer programs, thereby showing that this set is countably infinite.

Unfortunately, the set F of all possible functions is uncountably infinite. To see this, we need consider only a tiny subset of the set of all possible functions, the functions $f{:}Z \rightarrow \{0,1\}$ that map the integers Z either to zero or one. Assume by way of contradiction that the functions in the set F *can* be mapped one-to-one onto the integers. Let f_k be the function in the list mapped to the integer k. Now create a new function $g{:}Z \rightarrow \{0,1\}$ defined by

$$g(n) = \begin{cases} 0 & \text{if } f_n(n) = 1 \\ 1 & \text{otherwise} \end{cases}$$

This function cannot appear anywhere on the list, since if $g(n) = f_k(n)$ then $g(n) = f_k(k)$ and $g(n) \neq f_k(k)$, which is a contradiction. Thus, even this small subset of the set of all functions is uncountable. The inescapable conclusion is that there are far more functions than there are algorithms to compute them, so that the vast majority of conceivable functions are uncomputable.

Perhaps even more surprising is the fact that many functions having great practical interest are uncomputable, the most notorious being the halting function. To solve the famous Halting Problem, we seek a total recursive function $h(u,v)$ of two arguments u and v: the first argument u is an algorithm, the second v is the input that it is to operate on. The *halting function* is defined as

$$h(u,v) = \begin{cases} 1 & \text{if program } u \text{ halts on input } v \\ 0 & \text{otherwise} \end{cases}$$

Since $h(u,v)$ is total recursive, it should return the correct output for any algorithm u and any input v.

The cold truth is that the halting function, despite its potential usefulness if only one knew it, cannot be computable. Since both computer programs and their data can be represented by a finite sequence of bits, they both have a representation as integers. Thus, the halting function h can be thought of as mapping the set of all pairs of nonnegative integers to the set $\{0, 1\}$, h: $N \times N \to \{0, 1\}$. Now, by way of contradiction, assume that the halting function is computable. Define a partial recursive function Ψ by

$$\Psi(u) = \begin{cases} 1 \text{ if } h(u, u) = 0 \\ \text{does not halt otherwise} \end{cases}$$

Since by assumption h exists, Ψ is a perfectly valid computable partial recursive function. Furthermore, since h is a total recursive function, it can take as input all partial recursive functions, including Ψ. This means that $\Psi(\Psi)$ halts if and only if $h(\Psi, \Psi) = 0$. But by definition, $h(\Psi, \Psi) = 0$ if and only if $\Psi(\Psi)$ does not halt. So $\Psi(\Psi)$ halts if and only if it does not halt, a contradiction.

It is therefore impossible, even in principle, to construct a computable function that will take any given computer program and its data and reveal whether or not the program will ever halt. Let us reiterate the implications of this result for mathematical theory. Traditionally, physical theorists have been trained to believe that the most desirable result of a study of a mathematical model is an explicit, closed-form solution which exactly describes the behavior of the model for all time. If the model is sufficiently complex that it can be used to simulate a Turing machine, however, then an explicit solution would amount to a solution of the Halting Problem, which we now know is impossible. Furthermore, this level of complexity is not as high as one might suspect: As few as three continuous ordinary differential equations can simulate any TM (Moore 1991, Branicky 1994). The implication would seem to be that the dynamics of all systems above a certain level of complexity, a class including the vast majority of nonlinear systems, cannot be known without embedding them in the dynamics of a computer (Wolfram 1985, da Costa and Doria 1991).

2.2 Formal Grammars and the Chomsky Hierarchy

The study of the formal grammars is an integral part of theoretical computer science and bears an intimate relationship with the theory of computability discussed in the previous section. These grammars provide a means for capturing complex dynamical behavior in a concise form, and also for assigning a quantitative measure to a dynamical system's complexity based on the grammar that represents it. We will elaborate on this point further below.

A formal grammar G is a set $G = \{V, T, P, S\}$, where V is a finite set of *variable* symbols which are used transiently in the process of constructing strings; T is a finite set of *terminal* symbols comprising the alphabet from which all final strings are made; P is a finite set of *production rules* that are applied recursively to transform strings in successive stages of their construction; and $S \in V$ is a special variable called the *start* symbol, which is the initial seed from which the construction of all strings begins. The process of constructing a string, beginning with S and ending with a string comprised of symbols from T, is called a *derivation.* The set of all strings that can be produced by a given grammar G is called the *formal language* $L(G)$ generated by G.

As a simple example, consider the formal grammar G' with the set of variables $V = \{a, b\}$, terminals $T = \{0, 1\}$, and productions

$$P = \begin{cases} S \rightarrow a \\ a \rightarrow 1b \\ b \rightarrow 0a|1a|\varepsilon \end{cases}$$

The three production rules are interpreted as follows: (1) the start state S is replaced with an a; (2) an a may be replaced with $1b$; and (3) a b may be replaced with $0a, 1a$, or the empty string ε. The derivation is initiated with S, after which the production rules are applied recursively until all variables have disappeared. For this grammar, the last step is clearly always the transformation of b into the empty string ε. A sample derivation with this grammar appears below:

$$\begin{aligned} S &\rightarrow a \rightarrow 1b \rightarrow 10a \rightarrow 101b \\ &\rightarrow 1011a \rightarrow 10111b \rightarrow 101110a \\ &\rightarrow 1011101b \rightarrow 1011101 \end{aligned}$$

It is not difficult to see that the language $L(G')$ associated with the grammar G' contains all binary strings such that: (1) the number of digits is odd; and (2) every odd-numbered digit is a 1. The grammar G' thus represents this infinite set of strings in a very compact form.

In the space of all possible grammars, there exists a classification system called the *Chomsky hierarchy* which ranks grammars according to the form of their production rules. The Chomsky hierarchy contains four levels such that the grammars in each successively higher level subsume all of the grammars in the levels below. From lowest to highest, the levels are called *regular*, *context-free*, *context-sensitive*, and *unrestricted.* The production rules of regular grammars may contain only one variable on both the left- and right-hand sides, and a number of terminals on the right-hand side that must appear to the left of the variable (e.g. $a \rightarrow 01b$); context-free productions are the same as regular ones but may possess terminals on either side of the variable on the right-hand side (e.g. $a \rightarrow 01b1$); context-sensitive are the same as context-free but may possess additional variables and terminals on the left-hand side (e.g. $b1a01 \rightarrow b10a101$) such that the a may be replaced with $0a1$ only in the context of $b1\ldots01$, and with the additional constraint that the left-hand side must be shorter in length than the right-hand side; and unrestricted are the same as context-sensitive except that the left-hand side may now be longer than the right-hand side (e.g. $b01Aa00 \rightarrow 1b0$).

Associated with every possible grammar G there is a simple conceptual machine M similar to a Turing machine which defines exactly the same language L as does G. However, they do not generate strings from their language, but rather only *recognize* them, as follows. To start, the string x of symbols is placed on the leftmost cells of the machine's tape and its finite control is in the start state. The machine's transition function δ is then applied repeatedly. There are three possible outcomes: (1) the machine halts in one of potentially many *accepting* states (string x belongs to the language L); (2) the machine halts in a *nonaccepting* state (x is not in L); and (3) the machine never halts (membership of x in L cannot be determined). Like grammars, these devices fall into four broad classes.

The machine that recognizes regular languages is called a *finite automaton* (FA). The FA is similar to a Turing machine, except the head that scans the tape can only *read* symbols, and its read head can only move right one cell on each step. The end of the string must eventually be reached, so an FA always halts on a finite input. The states in the finite control of an FA correspond to the variable symbols in the grammar whose language the FA recognizes. There is a useful alternative representation of an FA or the regular grammar G whose language it recognizes, called a labeled directed graph or *l-digraph*. A labeled digraph is a pair (V, E) of two sets, a set of vertices V and a set of labeled directed edges E. Each of the edges e from E consists in a triple $e = (v, v', s)$, where v and v' belong to V and s is a symbol that *labels* this edge. An edge e is said to represent a transition on symbol s from vertex v to vertex v'. Each vertex $v \in V$ in a digraph corresponding to an FA represents one of the states in the finite control of the FA, while each edge $e \in E$ from the set of edges represents one of the possible transitions in the FA's transition function δ. The meaning of an edge $e = (v, v', s)$ is that if the finite control is in state v and the read head reads symbol s from the tape, the finite control will move to state v' (and of course the read head will move one cell to the right). The symbols s therefore correspond to the terminal symbols of the grammar G.

A simple digraph corresponding to the example grammar G' discussed above is shown in Figure 2. The state q_s represents the variable symbol S, q_2 represents a, q_3 represents b, and q_4 is the accepting halt state. A *path* through the digraph (V, E) is traversed by starting at vertex q_s, then traveling over a directed edge to another vertex, over another directed edge to another vertex,

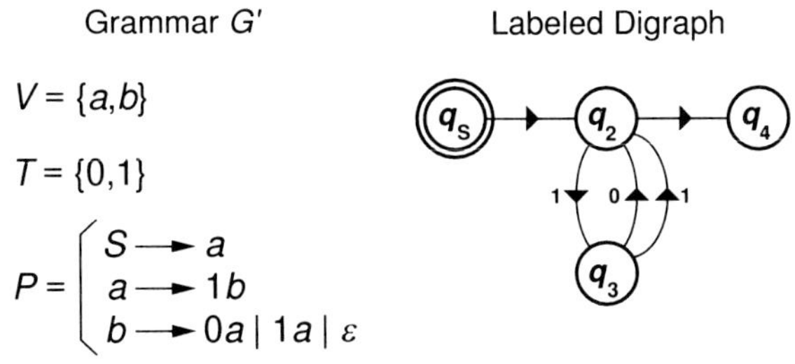

Figure 2: An ℓ-digraph corresponding to the example grammar discussed in the text.

and so on, until finally arriving in the halt state. Every possible string that can be generated by G' corresponds to a path through the digraph (V, E), and conversely. This is true not only for G', but for any regular grammar and the digraph corresponding to it. This duality is extremely useful in the theory of physics and computation, and we will discuss it in more detail below.

Ascending the Chomsky hierarchy, the machine that recognizes context-free languages is called a *push-down automaton* (PDA). The PDA is like an FA, except that now the finite control is augmented with a device called a *stack*, a data structure somewhat like a tape capable of holding a sequence of symbols from a finite alphabet. However, a stack is usually thought of as being vertical, with a top and a bottom. With a stack, one can only read and write symbols to the top position. All other symbols on the stack are displaced downward one cell, and the new symbol now occupies the top position. Conversely, when one reads a symbol from the stack, all other symbols move up one cell. The stack endows PDA with the ability to recognize context-free languages, while the simpler FA is limited to regular languages. Like the FA, the PDA is guaranteed to halt on all finite inputs.

The machine that recognizes context-sensitive languages, called a *linear bounded automaton* (LBA), is more powerful still. This machine is identical to a Turing machine, except that the length of its tape is finite and bounded by a linear function of its input string length. Since both the tape and the finite control are finite, there are only a finite number of states that the LBA can occupy. Thus, any computation that it performs in an attempt to recognize a given string must eventually halt, or loop back on itself endlessly. If one is careful to avoid the latter possibility, the LBA is always guaranteed to halt. The question of membership of a string in the language defined by the LBA can therefore always be determined.

Finally, as the reader may have already guessed, the machine that recognizes unrestricted languages is the Turing machine, whose definition includes

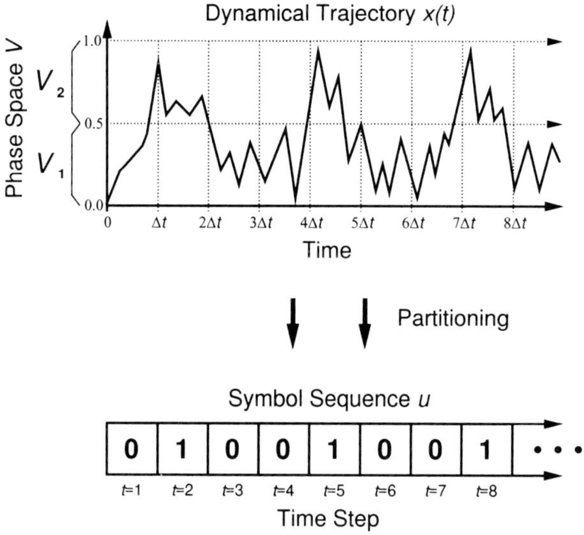

Figure 3: Constructing a phase space partition is the first step to formulating the symbolic dynamics of a dynamical system and connecting dynamics to computation.

an infinite tape. Since the TM's infinite tape endows it with an infinite number of states, it can compute for eternity without ever entering the same state twice. This means that, regardless of how careful the programmer is to avoid endless loops, the TM need never halt on a given input. For this reason, one cannot tell whether a given string is a member of the unrestricted language recognized by a given TM. Since there is no effective procedure to answer this membership problem for unrestricted languages, this question is called *undecidable*. Undecidability is in some sense the curse that comes with greater computational power: As one ascends the Chomsky hierarchy, the number of questions about languages in each level that are undecidable increases dramatically.

3 Symbolic Dynamics

The first step in translating the trajectories of a dynamical system $x(t)$ into the language of computation is to divide up or discretize the phase space volume V into a finite number n of smaller volumes V_i. Both V and V_i may have one or more dimensions. A special kind of discretization of V is needed, called a *partition* (Figure 3), for which: (1) the component volumes V_i span all of V, $\bigcup_{i=1}^{n} V_i = V$; and (2) none of the V_i overlap, $V_i \cap V_j = \phi \; \forall i, j | i \neq j$. A unique symbol s_i can then be assigned to each of the V_i, resulting in the conversion of the phase space V to a finite alphabet of symbols $S\{s_i | i = 1 \ldots n\}$. The next step is to choose a suitable time constant Δt to divide up the finite time interval $[0, \tau]$, over which the trajectory $x(t)$ is defined, into a finite number $N = \tau / \Delta t$ of equal-sized time steps. The constant Δt should be chosen small enough so that $x(t)$ only infrequently visits multiple V_i in the same time step. The behavior $x(t)$ can then be converted into a finite sequence of N of the symbols s_i by assigning one symbol for each time step (Figure 3).

3.1 Topological and Metric Entropies, Set Dimensions

With the trajectories $x(t)$ of the dynamical system of interest suitably converted into a sequence or string of symbols $u = s_i^{t=1} s_j^{t=2} s_k^{t=3} \ldots s_l^{t=N}$, there are a number of questions one can ask about u. For example, how often do the various possible symbols s_i appear in u? How often do all possible blocks of two symbols $s_i s_j$ appear? How often do all of the possible blocks $\ldots s_i s_j s_k \ldots$ of length L appear? Answering these questions about u can tell us a great deal about the original dynamical system $x(t)$. But first, we introduce a special data structure which is particularly suited to extracting the answers to the above questions from u.

Consider a labeled digraph shaped like an upside-down tree. At the top, or 0th level, there is a single vertex corresponding to the trunk of the tree. From this 0th-level vertex n edges emanate, each labeled uniquely by one of the n symbols s_i. At the end of each of these n labeled edges, there is a new vertex. These n new vertices form the 1st level of the tree. From each 1st-level vertex, n more edges emanate, each again labeled uniquely by one of the n symbols s_i, and arrive at n^2 new vertices which form the 2nd level of the tree. This procedure is repeated recursively to a desired level L, resulting in an object called a *parse tree* $T^{(L)}$ of depth L. The Lth level of the parse tree $T^{(L)}$ contains n^L nodes, each of which are called *leaves* since they have no edges emanating from them. The parse tree has exactly one path of length L from the 0th-level vertex down to one of the leaves for each possible string $w_j^{(L)}$ of length L, $j = 1$ to n^L, made up of symbols s_i from S. Thus, each leaf j corresponds uniquely to one possible string $w_j^{(L)}$. An example parse tree $T^{(L=4)}$ with $n = 2$, alphabet $S = \{0, 1\}$, and depth $L = 4$ is pictured in Figure 4. (Arrows are not shown on the tree's edges, as it is understood that all edges point downward.)

A parse tree $T^{(L)}$ used to analyze the sequence of symbols u is constructed by carrying out the following steps (Crutchfield 1989, Crutchfield and Young 1990, Crutchfield 1991):

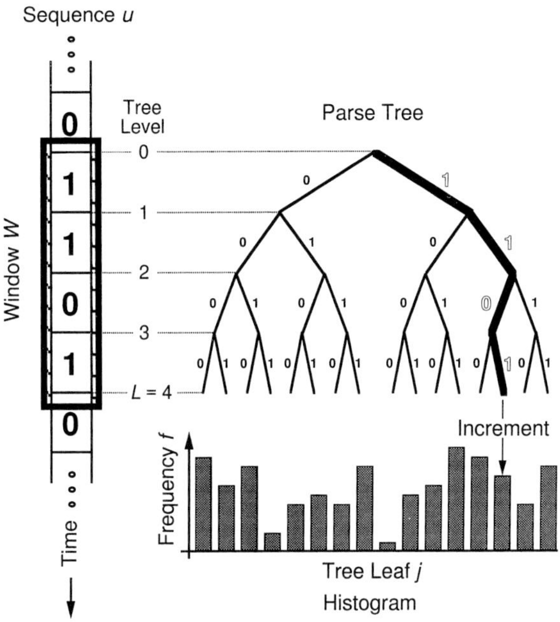

Figure 4: A parse tree $T^{(L=4)}$ and symbolic dynamics analysis for the alphabet $S = \{0, 1\}$ and depth $L = 4$. See text.

1. Assign a counter f_j to each leaf j of the Lth level, $j = 1$ to n^L, and assign it an initial value of $f_j = 0$.

2. Position a *window* W of length L over the sequence u such that only L consecutive symbols from u are visible at a time inside the window. The contents of W thus form a string $w_j^{(L)}$ of length L. Initially, the window is placed over the first L symbols of u.

3. For the L symbols inside the window forming the string $w_j^{(L)}$, trace the corresponding path down the parse tree to the appropriate terminal leaf j, and then increment the counter f_j corresponding to this leaf by one.

4. Slide the window W along the sequence u by one time step.

5. Go back to step 3, and repeat for all $N - L + 1$ possible positions of W on u.

This procedure causes every counter f_i to be incremented by one each time the string $w_j^{(L)}$ corresponding to it is found in u. The result is a histogram of frequency counts f_j, such that $\Sigma_{j=1}^{n^L} f_j = N-L+1$, tallying the number of times each possible string $w_j^{(L)}$ appears in the overall sequence u. A window of length $L = 4$ on a sample data sequence is shown in Figure 4. The corresponding path down the parse tree is highlighted, representing the string $w_{j=13}^{(L=4)} = 1101$, and the histogram of frequency counts is shown under the leaves at the bottom level of the parse tree.

The creation of a parse tree and histogram from the sequence u is the first step in analyzing its symbolic dynamics. However, even for modest values of n and L, for example $n = 4$ and $L = 8$, one is still left with a histogram containing 65,536 counts. Visual inspection of a histogram of this size is still unlikely to tell us very much about u. Fortunately, there are a number of important measures that one can easily compute using the counts f_j that reveal very important characteristics of u.

The first of these is the *topological entropy*, defined as (Crutchfield and Packard 1982, Crutchfield and Packard 1983, Wolfram 1984c)

$$h_0^{(L)} = \frac{1}{L} \log_n \left(\sum_{j=1}^{n^L} \theta(p_j) \right)$$

where $\theta(p_j) = 1$ for $p_j > 0$, $\theta(0) = 0$, and $p_j = f_j/(N-L+1)$ is the probability of observing string $w_j^{(L)}$ at a randomly chosen location in u. Essentially, $h_0^{(L)}$ is a measure of the number of different possible strings $w_j^{(L)}$ that actually appear in u, out of a possible n^L. If we think of the set of n^L strings $w_j^{(L)}$ as representing all of the possible dynamical fragments from $x(t)$ of length $L\Delta t$, then $h_0^{(L)}$ indicates how many of these fragments actually appear in $x(t)$ at one time or another during the whole interval $t \in [0, \tau]$. (Dynamical fragments defined by a sequence of symbols are often called *cylinders*.) If all possible dynamical fragments are encountered, then $h_0^{(L)} = 1$. If only one fragment appears repeatedly, then $h_0^{(L)} = 0$. Distributions p_j between these two extremes will yield $h_0^{(L)}$ between 0 and 1. As the length of the dynamical fragments L is increased, $h_0^{(L)}$ provides a progressively more refined analysis of $x(t)$. In fact, it has been shown that under suitable conditions, the limit as $L \to \infty$ yields the *topological set dimension* or *fractal dimension* of the attractor to which trajectory $x(t)$ is drawn (Wolfram 1984c),

$$d_0 = \lim_{L \to \infty} h_0^{(L)} .$$

While the topological entropy $h_0^{(L)}$ is useful for telling us *how much* of the space of dynamical fragments $w_j^{(L)}$ is actually traversed by $x(t)$, it does not give any indication of *how often* each of the possible fragments is encountered. For example, $h_0^{(L)}$ would have the same value if each fragment appears an equal number of times, or if some fragments appear much more frequently than others, since it is only sensitive to whether or not $p_j > 0$. The second useful measure that can be culled from the parse tree is called the *metric entropy* (or Kolmogorov-Sinai entropy) (Crutchfield and Packard 1982, Crutchfield and Packard 1983, Wolfram 1984c), and this measure *does* give us an indication of how evenly $x(t)$ is spread out over the space of possible fragments $w_j^{(L)}$. It is defined as

$$h_1^{(L)} = -\frac{1}{L}\sum_{j=1}^{n^L} p_j \log_n p_j \ .$$

As with $h_0^{(L)}$, if all possible $w_j^{(L)}$ appear an equal number of times, then $p_j = 1/n^L$ for all j and $h_1^{(L)} = 1$. If only one $w_j^{(L)}$ appears repeatedly, then $p_j = 1$, $p_{k\neq j} = 0$ and $h_1^{(L)} = 0$. However, $h_1^{(L)}$ is bounded from above by $h_0^{(L)}$, $h_1^{(L)} \leq h_0^{(L)}$, for all possible distributions of the p_j, where equality is attained only if all $w_j^{(L)}$ that appear with nonzero probability $p_j > 0$ appear with the *same* probability p for all j. $h_1^{(L)}$ is therefore lower than $h_0^{(L)}$ only if the p_j are skewed in the sense that some are larger than others. $h_1^{(L)}$ thus adds to the measurement made by $h_0^{(L)}$ by telling us how evenly spread $x(t)$ is over the space of all possible dynamical fragments $w_j^{(L)}$. As with $h_0^{(L)}$, in the limit $L \rightarrow h_1^{(L)}$ becomes the metric set dimension,

$$d_1 = \lim_{L\rightarrow\infty} h_1^{(L)}$$

The metric entropy $h_1^{(L)}$ has one other remarkable property that makes it extremely useful for the analysis of trajectories from nonlinear dynamical systems. A description of this property requires a brief digression into nonlinear dynamical systems theory. One of the hallmarks of nonlinear systems is their ability to display chaos, a type of extremely disordered dynamical behavior that appears random but nonetheless is entirely deterministic. Chaos is associated with a sensitive dependence on initial conditions, in the sense that no matter how close together two different trajectories $x_1(0)$ and $x_2(0)$ begin, they immediately diverge at an exponential rate, $|x_1(t)-x_2(t)| = e^{\lambda t}$. The exponent λ that measures the rate of separation of the trajectories is called a *Lyapunov exponent*, and the degree to which $\lambda > 0$ is used as a measure of the level of chaos in a system. Multidimensional systems are generally characterized by several such exponents λ_i.

The metric entropy of $x(t)$ is useful as a measure of the degree of chaos in $x(t)$ because, under certain conditions, it is equal to the sum of the positive Lyapunov exponents $\lambda_i^+ > 0$ describing $x(t)$ (Crutchfield and Packard 1983),

$$h_1^{(L)} = \sum_i \lambda_i^+ .$$

As we will discuss further below, this fact has caused $h_1^{(L)}$ to be used often as a scale between total order ($h_1^{(L)} = 0$) and fully developed chaos ($h_1^{(L)} = 1$). Intuitively, this means that chaotic systems traverse all possible dynamical fragments with equal probability as they evolve in time.

3.2 Generating Partitions

The first essential step in symbolic dynamics analysis is the partitioning of a continuous phase space V into a finite set of symbols s_i. In the example of Figure 3, we created a partition of two symbols, $s_1 = 0$ and $s_2 = 1$, to span the continuous space $V = \{x \in \mathbf{R} | 0 \leq x \leq 1\}$. There was only one variable that we could adjust in creating this partition, corresponding to the line separating the two symbols at $x = 0.5$. This line is called the *decision point.* The choice of 0.5 seemed suitable for the trajectory $x(t)$ in Figure 3. But what if, instead, we had chosen a decision point of 0.99 or 0.01? Clearly, $x(t)$ simply never would have entered one of the volumes V_1 or V_2, and we would have been left with a sequence u consisting in either all zeros or all ones, two situations that would tell us nothing about $x(t)$. This extreme example underlines the importance of choosing an appropriate partition.

To see how a suitable partition can be chosen, consider the phase space $V = \{(x, y) \in \mathbf{R}^2 | 0 \leq x, y \leq 1\}$ depicted in Figure 5, which has been separated into two volumes V_1 and V_2 corresponding to the symbols 0 and 1 at time $t = m$.

In what follows, we will simply call the two volumes 0 and 1. At a given time step $t = m$, the partition tells us which points (x_m, y_m) are in 0, and which are in 1. But what about the previous time step, $t = m - 1$? Which points (x_{m-1}, y_{m-1}) at time $t = m - 1$ were mapped to the set of points (x_m, y_m) in 1 at $t = m$ under the dynamics of the system? Some of the points (x_{m-1}, y_{m-1}) must have been in 0 at $t = m - 1$, and some must have been in 1. Let us label the set of all points (x_{m-1}, y_{m-1}) that were in 0 at $t = m - 1$ and that moved under the dynamics of the system to points (x_m, y_m) in 1 at $t = m$ by 01. Similarly, all points that were in 0 at $t = m - 1$ and that moved to another

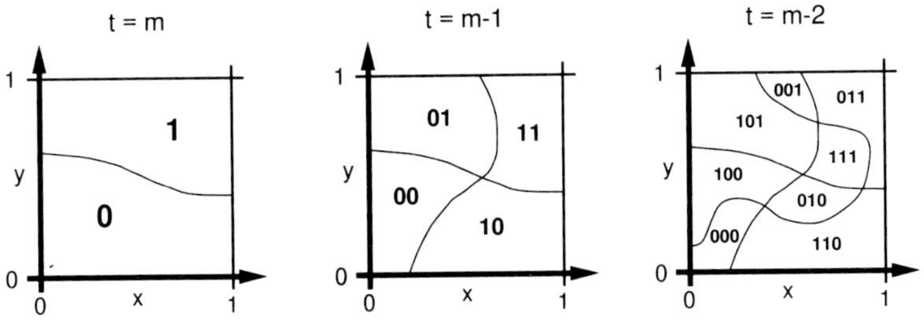

Figure 5: Choosing partitions for a phase space V.

point in 0 at $t = m$ will be labeled by 00, and so on. This is depicted in Figure 5, $t = m - 1$, where we have labeled all points that were in a at $t = m - 1$ and are in b at $t = m$ by ab, and a and b represent either 0 or 1.

By separating the points in V according to the trajectories that they will traverse over time steps $t = m - 1$ and $t = m$, we have been able to box them into smaller subsets than when we simply considered their position over the single time step $t = m$. We can apply this procedure again by asking where each point was on time step $t = m - 2$, which is also pictured in Figure 5, and so on backward as far as desired. The further back we go, the smaller the subsets containing points for each specific trajectory $\ldots s_i^{t=m-2} s_j^{t=m-1} s_k^{t=m}$ will become. The precise shape of the smaller and smaller subsets created through this procedure will of course depend on the shape of the original partition of V into subsets 0 and 1, and on the precise form of the dynamics.

Now here is the crucial point: For some special partitions, as we apply the above procedure, tracing longer and longer trajectories backward and subdividing V into smaller and smaller sets, in the limit $t \to -\infty$ each subset will contain only one unique point (x, y). These special partitions are called *generating partitions*, and they are precisely the ones that are best suited to analyses with the techniques of symbolic dynamics. Remarkably, it has been shown that generating partitions are precisely those partitions which maximize the metric entropy h_1.

It can be seen on an intuitive level why these partitions are the most suitable. If a dynamical trajectory $x(t)$ of a system has the ability through an appropriate choice of partition to generate many different symbol sequences $w_j^{(L)}$, but is prevented from doing so by choice of an unsuitable partition, then one will necessarily miss much of the most complex behavior displayed by $x(t)$. This would be equivalent to choosing a 0.01 or 0.99 decision point in Figure 3. The practical result of this discussion is that, when using symbolic dynamics to study a dynamical trajectory $x(t)$, one must experiment with different partitions to find one that maximizes h_1. This maximal h_1 will reveal the partition that captures the most detail from the trajectory $x(t)$.

4 Computation in Dynamical Systems

We introduced above the idea that every dynamical system experiencing a stimulus from the outside world can also be thought of as a computational device being fed information. But how are the dynamical systems concepts of attractors, basins and chaos related to the computational concepts of algorithms, computability and grammars? This section will describe three recent efforts to flesh out the duality between dynamical systems and computers. Their discoveries point to several bold conjectures regarding the relationship between our two fields of interest.

4.1 ϵ-Machines and Statistical Complexity

Imagine having a method to rigorously infer the smallest possible labeled digraph, corresponding to a regular grammar G, capable of generating the strings from a symbolic sequence u — and only those strings. This would allow us to concisely represent the sequence u corresponding to the dynamical trajectory $x(t)$ by a regular grammar, and also lead naturally to a measure of the complexity of any symbolic sequence u in terms of the grammar that represents it! James Crutchfield of the University of California at Berkeley has devised just such a method and named it *topological ϵ-machine reconstruction* because it involves reconstructing a finite automaton from a data series. After exploring the method itself, we will discuss its application to the logistic map $f(x) = \mu x(1-x)$, a simple nonlinear dynamic relevant to many biological phenomena.

There are two crucial concepts that must be understood. First, any parse tree $T^{(D)}$ of depth D can be broken into smaller subtrees $T^{(L)}$ of depth $L < D$: Pick any vertex V at some depth d such that $0 \leq d \leq D-L$; next, from V trace all n^L paths emanating downward L levels from V to level $d+L$. This gives subtree $T^{(L)}$. Define frequency counts f_i by starting with each individual leaf i of $T^{(L)}$ and adding up all f_j for leaves j of the tree $T^{(D)}$ that pass through leaf i. There will be n^{D-l-d} such counts f_j to add up for each leaf i. Given that the vertex V is at depth d in $T^{(D)}$, there must be a path of depth d, or equivalently a string $\omega^{(d)}$ of length d, leading from $T^{(D)}$'s source vertex to V. Then, emanating from V there are n^L length L strings $w_i^{(L)}$ with associated frequencies f_i. Thus the conditional probability of observing dynamical fragment $w_i^{(L)}$, given that fragment $\omega^{(d)}$ has just been observed, is

$$p(w_i^{(L)}|\omega^{(d)}) = f_i \Big/ \sum_{1=1}^{n^L} f_i \ .$$

Thus, the frequencies f_i give us a probability distribution over the possible future dynamical fragments $w_i^{(L)}$ given the past fragment $\omega^{(d)}$. Summing over the $D - L + 1$ possible depths d for a subtree $T^{(L)}$ inside $T^{(D)}$, there are a total of $\Sigma_{d=0}^{D-L} n^d = (1 - n^{D-L+1})/(1 - n)$ individual subtrees $T^{(L)}$ inside an overall tree $T^{(D)}$.

The second crucial notion is *topological subtree equivalence*. Consider subtrees $T_1^{(L)}$ and $T_2^{(L)}$ rooted at vertices V_1 and V_2 in $T^{(D)}$, not necessarily at the same depth. Both subtrees have depth L, and both must also have exactly n^L leaves i. Now consider each leaf i_1 from $T_1^{(L)}$ and its corresponding leaf i_2 in $T_2^{(L)}$ as a pair. The two trees are topologically equivalent if for each pair: (1) either $f_{i_1} = 0$ and $f_{i_2} = 0$; or (2) $f_{i_1} > 0$ and $f_{i_2} > 0$. Thus, two trees are said to be equivalent if their corresponding histograms are topologically conjugate, in the sense that each pair of corresponding counts f_{i_1} and f_{i_2} must be both zero or both nonzero.

By way of example, consider a data sequence u of length $N = 1001$ of the form 1X1X1X1X..., where X = 0 or 1, produced by the example grammar in Figure 2. Imagine, however, that we do not know the underlying grammar that produced this sequence, and that we are given the problem of trying to infer it. To reconstruct a finite automaton from this data sequence, we carry out the following steps (Crutchfield 1989, Crutchfield and Young 1990, Crutchfield 1991):

1. From u, construct a parse tree $T^{(D)}$ of depth D using the procedure described in section 3.1 above. This is pictured in Figure 6A for our example grammar.

2. From $T^{(D)}$, obtain all $(1 - n^{D-L+1})/(1 - n)$ possible subtrees $T^{(L)}$ of depth L, with $L = \lfloor D/2 \rfloor$ and $\lfloor k \rfloor$ meaning the largest integer not greater than k. For each subtree obtained, record the vertex V from $T^{(D)}$ at which the subtree starts. In our example, $L = \lfloor 5/2 \rfloor = 2$. The collection of subtrees $T^{(L=2)}$ from the tree $T^{(D=5)}$ for the example sequence is shown in Figure 6B.

3. Group all topologically equivalent subtrees into equivalence class M_k. Crutchfield calls each equivalence class a *morph*. Our example sequence

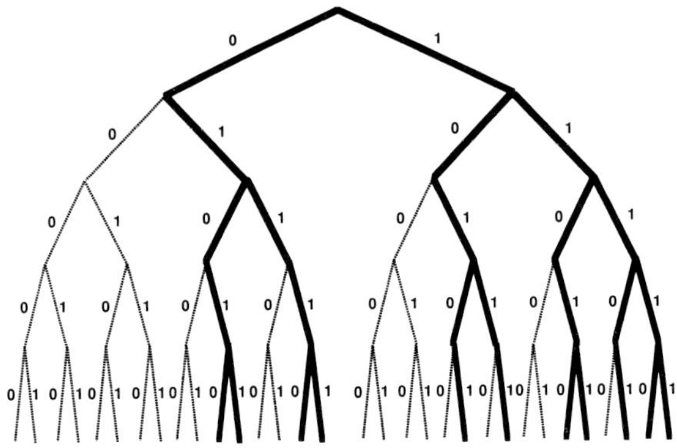

Figure 6: A. ϵ-machine parse tree for the grammar shown in Figure 2.

in Figure 6B gives three different morphs numbered $k = 1, 2$ and 3. Subtrees $T^{(L=2)}$ containing no frequency counts are not considered as a separate morph.

4. Construct a new subtree $T_M^{(D-L)}$ of depth $D - L$ whose source vertex coincides with the source vertex of $T^{(D)}$. This tree is easily obtained by simply clipping off the bottom L levels of $T^{(D)}$. Each vertex in $T_M^{(D-L)}$, including the leaves, is the source vertex V for one of the subtrees in steps 2 and 3. For each vertex V in $T_M^{(D-L)}$, look up the corresponding subtree $T^{(L)}$ that starts at this vertex. Take the number of the equivalence class M_k to which $T^{(L)}$ belongs, and write it down with the vertex V in $T_M^{(D-L)}$. The subtree $T_M^{(D-L=3)}$ for our example sequence is shown in Figure 6C. Each of the vertices in $T_M^{(D-L=3)}$ with a nonzero frequency count has been assigned to one of the three equivalence classes, whose number appears inside a circle.

5. Construct a labeled digraph: For each morph M_k in the subtree $T_M^{(D-L)}$, create a state (vertex) q_k in the digraph. Add labeled edges to the digraph according to the following rule: if anywhere in $T_M^{(D-L)}$ there appears an edge from a vertex of class M_k to another vertex of class $M_{k'}$ labeled with symbol s, then in the digraph draw an edge leading from state q_k to state $q_{k'}$ labeled with the symbol s. For our example sequence the resulting digraph appears in Figure 6D.

The grammar that we have analyzed to illustrate the reconstruction technique is admittedly not very complex, but one of the technique's principal

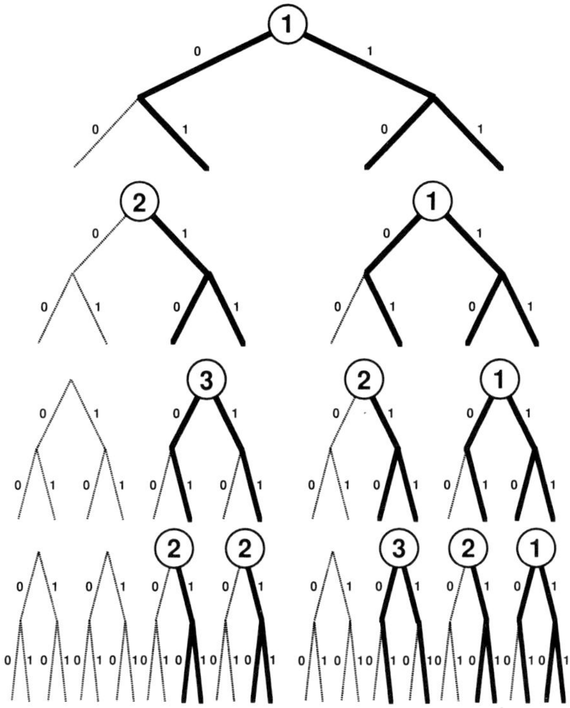

Figure 6: B. Subtree collection for the parse tree shown in Figure 6A.

assets is that it furnishes a means for precisely quantifying this impression. The measure of complexity defined on the ϵ-machines, which Crutchfield has called *statistical complexity*, has both a topological and a metric version as with the entropies for dynamical fragments (Crutchfield and Young 1990). In fact, the entropies and complexities are calculated in very much the same way. The topological complexity c_0 is just the base-n logarithm of the number of morphs, $c_0^{(L)} = L^{-1} \log_n(N_{morph}) = 0.5 \log_2(3)$ in our example, just as the topological entropy h_0 was the logarithm of the number of dynamical fragments with nonzero probability, divided by the tree depth L. The metric complexity, as with the corresponding entropy h_1, refines the topological measure by considering how much each morph is used.

To obtain a measure of the morph probabilities p_k, we go back to the morph subtree $T_M^{(D-L)}$ and create a frequency histogram f_i for the leaves of this subtree. We then calculate the probability for each morph M_k by summing

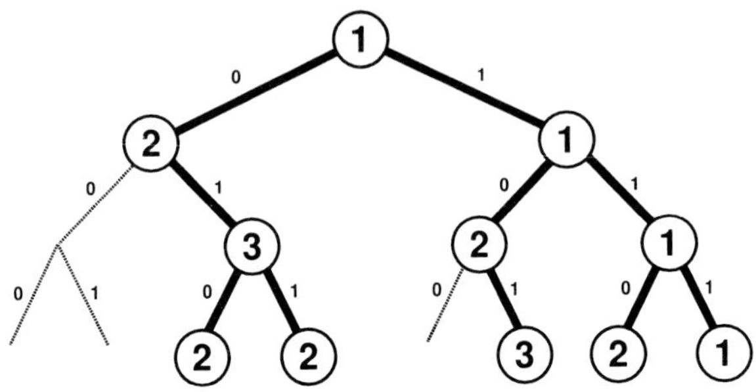

Figure 6: C. Associating tree vertices and morph equivalence classes. See text.

the frequencies of all the leaves i belonging to M_k and dividing by $N - D + 1$,

$$p_k = \frac{1}{N - D + 1} \sum_{i \in M_k} f_i \ .$$

The morph probabilities p_k are then used to calculate the metric complexity of the reconstructed digraph exactly as with the metric entropy,

$$c_1^{(L)} = \frac{1}{L} \sum_{k=1}^{N_{morph}} p_k \log_n(p_k) \ .$$

where L is the subtree depth. (It is noteworthy that there are unnormalized versions of the entropies and complexities in common use in the literature as well which are not divided by the tree depths, defined as $H_0 = Lh_0^{(L)}, H_1 = Lh_1^{(L)}, C_0 = Lc_0^{(L)}, C_1 = Lc_1^{(L)}$.)

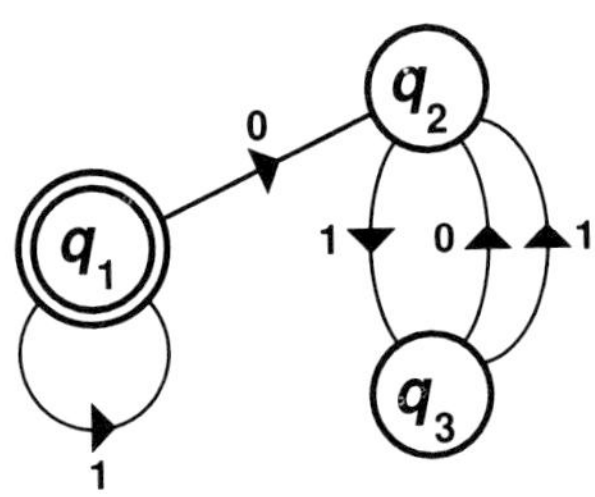

Figure 6: D. Final, labeled digraph for the Crutchfield ϵ-machine corresponding to the grammar in Figure 2.

The reconstructed digraph complexity is essentially a measure of the amount of memory in a process, or to put it in somewhat anthropomorphic terms, it is a measure of how much the process remembers where it has been and to what extent it uses this information to determine its future actions. To see how this interpretation comes about, let us recall for a moment what a morph subtree represents. For a given string $\omega^{(d)}$ leading up to its source vertex, the morph indicates which future strings $w_i^{(L)}$ have a nonzero probability of occurring. Thus, if there is a different morph, or equivalently a different set of possible future strings $w_i^{(L)}$ for every possible preceding string $\omega^{(d)}$, then obviously the process's future behavior depends very much on its past. If there is only one morph for all preceding strings $\omega^{(d)}$, then the future behavior of the process is independent of its past. These are the two extremes, and they are of course joined by a spectrum of intermediate possibilities between them. If only one morph exists, then $c_0 = c_1 = 0$, while at the opposite extreme if there is a different morph for every one of the n^{D-L} strings $\omega^{(d)}$ and they are all used equally, then $c_0 = c_1 = 1$.

Now here is the single most useful and important property of this measure of complexity: For purely constant processes which produce a homogeneous sequence of the same symbol, the complexity is zero; and for processes which simply emit a new symbol chosen at random at every time step, the complexity is also zero (since in both cases, the set of future possibilities is the same regardless of the past). The statistical complexity therefore conforms to a criterion met by virtually no other such measure. It coincides with our intuitive belief that purely homogeneous or purely random processes are not complex. Rather, complexity is to be found for some intermediate behavior between the two, and it has been quantified in the case of this so-called *statistical complexity* by the computational power of the regular grammar corresponding to a dynamical system. Thus, statistical complexity bears out our belief, mentioned in section 1., that metaphors from computer science can be fruitfully employed to understand the complexity of dynamical systems in a way not previously possible.

There is a catch, however. What we have given above is an algorithmic procedure for reconstructing the regular grammar corresponding to a dynamical system that can be almost completely automated in a concise computer program. But what if the dynamical system is so complex that no regular grammar can capture its dynamics? What if only grammars from a higher level in the Chomsky hierarchy, say context-sensitive or unrestricted, can accurately capture the system's dynamics? Then we are currently out of luck. There is no known automated procedure for inferring higher-level grammars from symbol sequences, and many investigators suspect that the problem is undecidable. And, as fate would have it, this higher level of complexity can arise even in relatively simple systems.

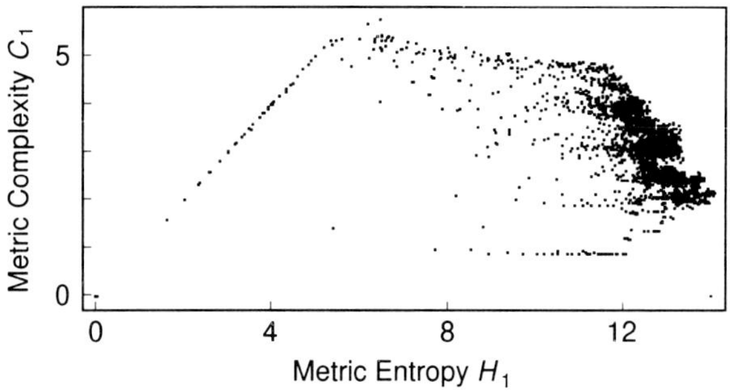

Figure 7: The metric complexity C_1 for the logistic map, considered as a function of the metric entropy H_1.

4.1.1 ϵ-Machines for the logistic map

To illustrate many of the points discussed in the above paragraphs, we turn to an example application of the ϵ-machine reconstruction technique first undertaken by Crutchfield, a study of the logistic map $f(x) = \mu x(1-x)$ (Crutchfield and Young 1989, 1990). Using the 2-symbol generating (metric entropy maximizing) partition $s_1 = 0 = \{x \in \mathbf{R} | 0 \leq x \leq 0.5\}$ and $s_2 = 1 = \{x \in \mathbf{R} | 0.5 < x \leq 1.0\}$, this map was studied by constructing parse tree and ϵ-machines for symbol sequences millions of characters long, for many different values of the parameter μ. For those readers not already acquainted with the dynamical behavior of this charmingly complex equation, we provide a brief summary here. For $0 < \mu < 3$, the map has one attracting fixed point at $x = 1 - \mu^{-1}$. Beyond this value of μ, there is a period-doubling bifurcation that replaces the fixed point with a period-2 attractor. As μ is increased further, more period-doubling bifurcations take place over successively smaller intervals of μ, until finally at $\mu = \mu_\infty = 3.569946$, the period becomes infinite and there is a transition to chaos. For $\mu_\infty < \mu < 4$ there are many interesting regimes showing varying degrees of chaos, until finally at $\mu = 4$ fully developed chaos is initiated and all symbol sequences of any given length appear with equal probability. This great diversity of behavior has made the logistic map an important tool in for understanding how nonlinear dynamics and complexity may be interrelated in biological systems.

A plot of the metric complexity C_1 versus the metric entropy H_1 for the logistic map appears in Figure 7, each point representing a different value of μ. In all, there are 5000 values of μ sampled at intervals of $\Delta\mu = 10^{-4}$ between $\mu = 3.5$ and $\mu = 4$. It is worth taking note of several of the features of this plot, as it is rapidly becoming a standard for establishing the computational

(C_1 or c_1) and informational (H_1 or h_1) behavior of complex dynamical systems relative to other such systems. Purely static systems which repeatedly produce the same symbol appear in the lower left corner of the plot, where $C_1 = H_1 = 0$. Purely random or chaotic systems which produce all symbols at random with the same probability appear in the lower right corner of the plot, where $C_1 = 0$ and $H_1 = L = 14$. The straight line emanating from the lower left corner of the plot represents purely periodic orbits, whose periodicity increases as the line extends upward and to the right. The slope of this line is one, since the number of recurrent symbol sequences and recurrent morphs for a periodic orbit are both equal to the period. Finally and most noteworthy, the maximum complexity occurs at intermediate values of the entropy, not at the most ordered ($H_1 = 0$) or chaotic ($H_1 = L$) values.

Crutchfield has found that the complexity of the logistic map has its maximal value at $\mu = \mu_\infty$, where the period doubling cascade diverges to an infinite period and the map becomes chaotic. At this point, in fact, the reconstructed machine size diverges, and there is no regular grammar that can capture the map's symbolic sequences. Through careful examination of the structure of the diverging regular grammars (Crutchfield and Young 1990), Crutchfield was able to establish that symbolic sequences at the period-doubling transition to chaos can be captured by an indexed context-free grammar, which lies between context-free and context-sensitive in the Chomsky hierarchy. While this is an extremely interesting result, it is noteworthy that establishing this indexed context-free grammar was a labor-intensive task for which there is no simple algorithm. Nevertheless, Crutchfield's work on the logistic map (and other work which we will now discuss) has led to the speculation that all nontrivially computing dynamical systems are situated at a phase transition between order and chaos. We will have more to say on this conjecture below.

4.2 Cellular Automata and Computation at the Edge of Chaos

While symbolic dynamics approaches have been very fruitful, there is yet another system for analyzing computation in dynamical systems that is far more popular, called the *cellular automaton* (CA). Invented by John von Neumann in the 1940s, these systems have been the subject of a vast amount of study, especially since having been popularized and scrutinized by Stephen Wolfram in the early 1980s (Wolfram 1984a–c). A cellular automaton consists in a 1–, 2–, or d–dimensional hypercubic lattice of *cells* $s_{1,2,\ldots,d}$, each of which is capable of being in a state q from a finite set of states Q. Typically, the one or more dimensions of the lattice are associated with directions in space, so that each cell is associated with a specific location in the region being modeled. Cellular automata are consequently said to be *spatially extended systems*.

Temporal dynamics are added to the cellular automaton via a transition function $\delta: Q^{2rd+1} \rightarrow Q$ somewhat like the one introduced in section 2.1 for the Turing machine. The transition function is identical for all cells s_i in the lattice, taking as input the cell s_i's current state and the states of a number of its neighbors, and mapping these to a new output state for s_i. The variable r in the definition of δ is called the *radius* of the transition function's input neighborhood and d is the dimensionality of the lattice, so that in addition to s_i's current state, δ takes as input the current states of the r cells in each of the $2d$ directions away from s_i, for a total of $2rd + 1$ states.

Aside from the fact that the spatiotemporal dynamics of even simple CA can look complex and pretty, there are essentially two reasons why CA have been the object of so much attention. First, cellular automata have the unique property of being discrete in all possible senses: (1) space is separated into discrete cells; (2) time is separated into discrete steps; and (3) state variables in each cell are discrete, in that they may only take on only one of the finite number of states q. As such, CA are already formulated in the language that is native to computation, and it is consequently much easier to assess their computational capabilities (particularly for 1-dimensional CA, where one simply associates the different lattice configurations produced by the CA with the different strings produced by a grammar).

The second reason has to do with the intimate relationship of CA to physics. Because the same transition function is applied to all cells in the CA lattice, the dynamical behavior is translationally invariant, very much like the laws of physics. Similarly, it is a simple matter to make the transition function rotationally invariant as well. Finally, a judicious choice of the transition function combined with the preceding two points can create a CA that mimics physical laws on a discrete spatiotemporal grid, leading to strong methods for simulating spatially extended physical systems.

One piece of work in particular that has attracted much attention was carried out in the late 1980s by Christopher Langton, then at University of Michigan (Langton 1990, 1992). Langton's goal was to search through the space of all possible one- and two-dimensional CA transition functions to locate regions associated with the different kinds of dynamical behavior, and in particular to locate those regions best able to support nontrivial computation.

A central problem faced by Langton was to develop a means for associating the different dynamical behaviors observed in his simulations with different regions in the space of possible transition functions. Transition functions are made up of discrete symbols that have no magnitude or order defined on them, so how can one determine whether two different functions are close together or far apart? Langton used the following method: (1) choose one of the c CA states q at random and call it the quiescent state q_q; (2) count the number of neighborhoods n in the transition function that map to q_q; and (3) calculate the parameter $\lambda = (c^{2rd+1} - n)/c^{2rd+1}$. If all states map to q_q, $\lambda = 0$, and if none map to q_q, $\lambda = 1$, so the λ parameter measures the level of bias in a given transition function towards the quiescent state. (Unfortunately, λ says nothing about how the remaining neighborhoods are distributed among the remaining non-quiescent states.)

Langton used this single parameter to sort all the CA transition functions in order of increasing λ. The dynamics of the simulated CA showed a strong correlation with λ. At low values, the CA quickly reached a fixed point or entered a periodic orbit, resulting in a very low entropy, while at high values, the CA quickly entered a state of apparent total disorder with a correspondingly high entropy, indicating a chaotic state. Approaching $\lambda = 0.5$ from below, the transient length displayed a sharp peak, the entropy rose sharply, and the correlations between cells peaked and fell gradually. Visual inspection of the evolving CA revealed that the most complex behavior, defined by virtue of its being the most qualitatively unpredictable, occurred in the vicinity of $\lambda = 0.5$.

In analyzing and interpreting his results, Langton argued that a system capable of universal computation could not be simply periodic or chaotic, but must instead occur at the "critical" value of $\lambda = 0.5$ between order and chaos. In support of this conjecture, the plot of the correlation between a cell and itself at the next time step (autocorrelation) versus the cell entropy revealed low autocorrelations at low entropy (order) and at high entropy (chaos), while at middle values of entropy the autocorrelation was higher. Suggesting an analogy between his cell autocorrelation and Crutchfield's regular language complexity, he pointed out that his autocorrelation-entropy plot looked very similar to Crutchfield's complexity-entropy plot for the logistic map (Figure 7). He then went on to offer the intriguing hypothesis that λ acts as an order parameter for the transition between order and chaos in CA the same way that temperature acts as an order parameter for the transition between solid and liquid states in physical matter. Finally, he speculated that, in light of the above arguments, computation in physical systems is most likely to occur at a phase transition between ordered and chaotic states.

Langton's arguments are extremely provocative and have produced extensive speculation about edge-of-chaos phenomena in biology (e.g. Kauffman 1993). The force of his suggestions is currently limited by several problems. First, Langton has not as yet supplied quantitative evidence to support the proposal that the dynamical behavior observed at "critical" values of λ can actually support computation. Second, he has as yet provided only heuristic arguments to justify his assumption that the dynamics of computation are inconsistent with, say, quasiperiodic or weakly chaotic dynamics, both of which would occur away from the "critical" values of the metric entropy. Third, it may be questionable to associate a single "average" type of dynamical behavior with each value of the λ parameter. Simple combinatorics shows that the number of possible transition functions increases exponentially with the value of λ, so that the vast majority of possible functions must occur at high λ values. Among this latter group there is certainly a wide variety of diverse possible behaviors, possibly blurring the order/chaos distinction. Finally, since a universal computer can simulate any kind of dynamical system, it is conceivable that every possible kind of dynamical behavior could be exhibited by the same transition function, just by starting it with different initial configurations. So for a computationally complex CA, there is no well-defined "average" behavior — instead, the behavior will change dramatically depending on the initial configuration. Indeed, there are CA which perform nontrivial computations but do not sit at the phase transition values of λ predicted by Langton (Mitchell, Crutchfield and Hraber 1994), so many mysteries of outstanding importance remain.

4.3 Sulis Machines: the Organism in Its World

Up to this point, the duality we have described between dynamical systems theory and theoretical computer science has played captive to the latter's definition of computation. Accordingly, there are three equivalences: (1) the data to be processed by a computer program is encoded in and represented by the initial condition of the corresponding dynamical system; (2) the algorithm to be executed by the computer corresponds to the rules for time evolution of the dynamical system; and (3) the process of running the program until it terminates corresponds to allowing the dynamical system to evolve in time through a basin of attraction until it reaches a fixed-point attractor.

It must be realized, however, that the vast majority of biological systems in which one might wish to look for computational behavior do not have this luxury. A society, a brain, or even a single cell cannot simply shut out input from the outside world while it allows its internal dynamics to unfold. Instead, it is constantly receiving new information from the outside world throughout the dynamical process or computation that it is performing.

This is one of the central themes in the recent work of William Sulis, a psychiatrist and mathematician at McMaster University. In his work (Sulis 1994a,b, 1995) on what he calls "naturally occurring computational systems" (as opposed to artificial computational systems like the modern digital computer), Sulis delimits the notion of natural versus artificial computing according to various constraints that are substantially different for the two different kinds of system and which, he argues, are responsible for the many differences between the two. We will focus here on only two of the many interesting arguments that he has made. The first is philosophical, striking at the heart of the very definition of computation, while the second is intended to give the reader a taste of the technical flavor of Sulis' work.

Paralleling trends in cognitive science and developmental psychology, Sulis has framed a challenge to the currently accepted definition of computation in terms of what he calls *computational competence* and *computational performance*. A system's computational competence is the set of all computations that it can carry out or functions that it can evaluate, while its computational performance is the actual method it employs to do so (Sulis 1994a). Sulis notes that the vast majority of research on computation that has been and continues to be done is devoted to computational performance. It is now common to assume that any computer can evaluate all computable functions, and focus on optimizing computational performance.

Sulis argues that this view is far too limiting. First of all, most biological systems that are computationally complex are very likely to be highly specialized in terms of their adaptations, and therefore very unlikely to be universal computers. So the issue of establishing computational competence for each individual system is extremely important. Second, the fact that there exists a Turing machine with the same computational competence as a given natural system does *not* mean that the system carries out its computations in the same way as the TM. In particular, Sulis claims that naturally occurring computational systems do not necessarily carry out their computations by following an algorithm of the familiar sort. There *may* be such an algorithm that computes the same function, but so what? The dynamics of the natural system do not necessarily resemble it in any way. An example of this is a simple analog computer, a network of resistors and capacitors that finds the solution of an ordinary differential equation. The solution to such an equation can also be found by digital computer using, say, a Runge-Kutta integration algorithm. The two systems may thus have the same computational competence, but to assert that the only way to understand the operation of the analog computer is in terms of the corresponding digital computer algorithm seems clearly in error.

If it is true that computation transcends its definition in terms of algorithms, then there must be some other way of representing computation in dynamical systems that does not involve the execution of an algorithm. What could it be? Sulis begins with the dynamics of a system taking input from the outside world, and builds a very natural and novel way of accommodating the attractor and basin metaphors of dynamical systems into a symbolic framework. On the surface, the formalism looks similar to the one that results from Crutchfield's ϵ-machine reconstruction technique in that both involve the use of a finite automaton (FA) to represent the dynamical evolution of a system, and the idea of attractor$\leftrightarrow$symbol metaphors is reminiscent of J.J. Hopfield's influential work on the attractor dynamics of spin-glass-like networks as models of memory formation and recall (for an introduction to Hopfield models see the essays by Busch and Trainor and by Sutton, this volume). However, the crucial difference involves the meanings that are assigned to the states and symbols of the FA. Recall that an FA consists in a set of b symbols a from a finite alphabet S, a set of c states q from a finite set Q, and a transition function $\delta{:}S \times Q \rightarrow Q$ that maps every symbol-state pair (a, q) to a new state q'. The symbols therefore act as input to the FA.

In Crutchfield's framework, the input symbols represent the configuration of the dynamical system being observed (which is isolated from the outside world), and the states of the FA represent the "morphs," abstract structures obtained from the dynamical system's parse tree for the purpose of optimal forecasting of the system's future behavior. In Sulis' framework, the system being modeled is not isolated from the outside world, but is rather constantly receiving input from the outside. Consequently, the input symbols to the FA represent the configuration of the outside world (at least the part of it that the system can detect) and the FA's states represent the system's internal configuration. Thus, on every time step in Sulis' framework, the transition function δ maps the current input and the current internal configuration to a new internal configuration.

This alternate framework has a number of consequences for the form of the finite automaton's corresponding digraph. Since a Sulis machine cannot make any *a priori* assumptions about the behavior of the outside world, it must be prepared to receive any input symbol no matter what its internal state. Thus, the digraph will possess c vertices representing the states q, but from every vertex there must emanate b edges labeled with each of the possible input symbols a. What makes each system unique is the destinations of the edges. Furthermore, since this framework is designed to model the behavior of naturally occurring systems, it has *no* fixed points or attractors from which it can never escape. How, then, is one to analyze the dynamics of such a system?

Sulis has introduced the following new and useful concepts, which we will describe by way of the example system shown in Figure 8A. It is expected that systems complex enough to warrant the use of this formalism will not fall into simple periodic behavior, but will instead traverse what is essentially

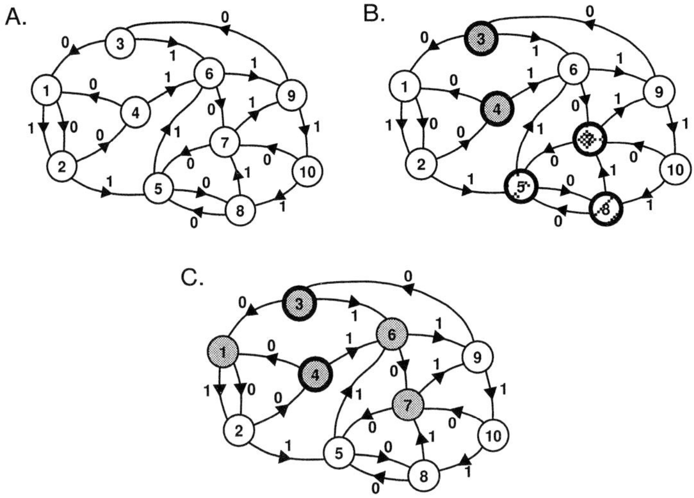

Figure 8: Never-ending transients and transient attractors in a simple Sulis machine.

a never-ending transient. Thus, each of the input symbols and each of the internal states actually represent transient fragments of dynamical behavior from the outside world and the system's internal motion, respectively. These are essentially identical to the dynamical fragments $w_j^{(L)}$ that we introduced in the context of parse-tree construction, in section 3.1. The example system in Figure 8A has 10 possible internal states or transients that it can traverse, while its input comes in the form of 2 possible symbols 0 and 1, which represent transient behavior in the outside world.

The first of the concepts introduced by Sulis is that of *dynamical equivalence.* Two states are said to be dynamically equivalent if they map to the same next state no matter what the input. For example, states 3 and 4 in Figure 8A are dynamically equivalent, because under input 0 they will both map to state 1, while under input 1 they will both map to state 6. The system will thus behave identically whether in state 3 or 4.

The next two concepts are analogs of the attractors and attractor basins common to ordinary dynamical systems, called *transient attractors* and *transient attractor basins.* The difference between these latter structures and their more common cousins is that they are defined relative to some sequence of input symbols ρ. Each state which is reached from any other state under input ρ is defined as an attractor for ρ. Dynamically equivalent states are lumped together into the same attractor, since they are indistinguishable. For example, in Figure 8B the transient attractors for input $\rho = 10$ are states 3 and 4, 5, 7, and 8, each of which is bolded and shaded differently to indicate that each is a separate attractor (states 3 and 4 are shaded the same since they are dynamically equivalent and therefore form only one attractor). The transient attractor basins are all those states which feed into each of the attractors under input ρ. In Figure 8C, the attractor basin states for the attractor composed of states 3 and 4 are shown shaded the same as the attractor, but not bolded to differentiate them from the attractor itself.

Sulis defines the computational competence of his systems as the set of all possible sequences of pairs (a, q) that they can produce, where a is the input symbol and q the system's internal state. The input sequence can of course consist of all possible words of all lengths made up of symbols from the alphabet S, but the sequence of corresponding internal states is very much a function of the particular system's transition function. Each system with a different transition function will therefore have a different competence, that is, a different set of computations that it can perform.

Although Sulis' framework possesses attractor and attractor basin structures similar to those for an isolated dynamical system, it is substantially different from the latter in that its attractor-basin structure is constantly changing, remodeled by events in the outside world. Without resorting to the complex and abstract production rules of a context-sensitive grammar, this allows the dynamical system to act in the context of its environment and its past experience. The same stimulus will not always elicit the same response — instead, the system will react according to the present input and its current internal state, itself a representation of the system's past. Most importantly, this is accomplished without having to redefine the system in terms of traditional algorithmic constructs.

5 More Questions

Complexity, so commonplace in biological things, leads quickly to mysteries that no amount of formalism can as yet dispel. There *is* a search for the computational Rosetta Stone of nonlinear dynamics, and we have sampled here a few of its more salient forms. Most of the important questions are open, so we close with three observations that seem to us crucial to resolving how dynamics, computation, and complexity are tied together in living matter:

- **No amount of parallelism can do the job of recursion.** While massive parallelism is obviously very efficient for some problems, there are many other problems where results must be built upon, iteration by iteration. Any theory of computation in distributed networks, such as those now so popular in CA and neural-net modeling, must face this issue head on.

- **Computation exists by design.** We don't look at something and say that it is computing; we design something that computes. Can natural selection do the same (e.g. Crutchfield and Mitchell 1995)? Computation is a representation of a stream of logic and reasoning, and how does one make sense of these concepts in a dynamical network? Each variable and conditional statement in a computer program has a specific meaning. How is one to assign meaning to the various portions of the phase space in a high-dimensional dynamical system?

- **Concepts from low-dimensional systems are limiting.** Attractors, basins and chaos are phenomena discovered and most properly understood in low-dimensional systems. In analyzing systems with higher dimensionality, are there striking new features of computational significance?

Acknowledgements

We gratefully acknowledge the grant support to CJL by the Natural Sciences and Engineering Research Council of Canada, and the support of the Medical Research Council of Canada, in the form of operating grants and career awards to CJL and a Postgraduate Scholarship to PAD.

References

Barwise, J. and Etchemendy, J. (1993) *Turing's World 3.0: An Introduction to Computability Theory.* Center for the Study of Language and Information (CLSI) Lecture Notes, Number 35. Stanford, CA: CSLI Publications.

Blum, L., Shub, M. and Smale, S. (1989) On a theory of computation and complexity over the real numbers. *Bulletin of the American Mathematical Society* **21**, 1–46.

Branicky, M.S. (1994) Analog computation with continuous ODEs. *PhysComp '94.* Proceedings of the Workshop on Physics and Computation 1994. Los Alamitos, CA: IEEE Computer Society Press, pp. 265–274.

Crutchfield, J.P. (1989) Inferring the dynamic, quantifying physical complexity. In N.B. Abraham, ed., *Measures of Complexity and Chaos.* New York: Plenum Press, pp. 327–338.

Crutchfield, J.P. (1991) Reconstructing language hierarchies. In H. Atmanspacher and H. Scheingraber, eds., *Information Dynamics.* New York: Plenum Press, pp. 45–60.

Crutchfield, J.P. and Mitchell, M. (1995) The evolution of emergent computation. *Proceedings of the U.S. National Academy of Sciences* **92**, 10742–10746.

Crutchfield, J.P. and Packard, N.H. (1982) Symbolic dynamics of one-dimensional maps: entropies, finite precision, and noise. *International Journal of Theoretical Physics* **6/7**, 433–466.

Crutchfield, J.P. and Packard, N.H. (1983) Symbolic dynamics of noisy chaos. *Physica D* **7**, 201–223.

Crutchfield, J.P. and Young, K. (1989) Inferring statistical complexity. *Physical Review Letters* **63**, 105–108.

Crutchfield, J.P. and Young, K. (1990) Computation at the onset of chaos. In W.H. Zurek, ed., *Complexity, Entropy, and the Physics of Information.* Proceedings of the 1988 Workshop on Complexity, Entropy, and the Physics of Information. Santa Fe Institute Studies in the Sciences of Complexity, Volume VIII. Reading, MA: Addison-Wesley, pp. 223–260.

da Costa, N.C.A. and Doria, F.A. (1991) Undecidability and incompleteness in classical mechanics. *International Journal of Theoretical Physics* **30**, 1041–1073.

Hopcroft, J.E. and Ullman, J.D. (1979) *Introduction to Automata Theory, Languages, and Computation.* Reading, MA: Addison-Wesley.

Kauffman, S.A. (1993) *The Origins of Order: Self-Organization and Selection in Evolution.* New York: Oxford University Press.

Kfoury, A.J., Moll, R.N. and Arbib, M.A. (1982) *A Programming Approach to Computability.* New York: Springer-Verlag.

Langton, C.G. (1990) Computation at the edge of chaos: phase transitions and emergent computation. *Physica D* **42**, 12–37.

Langton, C.G. (1992) Life at the edge of chaos. In C.G. Langton, C. Taylor, J.D. Farmer and S. Rasmussen, eds., *Artificial Life II.*

Proceedings of the Workshop on Artificial Life, Santa Fe, New Mexico, February 1990. Santa Fe Institute Studies in the Sciences of Complexity, Volume X. Reading, MA: Addison-Wesley, pp. 41–91.

Minsky, M.L. (1967) *Computation: Finite and Infinite Machines.* Englewood Cliffs, NJ: Prentice-Hall.

Mitchell, M., Crutchfield, J.P. and Hraber, P.T. (1994) Dynamics, computation, and the "edge of chaos": a re-examination. In G. Cowan, D. Pines and D. Meltzer, eds., *Complexity: Metaphors, Models and Reality.* Santa Fe Institute Studies in the Sciences of Complexity, Volume XIX. Reading, MA: Addison-Wesley, pp. 497–513.

Moore, C. (1990) Unpredictability and undecidability in dynamical systems. *Physical Review Letters* **64**, 2354–2357.

Moore, C. (1991) Generalized shifts: unpredictability and undecidability in dynamical systems. *Nonlinearity* **4**, 199–230.

Nicolis, G. and Prigogine, I. (1977) *Self-Organization in Nonequilibrium Systems.* New York: Wiley.

Sulis, W. (1994a) Naturally occurring computational systems. *World Futures* **39**, 225-221.

Sulis, W. (1994b) Synchronization in NOCS: transients, disorder and irreducibility. Preprint.

Sulis, W. (1995) Driven cellular automata, adaptation and the binding problem. In F. Moran, A. Moreno, J.J. Merelo and P. Chacon, eds., *Advances in Artifical Life.* Proceedings of the Third European Conference on Artifical Life. Berlin: Springer-Verlag, pp. 824–840.

Wolfram, S. (1984a) Algebraic properties of cellular automata. *Communications in Mathematical Physics* **93**, 219-258.

Wolfram, S. (1984b) Computation theory of cellular automata. *Communications in Mathematical Physics* **96**, 15–57.

Wolfram, S. (1984c) Universality and complexity in cellular automata. *Physica D* **10**, 1–35.

Wolfram, S. (1985) Undecidability and intractability in theoretical physics. *Physical Review Letters* **54**, 735–738.

II

Development:

Field Approaches to Pattern Formation

Vector Field Models of Morphogenesis

Wendy A. M. Brandts and John Totafurno

1 Introduction

Pattern formation in biological morphogenesis is a crucial process that contributes to the complex organization of life. Biological development epitomizes gloriously the ordered processes whereby structural and functional organization of a system is successively created and expressed, thus leading to the most complex forms of highly structured, dynamical matter.

The challenge of developing models that capture the essence of pattern and form generating mechanisms is particularly great in biological systems because both the fundamental "constituents" contributing to a pattern or form, and the underlying processes that govern "interactions" are usually unknown. In addition, the great complexity of biological systems makes it difficult to immediately ascertain the most appropriate simplifications at any of several levels of description (e.g. molecular, cellular, tissue). In this essay we will treat pattern and form generation at a global or "whole system" level, so that we can compare our results directly with the numerous and varied experimental data and organise these results relevant to the taxonomy of the organism. This sort of approach or theory does not exclude bridging to the familiar reductionistic microscopic mechanisms of physics and chemistry, but its primary purpose is to describe the large-scale pattern and form generation actually observed. Indeed, organisms comprised of the same macromolecules, and even the same genetic information, can demonstrate different yet stable configurations of their body structures. How are these diverse patterns generated and controlled?

One can gain insight into pattern and form generation and regulation in biological organisms by studying systems which display unusual symmetries, and by observing the effects of imposed perturbations. Symmetry is a particularly valuable concept in that it tends to reflect global properties characterizing particular biological organisms or groups of organisms. In this sense they contribute to the discovery of universal laws of pattern and form generation in developmental biology.

As an example of the power of symmetry principles, consider the occurrence of triplications and mirror reversals for a wide variety of biological organisms. In a triplication, three structures form instead of the single which is expected in the normal case, and the middle structure has reversed symmetry (handedness) with respect to the other two. In a mirror reversal, structures are arranged mirror-symmetrically. Many varieties of triplications and mirror reversals, occuring naturally or experimentally induced, have been examined (e.g. Goss 1969, Kumazawa 1979, Maden 1982, Russell 1985, Shi et al. 1991). What is it about the development of biological order, the unfolding of body plans and the processes of regulation, that makes such symmetries predominate? Are there universal generative rules which orchestrate similar developmental patterns in a wide variety of biological organisms?

Much of the work on pattern formation in living systems has focused on multicellular organisms, such as arthropods, insects and amphibians; however, even "simple" single cells can show elaborate pattern formation capabilities reminiscent of the complexity found in these multicellular systems (e.g. Kumazawa 1979, Suhama 1990, Shi et al. 1991). By describing unicellular organisms within a conceptual framework, such as our dynamic vector field, that also applies to multicellular organisms, a unified interpretation of pattern formation in morphogenesis is achieved.

The mechanisms for the expression of morphogenetic information, and for coordinating and controlling activity in a biological system, have been interpreted in several ways: positional information, prepattern, mechanical forces. Positional information in its most literal interpretation, suggests that cells have their position specified with respect to some natural coordinate system in the body. Differentiation is then decoupled from position: positional values are independent of the structures that they form. This idea is exemplified in transplant experiments where the transplanted material differentiates into a structure appropriate for the position specified by the host environment, though still characteristic of its origin.

Positional information, in particular in the form of the polar coordinate model (French, Bryant and Bryant 1976) and related works (Bryant, French and Bryant 1981; Lewis 1981; Mitthenthal 1981) implies a set of phenomenological rules that codify cellular responses to surgical intervention. It has provided a unifying explanation for some aspects of epimorphic regulation which involves localized growth and pattern formation at a discontinuity (e.g. a cut surface or grafting of cells which are not normal neighbors). Intercalation, the insertion or interpolation of new positional values, is attributed to the occurrence of this discontinuity. However, these models lack the formal structure capable of dealing *quantitatively* and *dynamically* with regulation. In particular, they are inadequate to treat morphalactic regulation, in which reorganization of the biological system results from long range interactions, such as a change in its size, rather than a distinct discontinuity. Our dynamic vector field approach overcomes this limitation and rearrangement of patterns may occur over a large part of the system, as a result of local interactions, even when there is no discontinuity. This demonstrates intercalation within a morphalactic as well as an epimorphic system.

Quantitative models following the so-called prepattern scheme, such as the classic reaction-diffusion mechanism envisioned by Turing (1952), involve biological systems differentiating in response to threshold values of a nonhomogeneously distributed quantity. While the system then does not need to carry an internal geometric representation of its position as in the case of positional information, other difficulties are encountered, such as size regulation (the patterns depend on the size of the domain), and the fact that the putative morphogens have not yet been identified experimentally. More recently it has also been noted that explicitly physical interpretations of morphogenetic information, such as via mechanical forces of stress and strain, can also generate rich repertories of spatial patterns (e.g. Goodwin and Trainor 1985).

In this essay, we shall describe two models of a morphogenetic field (Totafurno and Trainor 1987; Brandts and Trainor 1990a,b; Brandts 1993) that replace the traditional scalar morphogenetic field with vector fields. Although still a minimal description of biological patterns, vector field theories appear to offer many advantages for expressing and analyzing the dynamics of pattern formation, in particular the role of underlying symmetries and symmetry breaking. The dynamics can be based either on general mathematical considerations (Totafurno and Trainor 1987) or on phenomenological biological rules (Brandts and Trainor 1990a, Brandts 1993), but in either case they lead to configurations (solutions) that correspond sufficiently closely to the data that the formalism can be used to organize the plethora of experimentally observed forms and to account for their pathways of regulation. We have found this to be the case for both unicellular and multicellular organisms. We will review the dynamic vector model approach applied to two diverse organisms and different biological contexts: morphallaxis in the single-celled *Tetrahymena*, and epimorphosis in multicellular salamander limb regeneration.

2 Morphogenetic Fields

The concept of a morphogenetic field in biology captures the idea that there is a region with coordinated activity controlled by the differential distribution of some property or quantity. The field concept is appropriate not only because the mathematics is convenient, but because biological systems actually exhibit behaviors characteristic of physical fields, such as smoothness and continuity. The holistic nature of physical fields which allows, for example, a unique electric field to be determined everywhere in a region given sufficient information about the field on the boundary, is reminiscent of the ability of a morphogenetic field in biology to interpolate in regeneration, and thus to reform a whole organism from the information contained in some of the parts.

Generally speaking the morphogenetic fields referred to in the biological literature have not been fully explained on a physical basis at the present time. Interpretations include concentrations of morphogens or other cell-surface structures, mechanical effects such as elasticity, or more abstractly, order parameter fields (Brandts 1995). In this essay, the field concept is used as an effective descriptor of observed phenomena without trying to identify the underlying plethora of contributing processes, such as numerous chemical reactions, and physical properties of macromolecular assemblies. The point of view is to identify and emphasize such basic properties as smoothness, symmetry and regulation in order to establish a dynamic field theory capable of describing the broad features of pattern and form generation in diverse biological organisms. This approach does not preclude the eventual bridging between such a field description and the underlying microscopic processes contributing to it (Brandts 1995).

3 Biology of *Tetrahymena*

An impressive variety of cell configurations has been obtained experimentally (Nelsen and Frankel 1986, 1989). We will briefly describe the cell features and the experimental manipulations that lead to the plethora of forms during the subsequent regulation (see also Frankel, this volume).

3.1 Cytogeometry

Tetrahymena has a number of cell surface features which maintain a stable characteristic organization over generations, both in intrinsic structure and in location. Longitudinal bands, typically 18 to 21 of them, composed of basal bodies, run from anterior to posterior of the cell. We can number and use these rows to measure circumferential position around the cell. A normal cell has one oral apparatus, OA, composed of four membranelles. Asymmetry in the organization of the membranelles allows us to assign a handedness to the OA, normally defined as right-handed (RH). Located in a normal cell, to the cell's right of the right-handed OA and near the posterior end of one, two or three adjacent ciliary rows, are the contractile vacuole pores, CVP, used for osmotic regulation. As evidenced in situations where regions of pattern reversal occur, there are many interacting levels of patterning operating in the cell, some localized, others more global.

3.2 Tetrahymena Doublets

Experimental manipulation (Nelsen and Frankel 1986) has produced homopolar doublets, fusions of two cells, joined side by side. Doublet configurations are not stable, however; they regulate down in size from one generation to the next. A plethora of forms can be observed as doublets "shrink" back to singlet forms via several different pathways. To return to the singlet state, the doublet must lose an entire set of cell features: ciliary rows, an OA and a CVP set. Surprisingly, these superfluous cell structures are not simply resorbed. Rather, the OAs shift position and transient states with additional and reversed OA's can be formed. In the conversion of two OAs back to one OA, a triplication with three OAs, the middle one of reversed symmetry, sometimes appears. These additional OAs have a highly variable location but are found only in the narrower arc spanned by the two RH OAs. In addition to such triplication forms, many configurations of doublets and singlets appear, with varying numbers of CVP sets and positioning (Brandts and Trainor 1990b).

3.3 Left-Handed Cells

A left-handed *Tetrahymena* cell has the global position of all structures that are asymmetrically arranged with respect to one another around the circumference reversed. For example, the CVP set appears to the cell's right of a RH OA, but to the cell's left of a LH OA (Figure 1). The control and specification of patterning can be explored by performing experiments on LH cells analogous to the doublet regulation described above for RH cells. Symmetry reversal invariance can also be probed. In these complex biological forms we might not expect perfect symmetry invariance, but many of the features of cellular organization are in fact found to satisfy symmetry reversal invariance. As we will see, this can be understood in terms of the field, global organizing the cell.

3.4 Regulation Pathways

The variety of surface patterns appearing during doublet regulation can be ordered according to two pathways, the reverse intercalated RI pathway, and the mirror-reversed, MR, pathway (Figure 2). Presumptive LH cell regulation pathways are mirror-image counterparts to the RH pathways. Surprisingly, the pathway most frequently followed for LH cells is not simply the mirror-image counterpart to the RH pathway.

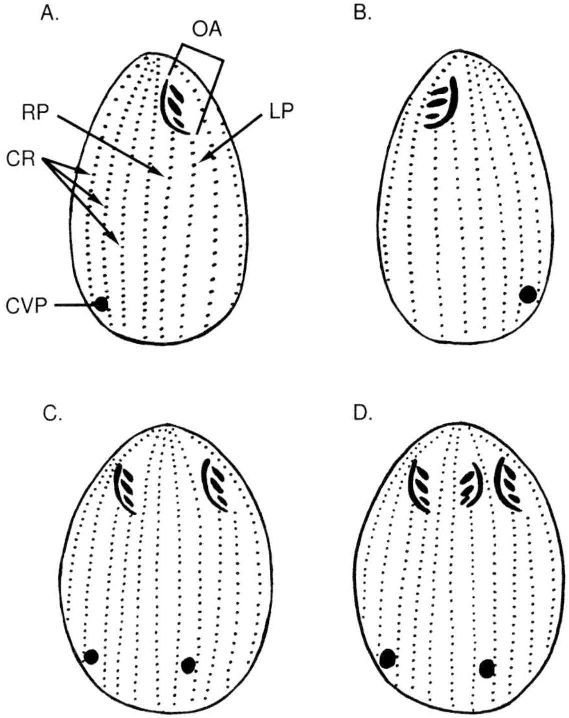

Figure 1: Configurations of *Tetrahymena* A. Right-handed singlet. The surface of *Tetrahymena* has several distinctive features which can be used as markers in describing the regulation of cell configurations. Shown here are: OA, oral apparatus whose asymmetry allows the definition of a "handedness" of the OA; CR, longitudinal ciliary rows; RP, LP, right and left postoral ciliary rows, respectively, extending from the oral area; CVP, contractile vacuole pores, which appear to the cell's right of the OA in the right-handed cell. B. Left-handed singlet. Left-handed cells are "wound" in the opposite direction to right-handed cells: The global positions of structures that are asymmetrically arranged with respect to one another (e.g. the CVP location with respect to the OA) around the circumference are reversed, and the symmetry of the OA itself (normally right-handed) can be partially reversed. C. Right-handed doublet. Right-handed doublets are created experimentally by the fusion of two right-handed singlet cells and thus have a double complement of all cell features. The doublet configuration is not stable and tends to down-regulate in size, losing ciliary rows, one of the OAs and one of the CVP sets. D. Right-handed triplet. During the regulation of doublet back to singlet, a triplet configuration sometimes appears, with a third OA having opposite (left) handedness. From W.A.M. Brandts (1993) *IMA Journal of Mathematics Applied in Medicine & Biology* **10**, 31–50, Oxford University Press, with permission.

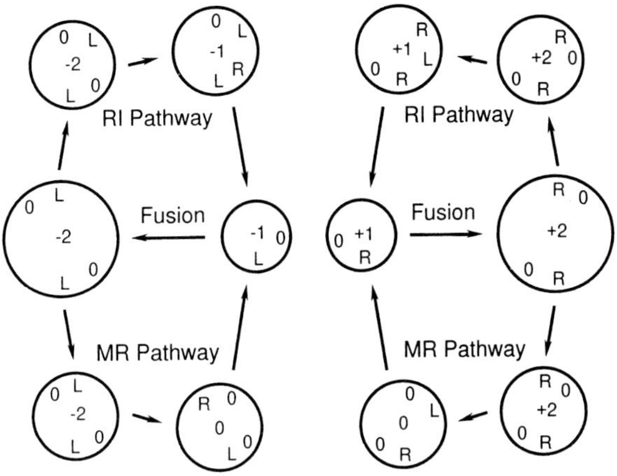

Figure 2: Pathways of right- and left-handed doublet regulation: The most distinctive features of the cell surface of *Tetrahymena*, the OA and the CVP set, can be used to follow regulation in populations of doublet cells. As doublet cells revert to singlet cells, a wide variety of intermediate forms appears and these can be related to each other via pathways of regulation. The strategies that result in symmetry reversals can be explored in experiments on left–handed cell doublets analogous to right–handed cell doublet regulation. Each circle represents a cell configuration, with the cell features displayed about the circumference in a polar view. R and L are the right and left OA's respectively. 0 represents a CVP set. The winding number of the configuration is shown in the center. The forms on the right originate from right-handed cells, those on the left represent the counterpart pathways originating from left-handed cells. For right-handed cell regulation, the pathway through a triplication state (RI) is preferred. For left-handed regulation, the pathway through a mirror-reversal configuration (MR) is preferred. Our model can account for why these cells, which appear as mirror symmetric counterparts, favor different developmental pathways. From W.A.M. Brandts (1993) *IMA Journal of Mathematics Applied in Medicine & Biology* **10**, 31–50, Oxford University Press, with permission.

4 The *Tetrahymena* Field Model

To describe the patterns on the surface of *Tetrahymena* in terms of a morphogenetic field and its dynamics, we shall interpret different values of the field as different cell structures. Thus the field values could be thought of as representing positions characteristically associated with the formation of certain structures (positional information interpretation) or, alternatively, the field values themselves could directly represent or result in the formation of structures (prepattern interpretation). Mathematically, we need a field which is single-valued, smooth, and continuous about a closed circumference, so that there will be a unique value assigned to each cell feature. We choose the simplest such field, the two-dimensional vector field, where different values of the vector's orientation, $\theta(x)$, in morphogenetic space represent different cell structures. (Two scalar fields could, in principle, also have been used with considerable sacrifice in the conceptual clarity of the formulation.) To obtain the field structure we suppose that biological regulation is guided by some sort of optimization. In our case it will not be a neodarwinian sort of optimization involving genes (see the chapter by Rowe, this volume), but rather a dynamical process. This point of view is natural because in the clonal cell situation here, genetics cannot be the determinant of pattern changes. Moreover, the regulation strategy leads to the formation of strange and superfluous structures and can even break down to the extent that unviable organisms are formed (e.g. astomatous cells which cannot feed so die). Clearly, a physical principle is needed, and we have considered the existence of a phenomenological, energy-like function related to the global distribution of the morphogenetic field across *Tetrahymena*'s surface. This "energy" is a dynamical quantity which the system minimizes. The energy as a functional of the field incorporates two competing tendencies in the field behavior:

1. There is an *optimal gradient* of field winding. This reflects the idea that there is an optimal spacing of pattern features for the cell. $\phi(x) = \partial\theta/\partial x$ has an optimum value, where θ is the angle of the morphogenetic field vector, x is the circumferential position and ϕ is defined as the gradient field.

2. The field is *smooth*. Changes in the direction of winding of the field cost energy.

The simplest form for an energy functional which satisfies these criteria is a double-well quartic in the field gradient, combined with a higher derivative smoothing influence. Our original symmetric model for *Tetrahymena* (Brandts and Trainor 1990a) used a square power for the smoothing term. In the asymmetrical model (Brandts 1993), a modification designed to permit a small intrinsic asymmetry between left- and right-handed winding, an asymmetrical double-well combined with a fourth-power smoothing was used. These solutions can be matched to the symmetric model by appropriate choice of the one free parameter. The energy functional for the symmetric model is:

$$E(\theta(x)) = \int_0^L \{ \overbrace{\left[\left(\frac{\partial \theta}{\partial x} \right)^2 - 1 \right]^2}^{\text{optimal gradient}} + \overbrace{\beta^2 \left(\frac{\partial^2 \theta}{\partial x^2} \right)^2}^{\text{smoothness}} \} dx \tag{1}$$

where the boundary conditions are $\theta(L) = \theta(0) + 2\pi W$, where W is the winding number = 1,2, L is the circumference, and β^2 is the relative-weight-of-smoothness-term parameter. In terms of the gradient field $\phi(x) = \partial\theta/\partial x$,

$$E = \int_0^L \left\{ \left[\phi^2 - 1\right]^2 + \beta^2 \left(\frac{\partial \phi}{\partial x}\right)^2 \right\} dx \tag{2}$$

where the boundary conditions are $\phi(L) = \phi(0)$, $\int_0^L \phi dx = 2\pi W$.

Having such an explicit quantitative form for the dynamical rule allows us to test rigorously the biological rules of organization and to provide precise symmetry categories or taxonomy of patterns.

4.1 Solutions

From the minimization of E in the above equation we find three symmetry classes of solutions (Figure 3). In Figure 3, each cell configuration is read along a horizontal line. The handedness of OA's are determined by the direction of winding of the field at the field value for OA; right-handed OA's are denoted $\mathcal{R}$, left-handed ones $\mathcal{L}$. An affixed numeral 1 or 2 indicates the semicell (sc), when sc's are of unequal size. The domain of CVP formation is indicated by solid and dashed lines. This series of graphs is used both to follow the dynamical paths of individual configurations as L is reduced, and to generate statistical samples of configurations. Figure 3A shows the symmetric solution (W=2) for $L = 3.500\pi$. The SYM solutions are linear. All solutions correspond to doublets with 2 CVP sets. Figure 3B shows the RI solution for $L = 3.250\pi$. Reading from the top down the cell configurations are: (2×) singlet with 3 CVP sets; (1×) singlet with 2 CVP sets, one set broad; (1×) singlet with 2 CVP sets, one set narrow; (8×) singlet with 1 CVP set; (3×) triplet with 1 CVP set; (1×) fusing triplet with 2 CVP sets. Figure 3C shows the symmetric solution for (W=1) for $L = 2.750\pi$. The SYM solutions are linear. All solutions correspond to singlets with 1 CVP set.

While there are many different solutions allowed by the model, corresponding to a wide variety of cell configurations, it is nevertheless a grossly restricted set: Not all imaginable configurations are obtained. Most significantly, all the forms given by the model have been seen experimentally, including the 3-CVP-set configuration which was identified experimentally after the model first predicted it. The three basic symmetry groups or classes of solution in the model are called symmetric (SYM), reverse-intercalated (RI), and mirror-reversed (MR). The SYM solutions (see Figure 3) are the simplest. The field angle θ winds at a constant rate $\partial\theta/\partial x$ and corresponds to singlet (W=1) or

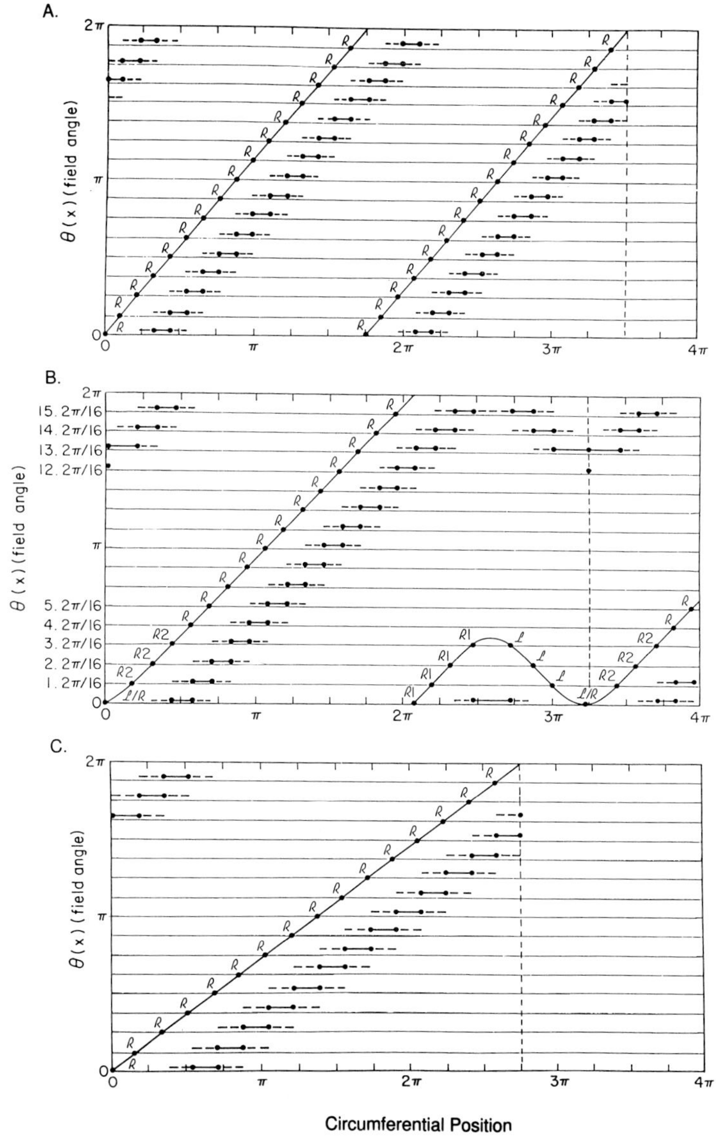

Figure 3: Minimum energy configurations: An illustrative sample of the minimum energy solutions for the symmetric model at several values of L. From W.A.M. Brandts and L.E.H. Trainor (1990) *Journal of Theoretical Biology* **146**, 57–86, Academic Press, with permission.

to doublet (W=2) states. The RI configurations contain a region of the field which winds in the opposite sense to the rest of the field. RI solutions can correspond to a variety of actual cell types, including singlets with three, two or one CVP sets, triplets and fusing triplets. The analysis of these solutions in detail allows us to compare with many documented and measured experimental configurations (Brandts and Trainor 1990b). There is a richness of successful correspondences, both for configurations representing individual cells in a pathway and for members of a statistical population of cells in different stages.

4.2 Pathways

Which of the locally minimum energy solutions has the globally lowest energy, and hence what form we would predict in *Tetrahymena* samples, depends on the cell's circumference L. Figure 4 traces the sequence of minimum energy solutions, beginning with large values of L (doublet size) and decreasing to smaller L (singlet size), simulating the regulation of doublets downsizing to singlets. The symmetric model is adequate to describe in detail the pathways and transitions of RH doublets. The asymmetric model, incorporating a small biological asymmetry between left and right, can also account for why the pathways of regulation in left-handed doublets differs from that of right-handed doublets. In particular, the RH doublets pass through triplications while LH doublets prefer transitional MR configurations.

These models suggest that the fundamental classifications of *Tetrahymena* cell types be based on symmetry (SYM, RI, MR) rather than on the directly observable "doublet" and "singlet" state. Similar cell configurations may arise from fundamentally different field solutions, and conversely, one type of field solution can give rise to very different biological patterns. For example, the observed 1-CVP-set singlet can result from either a SYM solution or an RI solution; conversely, the RI solution can give rise to triplets, 1-CVP-set singlets or 3-CVP-set singlets.

5 Biology of Salamander Limb Regeneration

We now turn to the fascinating and rather extensive phenomenology regarding the production of additional (supernumerary) limbs following various types of grafts in salamanders and insects (Maden 1982) that bring into apposition normally nonadjacent tissues. Attempting to account for the number, position and structure of such supernumerary limbs in regenerating systems has provided both fertile soil for a theoretical understanding of biological development and a proving ground for models of development. In particular the polar coor-

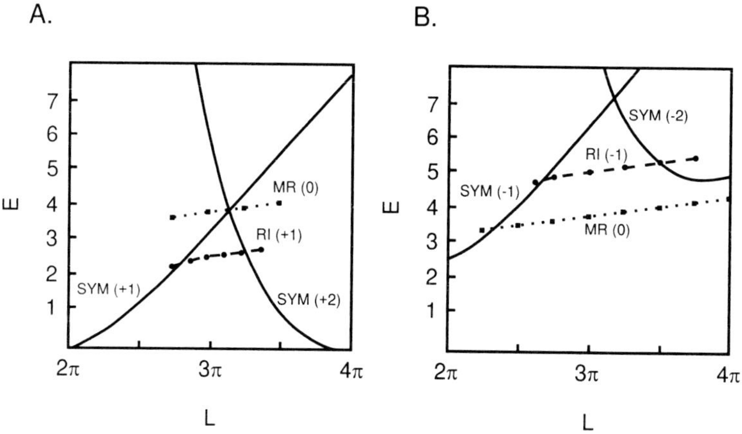

Figure 4: Energy E of solution versus total circumference L. Which cell configuration is predicted for any given cell size (stage of regulation) is determined by the energy of the solutions. Thus as the cell shrinks from doublet size (4π) to singlet size (2π), the sequence of minimum energy solutions can be followed. A. Asymmetric model for right-handed SYM, RI, MR solutions. The energy of right-handed field solutions SYM(+2), SYM(+1), and RI(+1), and mirror symmetric MR(0), over the whole range of cell sizes from doublet to singlet size are shown. The RI(+1) solutions have a lower energy than MR(0) for all cell circumferences L and thus the RI solutions would be favored over MR in regulation pathways. This corresponds precisely with the experimental observation that mirror-symmetric configurations are very rare compared to triplets. B. Asymmetric model for left-handed SYM, RI, MR solutions. The energy of left-handed field solutions SYM(−2), SYM(−1), RI(−1), and MR(0) over the whole range of cell sizes from doublet to singlet size are shown. In contrast to the right-handed solutions of Figure 3A the MR(0) solutions now have a lower energy than RI(−1) for all cell circumferences L and thus MR solutions would be favored over RI in regulation pathways. This corresponds precisely with the experimental observation that mirror symmetric configurations predominate over triplets in left-handed regulation samples. From W.A.M. Brandts (1993) *IMA Journal of Mathematics Applied in Medicine & Biology* **10**, 31–50, Oxford University Press, with permission.

dinate model of French et al. (1976), as already discussed, showed how much of the data could be understood in terms of a phenomenological rule governing cellular interactions.

While the polar coordinate model gives good insight into the results of grafts involving the apposition of tissues from limbs of opposite handedness (the so-called contralateral grafts as will be described below), it does not adequately account for the results of grafts involving the apposition of tissues from limbs of the same handedness (the so-called ipsilateral grafts as will also be described below). Moreover, in more exotic circumstances such as the regeneration of mirror symmetric limbs, the model fails entirely.

In this part of our essay we will briefly review a model that more adequately accounts for the experimental data that now exists (see Totafurno and Trainor (1987), and Ding, Totafurno and Trainor (1992) for further details). In constructing the model, we have found it necessary to both reinterpret and extend the conceptual framework of the polar coordinate model and to introduce a more sophisticated formal structure. The greater difficulty of modeling the ipsilateral graft as compared to the contralateral is explained by the model in terms of the difference in the boundary conditions of the grafts as regards their "rotational" symmetry (to be defined below) and as regards their "compatability" for the production of supernumeraries having normal structure (Tevlin and Trainor 1985). These differences in turn provide a justification for the greater variability observed in the ipsilateral data as compared to the contralateral with regard to the number, location and structure of the induced supernumeraries. For example, contralateral grafts generate up to two supernumeraries, both having normal structure; in contrast, ipsilateral grafts have been found to generate up to three supernumeraries, each having one of four distinct structures, with these structures occurring in various combinations (Tank 1981, Maden 1982, Maden and Mustafa 1982, Papageorgiou and Holder 1983). The field approach used here provides a basis intrinsic to the pattern formation mechanism for the greater variability in the ipsilateral data. The model is consistent with the size-independent nature of pattern formation found in normal regeneration after amputation (Maden 1981). It also decouples the location at which ipsilateral supernumeraries form from the angle of rotation of the apposed tissues (see below), allowing supernumeraries to form anywhere about the limb and at various angles of rotation, as is observed experimentally (the polar coordinate model requires ipsilateral supernumeraries to form at definite positions about the limb if they are also to form at angles of rotation other than 180°; see Stock, Krasner, Holder and Bryant 1980, and Turner 1981).

5.1 Limb Graft Experiments

5.1.1 Ordinary limb regeneration

In ordinary limb regeneration the more distal portions of a limb are amputated by a cut roughly perpendicular to the limb's proximo-distal axis, which runs from the body wall to the tip of the limb. The point along the proximo-distal axis at which the cut is made is known as the level of the amputation. Salamander limbs respond to such amputation anywhere along the proximo-distal axis by replacing those structures lost by amputation, i.e., just those structures distal to the level of the amputation. The process involves first a rapid wound healing phase in which epithelial tissue spreads over the cut surface of the stump to close the wound, followed within just a few days by the formation at the cut surface of the stump of a mound of relatively undifferentiated tissue called a regeneration blastema, which in turn is followed by a stage of active cell proliferation and differentiation lasting several weeks in which the blastema appears to move progressively distal as the amputated limb structure is redifferentiated from the blastema in a proximal to distal sequence. Finally a complete miniature copy of the amputated portion of the limb is produced which over the course of a few months grows to its original size. We note that the regenerate is of the appropriate limb type, whether forelimb or hindlimb; is of the appropriate handedness, whether left or right; is oriented with respect to the stump in the normal orientation, both along and about the proximo-distal axis; is usually identical or almost identical anatomically to the lost portion of the limb; and is functionally normal.

5.1.2 Contralateral and ipsilateral grafts

Ordinary regeneration in salamander limbs may be perturbed by grafts of regeneration blastemas onto cut stumps that bring into apposition normally non-adjacent tissues. In the so-called contralateral graft a regeneration blastema from a limb undergoing ordinary regeneration is removed and grafted onto the stump of a freshly amputated limb of opposite handedness. In the so-called ipsilateral graft the regeneration blastema is instead grafted, after rotation by some angle (usually 180°), onto a stump of the same handedness. In either case, in addition to the primary regenerate that forms from the grafted regeneration blastema, extra limbs called supernumerary limbs may arise from the region where the tissues of the stump and grafted regeneration blastema are apposed. This region of transition between the structure of the stump and the structure of the primary regenerate is called the intercalary region.

In keeping with ordinary limb regeneration, the pattern of the limb stump remains unchanged in such graft experiments. Somewhat surprisingly owing to its relatively undifferentiated state, the same may also be said of the grafted regeneration blastema since the primary regenerate conforms in its structure to that of the site of origin of the grafted regeneration blastema. Thus we may evidently associate with the regeneration blastema a *latent structure* both in terms of its proximo-distal level and in terms of its pattern about the proximo-distal axis that corresponds to the normal structure of the limb from which it was removed.

The pattern of the limb about the proximo-distal axis determines the handedness of the limb. This axial pattern is conventionally described in terms of its polarization along two orthogonal axes: the antero-posterior axis (AP) and the dorso-ventral axis (DV). As explained above, we may associate such axial structure both with the stump and with the grafted blastema. Now in the contralateral graft, if we align, for example, the AP axes of the grafted blastema and stump, then of necessity the DV axes must be inverted with respect to one another. Similarly if we align the DV axes, the AP axes will be inverted. In general the contralateral graft results in the inversion of some axis through the limb. In such grafts two supernumeraries may arise from the intercalary region, one centered at each of the points of inversion of the stump and blastema axes and both having the handedness of the stump.

The ipsilateral graft is characterized by the angle of rotation θ with which the grafted blastema is rotated relative to the stump. When $\theta = 180°$, we have the case of APDV inversion since both the AP and DV axes are inverted. In fact any axis through the limb is inverted by this graft. In this sense no particular points about the intercalary region are distinguished by the ipsilateral graft where supernumeraries might form in contrast to the contralateral graft. Nonetheless up to three supernumeraries have been found to arise as a result of such grafts. In contrast to the contralateral case where the supernumeraries have normal structure, after APDV inversion the supernumeraries have been found to have, in addition to normal structure, three other distinct structures including mirror imaged structures such as double ventral, and structures with discontinuities in the axial pattern. These structures occur in various combinations with differing frequencies. When two normal limbs are formed by such grafts they have opposite handedness to one another.

5.1.3 The proximo-distal shift graft

A proximo-distal shift graft involves grafting either a distal blastema onto a proximal stump or vice versa keeping the AP and DV transverse axes of both the stump and grafted blastema aligned. Supernumeraries are not generated by such grafts; apparently the misalignment of the transverse axes of stump and grafted blastema is a requirement for the generation of supernumeraries (Wallace and Watson 1979, and Table 1 of Bryant and Iten 1977). Instead only the missing limb structure between the proximo-distal levels of the stump, and grafted blastema are replaced. In the case of the graft of a distal blastema onto a proximal stump, the result is a single normal limb; however, the intercalated limb structure is composed only of tissue of stump origin (Pesticelli and Stocum 1980). In the case of the graft of a proximal blastema onto a distal stump, any intercalated structure between the levels of the stump and grafted blastema is only of blastema origin and is laid down in the reverse orientation to a normal limb, i.e., the intercalated structure goes from distal to proximal as one moves down the limb away from the body (Maden 1980).

6 An Interpretive Look at the Limb Graft Experiments

6.1 The Pattern Specification Field

We have already noted that in the grafting experiments discussed in section 5.1, the grafted regeneration blastema produces structure appropriate to its site of origin despite its altered relationship to the adjacent limb tissue. This suggests that the process of biological pattern determination in this system can be abstracted from the underlying molecular and cellular details and analyzed and understood in its own right. As such we introduce an as yet unspecified "pattern specification field" to which we attribute the local determination of the limb regeneration pattern. As in the notion of a prepattern, the complexity of the predicted patterns is commensurate with the complexity of the solutions of the field equations. From this point of view the parts of the limb pattern are seen not as separate entities with an independent identity but as the expression of the overall unity of the regenerative process (cf. Goodwin and Trainor 1983, especially pp. 75–76, p. 82, pp. 85–86).

Although the biological values of the field such as the proximo-distal level or the axial value (A, P, D, V) are conceptually distinct from position per se, position may be used to parameterize these values. The dimensionality of the tissue associated with the field therefore has a bearing on the nature of the field itself. French et al. (1976) have argued that biological patterns generally are established in two dimensions and have given evidence suggestive of this in the particular case of the limb pattern. This is a significant simplification, which we adopt until experiment indicates otherwise. We therefore assume that the values of the field are completely parameterized by the two-dimensional surface of the limb, so that the field specifies just the proximo-distal and axial values of the pattern. The internal structure of the limb is not directly specified by the field being considered here, but rather is assumed to follow the setting up of the proximo-distal and axial values of the tissue as a subsequent step associated with redifferentiation.

6.2 The Intercalary Region

For supernumerary production, the so-called *intercalary region* is the arena for the pattern generation mechanism. As described earlier this region is flanked proximally and distally by structures which remain relatively unchanged by the interaction that gives rise to supernumeraries. A similar situation occurs for the proximo-distal shift grafts as well. In other words, in these grafts the intercalary region is bounded along its proximal and distal boundaries by stump and blastemal tissue, respectively, having stable field values.

It is observed that the field values adopted by the tissue of the intercalary region vary continuously, to as great a degree as possible, between the values of the stump and grafted blastema boundaries. For example when two normal supernumeraries are produced in the contralateral graft, the orientation of these limbs is such that their axial field values (A,P,D,V) are continuous with the neighboring axial values of the stump and grafted blastema (see Figure 5). Maintaining continuity with the axial values of the stump and grafted

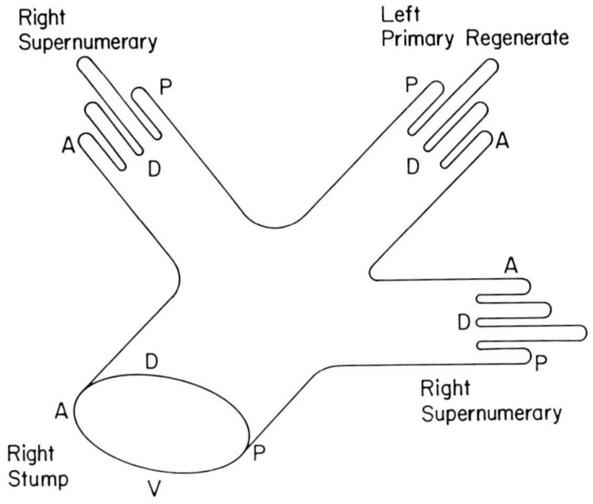

Figure 5: Results of grafting a left blastema onto a right stump (contralateral graft) with the antero-posterior (AP) axis of the blastema inverted with respect to the stump. Two normal supernumerary limbs of stump handedness have been produced with location and orientation as shown. A = anterior, P = posterior, D = dorsal, V = ventral. From J. Totafurno and L.E.H. Trainor (1987) *Journal of Theoretical Biology* **124**, 415–454, Academic Press, with permission

blastema in the contralateral graft results in two points where nonadjacent axial values converge; however, these are just the points where supernumeraries are centered. It would appear then that some kind of smoothing principle governs the development of the field of the intercalary region with respect to both its proximo-distal and axial values. Most models of supernumerary production have assumed a smoothing principle (Goodwin 1976, Papageorgiou and Holder 1983). Lewis (1981; see also Mittenthal 1981 and Winfree 1984) has argued that the similarity of the regenerative response observed in widely different organisms (amphibians, insects) is a consequence of a general principle that cells adopt field values which make the field continuous.

The intercalary region is a site of active cell proliferation. Since this mitotic activity is associated with local disparities between the proximo-distal or axial field values of the juxtaposed stump and grafted blastema tissues, and since a smoothing principle is evident in the development of the field values of the intercalary region, it has been suggested that cell division continues until local differences in the field values are smoothed out and the field becomes continuous (French et al. 1976, Bryant and Iten 1976, Tank 1978, Bryant et al. 1981, Stocum 1982). In this way an effective coupling of the process of pattern formation within the intercalary region and the control of the growth of the intercalary region is achieved (Maden 1981, Bryant and Simpson 1984). However, as we have argued above, the determination of the limb pattern occurs over a shorter time scale than that required for the expression of the regenerated pattern, so that the process of pattern formation may be seen as distinct from any subsequent growth. Furthermore it is found that supernumeraries with discontinuous axial structure (which are often generated by ipsilateral grafts) faithfully regenerate this discontinuous structure upon amputation (Maden 1982, Maden and Mustafa 1982). This suggests that discontinuities in the field values of neighboring tissue do not of themselves stimulate cell division and smoothing out of the field.

It would therefore appear that the growth of the intercalary region may be decoupled, at least to a first approximation, from the pattern formation mechanism. This means in particular that variability in the length of the intercalary region at the time the limb pattern is determined is to be expected since its size is dependent upon several factors including the extent of the release of cells to participate in pattern formation initiated by wounding, the rates of cell division and movement, and the time before redifferentiation sets in. (This is not to say that subsequent size regulation does not occur after pattern determination.) We have taken the view that the variety of structures generated by ipsilateral grafts can be seen as nonequilibrium or non-steady state structures which are "frozen in" when redifferentiation and outgrowth commence. When sufficient time is available for the development of the field, the limb patterns approach equilibrium or steady-state structures, i.e., the normal and probably the mirror-imaged structures.

6.3 Contralateral versus Ipsilateral Blastema Grafts

A striking difference between ipsilateral and contralateral supernumerary production is the greater variability of the ipsilateral data with regards to the number, location and structure of the induced supernumeraries. An important aspect of modelling supernumerary production is to incorporate this difference in a natural way.

The contralateral graft inherently defines two special points about the intercalary region where, as described above, the axial values of the juxtaposed stump and grafted blastema tissue are inverted with respect to one another. It is therefore not entirely surprising that just two supernumeraries are generated and that they are located at these points of inversion. Although the production of fewer than two supernumeraries is usually observed for these grafts, we assume this may be attributed to the suppression of the development of one or both of the supernumeraries actually determined by the pattern formation mechanism. This suppression is due to factors that are not intrinsic to the pattern formation mechanism itself such as variations in the degree of innervation about the stump.

The ipsilateral graft, in contrast, does not define any points about the intercalary region that may be distinguished from the others. We noted that when the grafted blastema is rotated by $\theta = 180°$, the axial values of the juxtaposed stump and grafted blastema tissue at every point about the intercalary region are inverted with respect to one another. This symmetry of the intercalary region of the ipsilateral graft is an instance of what we will call "rotational symmetry" (made more precise in Totafurno and Trainor, this volume). Since no special positions are defined by the boundaries at which to expect supernumeraries, one might anticipate more variable results in terms of the position, spacing and number of supernumeraries. In fact one might expect that supernumeraries could form anywhere about the intercalary region as is observed (Maden and Turner 1978; Turner 1981, Table 2).

An underlying rotational symmetry is also seen in the contralateral graft in that inversion of any transverse axis results in the production of a pair of supernumeraries, one at each of the poles of the inverted axis (Wallace and Watson 1979, Table 3, restricting to supernumeraries separated from the graft and taking derotation into account; Maden 1982, pp. 132–134). In the contralateral graft, however, a single axis is inverted as a consequence of the asymmetry between the axial field values of the stump and grafted blastema boundaries. In contrast, the production of supernumeraries in ipsilateral grafts corresponds to spontaneous symmetry breaking: the rotational symmetry of the boundary conditions is broken as the centers of the supernumeraries become distinguished from other points about the intercalary region.

7 The Regeneration Field Model

The notions discussed in our interpretive look at the limb graft experiments may be incorporated into an explicit model of intercalary regeneration as follows:

1. We associate with the two-dimensional surface of both the intercalary region and the normal limb a field vector $\vec{\sigma} = (u, v)$ which specifies the local pattern state of the tissue as a function of geometrical position. Here u and v are real, continuous functions of position which are used to assign a unique value of the field to every part of the pattern of a normal limb. Interpreting (u, v) as a vector incorporates the apparent equivalence of all transverse axes through the limb suggested by the underlying rotational symmetry of the contralateral graft. A useful analogy for the field $\vec{\sigma}$ is to think of it as a two-dimensional local order parameter similar to the average spin of an Ising spin model (Ma 1976, Amit 1978, Goodwin and Trainor 1980).

   The field values of $\vec{\sigma}$ are identified with the proximo-distal and axial elements of the normal limb pattern. The field magnitude $|\vec{\sigma}|$ specifies the pattern of the limb with respect to the proximo-distal axis, while the direction of $\vec{\sigma}$ specifies the axial field values about the limb from anterior (A), to dorsal (D), to posterior (P), to ventral (V), and back to anterior. The distal tip of a normal limb, or prospective normal limb, corresponds to $\vec{\sigma} = (0, 0)$, with $|\vec{\sigma}|$ increasing proximally. We have chosen $\vec{\sigma}$ parallel to $u = \pm 1$, $v = 0$ to correspond, respectively, to the anterior and posterior extremes of the limb, while $\vec{\sigma}$ parallel to $u = 0$, $v = \pm 1$ corresponds, respectively, to the dorsal and ventral extremes.

   Over the surface of the intercalary region, the field values of $\vec{\sigma}$ are interpreted as described above for the normal limb. In particular, the presence of isolated points where $\vec{\sigma} = (0, 0)$ over the intercalary region indicates the presumptive formation of supernumeraries (i.e., distal transformation) with complete proximo-distal structure. For simplicity, we take the intercalary region to be a cylinder of radius $R = 1$ and length L. The field $\vec{\sigma}$ is then a function of the cylindrical coordinates (z, ϕ) specifying geometrical position over the surface of the intercalary region.

2. The dynamical equations governing the development in time of the field $\vec{\sigma}(z, \phi)$ over the intercalary region were chosen on the basis of the following criteria:

- The fields generated in the intercalary region should be as smooth as possible; hence, the equations for u and v should contain Laplacian terms.
- The form of the equations should ensure that the model is rotationally symmetric.
- The equations should be nonlinear to facilitate the spontaneous breaking of the rotational symmetry of the model associated with the generation of supernumeraries in the ipsilateral graft.
- The appropriate solution of the field equations in the intercalary region is determined by the stable field values of the stump and grafted blastema.

With these guidelines, the following field equations were adopted:

$$\tau^2 \frac{\partial u}{\partial t} = \nabla^2 u + \alpha u - \beta u(u^2 + v^2) \tag{3}$$

and

$$\tau^2 \frac{\partial v}{\partial t} = \nabla^2 v + \alpha v - \beta v(u^2 + v^2) \ , \tag{4}$$

where τ^2, α and β are all positive constants, ∇^2 is the two-dimensional Laplacian for the cylindrical surface (with $R = 1$), and t is time.

It can be shown that solutions of equations (3) and (4) are such that an effective field energy

$$E = |\nabla u|^2 + |\nabla v|^2 - \alpha(u^2 + v^2) + \frac{\beta}{2}(u^2 + v^2)^2 \tag{5}$$

(where ∇ is the gradient operator for the cylindrical surface) decreases monotonically in time. The requirement of rotational symmetry of the model is ensured by the rotational invariance of the effective energy E. In fact E has additional symmetries which will be discussed in the next chapter. The rotational symmetry of the field equations (together with the boundary conditions to be described below) has the practical consequence that for the ipsilateral graft the supernumeraries predicted by the model may form anywhere about the stump, in keeping with experiment. The nonlinearity of the field equations is achieved in a simple way by including the fourth-order term in E. The linear and fourth-order terms in E may be seen as the first two terms of a series expansion in $|\vec{\sigma}|^2$. A practical consequence of including the fourth-order term in addition to the linear term is that spontaneous symmetry breaking occurs in the model over a range of values of the length of the intercalary region L, rather than only at special discrete values of L (Ding et al.1992, see also Totafurno and Trainor, this volume).

3. The field values along the proximal boundary of the intercalary region, which we take to be at $z = -L/2$, correspond to those of the stump, which for definiteness is assumed to be right-handed. We choose the positive sense of the cylindrical coordinate ϕ so that for a right-handed limb ϕ increases from anterior (A) to dorsal (D) to posterior (P) to ventral (V). The field values at the stump boundary are then assumed to be, using a two-component vector matrix notation,

$$\vec{\sigma} = m \begin{pmatrix} \cos\phi \\ \sin\phi \end{pmatrix} = \begin{pmatrix} u(\phi) \\ v(\phi) \end{pmatrix} \tag{6}$$

where m is the magnitude of $\vec{\sigma}$ at the boundary.

The field values along the distal boundary of the intercalary region at $z = +L/2$ correspond to the field values of the grafted blastema. Given a right-handed stump, the blastema for an ipsilateral graft has the field values of a right-handed limb, while for the contralateral graft it has the field values of a left-handed limb. The general form of the field along the blastema boundary for an ipsilateral graft is then

$$\vec{\sigma} = m \begin{pmatrix} \cos(\phi - \theta) \\ \sin(\phi - \theta) \end{pmatrix}, \tag{7}$$

where θ is the angle of rotation of the grafted blastema relative to the stump. When $\theta = 180°$ we have the case of APDV inversion for which

$$\vec{\sigma} = m \begin{pmatrix} -\cos\phi \\ -\sin\phi \end{pmatrix}. \tag{8}$$

Similarly for the contralateral graft the blastema boundary has the field values

$$\vec{\sigma} = m \begin{pmatrix} \cos(\phi - \theta) \\ -\sin(\phi - \theta) \end{pmatrix} \tag{9}$$

where θ is the angle of rotation of the grafted blastema relative to DV inversion (which corresponds to $\theta = 0$).

The proximo-distal shift grafts are ipsilateral grafts with blastema boundary of the form

$$\vec{\sigma} = m' \begin{pmatrix} \cos\phi \\ \sin\phi \end{pmatrix} \tag{10}$$

where $m' < m$ for a distal to proximal shift of the grafted blastema or $m' > m$ for a proximal to distal shift of the grafted blastema.

4. The proximo-distal shift grafts indicate that the tissue of the intercalary region having a particular proximo-distal value can only arise from tissue having a more proximal value, i.e., tissue can only transform distally. This means that, in the contralateral and ipsilateral grafts, the tissue of the intercalary region should have proximo-distal values more distal than the level of the stump and grafted blastema. For simplicity we have not attempted to incorporate this constraint into the model.

7.1 *Solutions*

We briefly review here the results of both numerical and bifurcation analysis of the model, concentrating particularly on supernumerary production in the ipsilateral graft since previous models have had difficulty in handling this case. For simplicity we have restricted ourselves to steady-state solutions of the field equations and their behavior as the length of the intercalary region L is varied. Such solutions will be especially relevant to the experimental situation if the rate of pattern formation is fast enough to ensure that as the intercalary region grows, the field maintains a quasi-steady-state up until the onset of re-differentiation and outgrowth. That L may vary considerably in blastema graft experiments has already been noted, so that the analysis of the model using L as the control parameter should have significance for the experimental situation. We are interested in the sequence of steady-state solutions of the field equations, each corresponding to an intercalary region of given size with fixed stump and grafted blastema boundaries and parameterized by the length L. We have found that the solutions in the contralateral graft correspond to the production of a pair of supernumeraries of stump handedness. In the ipsilateral graft with APDV inversion, we have found four distinct steady-state solutions which we refer to as the symmetric solution, the twist solution, the 2-centered solution (representing two supernumeraries), and the 4-centered solution (representing four supernumeraries). The solutions representing supernumeraries as well as the twist solution all arise as symmetry breaking bifurcations from the symmetric solution as the control parameter L is varied.

7.1.1 *The contralateral graft*

Numerical solutions of the model for the contralateral graft with DV inversion are shown in Figures 6 and 7. Both contain a pair of singular points (i.e.,

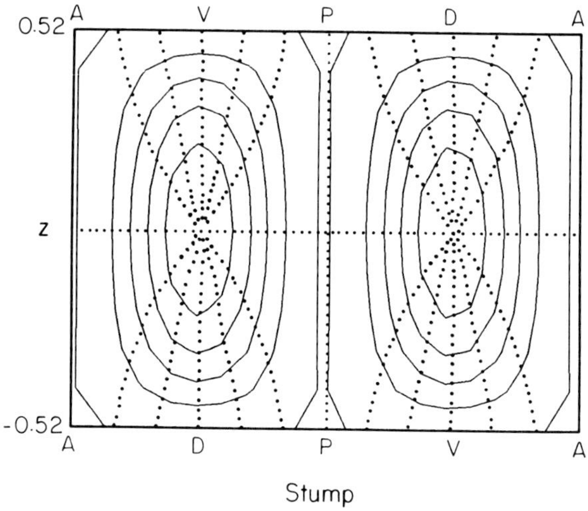

Figure 6: Two-dimensional representation of the field $\vec{\sigma}/m$ for a contralateral graft in which the DV axis of the blastema is inverted with respect to the stump. Here the surface of the cylindrical intercalary region has for convenience been represented by a rectangle with the left and right edges of the rectangle corresponding to the same line on the surface parallel to the z-axis. Out of the page corresponds to out of the limb. This numerical solution of the field equations for $\alpha = 2$, $\beta = m^{-2}$ and $L = 1.03$ represents 2 normal supernumeraries of stump handedness. The solid lines are contours of constant $|\vec{\sigma}|^2/m^2$ and have the values 0.2, 0.4 and so on increasing from 0 at the center of each of the two regions of concentric contours. Only over the relatively small area between these two regions is $|\vec{\sigma}|^2/m^2 > 1$ (maximum value of 1.02). The dotted lines are contours of constant direction of $\vec{\sigma}$. The biological values represented by $\vec{\sigma}$ along the stump and blastema boundaries are as indicated, where A, P, D and V are as in Figure 5. From J. Totafurno and L.E.H. Trainor (1987) *Journal of Theoretical Biology* **124**, 415–454, Academic Press, with permission

points at which $\vec{\sigma} = 0$) and represent the production of a pair of normal supernumeraries both of stump handedness. In Figure 6 however, the singularities in the field occur at $z = 0$, whereas in Figure 7 they occur displaced from $z = 0$ so that one of the supernumeraries is nearest the stump and one is nearest the grafted blastema.

The solution illustrated in Figure 6 occurs for relatively smaller values of L. The continuity of the axial values of the stump boundary with the axial values of the $z < 0$ parts of either supernumerary, and of the axial values of the blastema boundary with the axial values of the $z > 0$ parts of either supernumerary, suggests that in this case the supernumeraries are composed of both stump and grafted blastema tissue. In contrast, the solution illustrated in Figure 7 occurs for relatively larger values of L. The continuity of (1) the

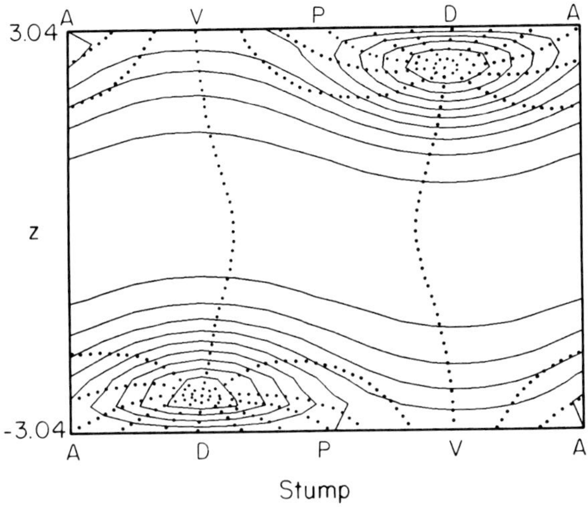

Figure 7: Contralateral graft as in Figure 6 except here $L = 6.08$. Again the solution represents two normal supernumeraries of stump handedness. Maximum value of $|\vec{\sigma}|^2/m^2 = 1.95$ in the vicinity of $z = 0$; minimum value of $|\vec{\sigma}|^2/m^2 = 0$ at the center of each region of concentric contours of $|\vec{\sigma}|^2$. The solid contours have the values 0.2, 0.4 and so on. From J. Totafurno and L.E.H. Trainor (1987) *Journal of Theoretical Biology* **124**, 415–454, Academic Press, with permission

axial values of the supernumerary nearest the stump with the axial values of the stump boundary, and (2) the axial values of the supernumerary nearest the blastema with the axial values of the blastema boundary, suggests that displaced supernumeraries are primarily composed of the tissue of the nearest boundary.

Although Figures 6 and 7 illustrate the solutions found after DV inversion, similar solutions of the model occur after inversion of the AP axis or any other transverse axis. The prediction of the model of both nondisplaced and displaced supernumeraries, the cellular composition of the supernumeraries as suggested by the model, and the production of supernumeraries after the inversion of any axis, are confirmed by the experimental data (Wallace and Watson 1979, Table 3, restricting to supernumeraries separated from the graft and taking derotation into account; Maden 1982, 1983; Stocum 1982; Muneoka and Bryant 1984).

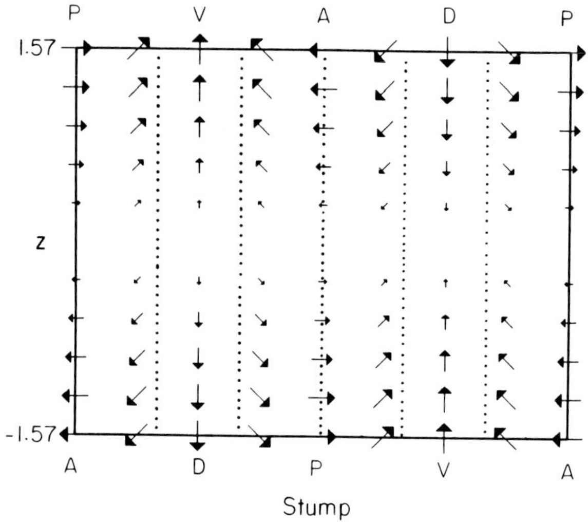

Figure 8: Symmetric solution of the field equations for the 180° ipsilateral graft with $\alpha = 2, \beta = m^{-2}$ and $L = 3.14$. The vectors $\vec{\sigma}/m$ at selected points over the intercalary region are shown. Dotted contours as in Figure 6. The magnitude of $\vec{\sigma}/m$, here indicated by the length of the vectors, decreases to 0 at $z = 0$. From J. Totafurno and L.E.H. Trainor (1987) *Journal of Theoretical Biology* **124**, 415–454, Academic Press, with permission

7.1.2 The ipsilateral graft

The four steady-state solutions we have found for the 180° ipsilateral graft, i.e., for the ipsilateral graft with APDV inversion, are illustrated in Figures 8, 9, 10, and 11.

The symmetric solution (so named because it displays the same symmetries as the model itself) does not represent supernumerary outgrowth; rather the field corresponds to the production of the parts of the limb normally distal to the level of amputation in normal sequence from $z = -L/2$ to $z = 0$, and in reverse sequence from $z = 0$ to $z = +L/2$. Such a pattern has not been explicitly identified in ipsilateral grafts of salamander limbs. This observation is not necessarily inconsistent with the model, however, since numerical solution of the model indicates that this solution is in fact unstable at larger values of L. Nonetheless we find, restricting ourselves here and throughout to the parameter values $\alpha = 2$, $\beta = m^{-2}$, that the symmetric solution exists for all L.

Bifurcation analysis shows that as L is progressively increased, first the 2-centered solution, then the twist solution, and finally the 4-centered solution arise via symmetry breaking bifurcation from the symmetric solution (Ding et al. 1992). The lengths L at which these solution branches bifurcate from the symmetric branch are labelled, respectively, L_2, L_T and L_4, where as explained, $L_2 < L_T < L_4$. As L is increased to L_2, L_T and L_4, we find that the symmetric solution becomes unstable to perturbations of the field having the symmetries corresponding to the 2-centered, twist, and 4-centered solutions, respectively. Numerical analysis indicates that the 2-centered solution is stable for $L > L_2$.

Like the symmetric solution, the twist solution illustrated in Figure 9 is rotationally symmetric. No supernumerary outgrowth is indicated; rather the axial pattern of the limb in the intercalary region rotates about the z axis either to the right or left until the values match up with those of the grafted blastema boundary. For the values of α and β considered here, the twist solution occurs

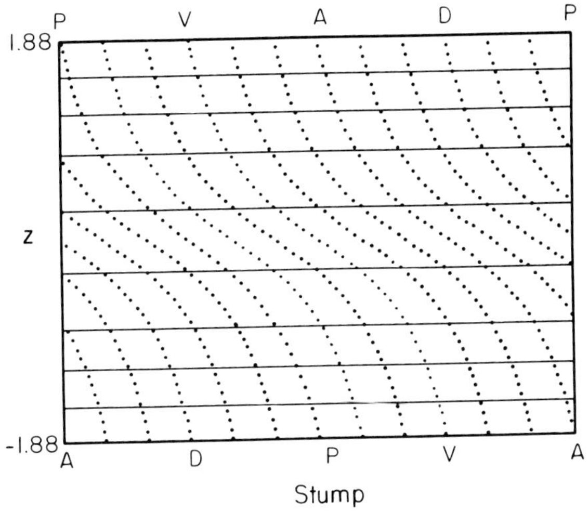

Figure 9: Twist solution of the field equations for the 180° ipsilateral graft. Here $\alpha = 2$, $\beta = m^{-2}$ and $L = 3.77$. Solid and dotted contours as in Figure 6. $|\vec{\sigma}|^2/m^2$ has a minimum value of 0.16 at $z = 0$ and increases to 1 at the boundaries. The solid contours have the values 0.2, 0.4 and so on. From J. Totafurno and L.E.H. Trainor (1987) *Journal of Theoretical Biology* **124**, 415–454, Academic Press, with permission

for $L > L_T = 3.5$. We note that the energy per unit area, $E/2\pi L$ where E is given by 5 for the twist solution (at least when $\alpha = 2$ and $\beta = m^{-2}$), is greater for rotation of the grafted blastema by $\theta = 180°$ than it is for rotation of the grafted blastema by $\theta = 0°$. This suggests that the model favors derotation of the grafted blastema over the development of the twist solution.

For $L > L_2$, a branch of solutions is found in which the field has two

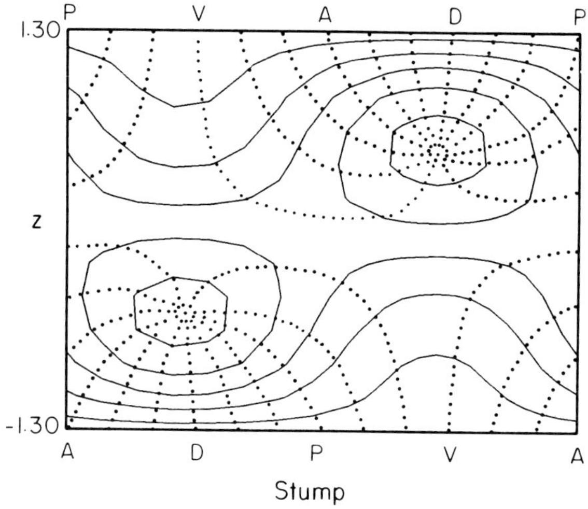

Figure 10: Two-centered solution of the field equations for the 180° ipsilateral graft. Here $\alpha = 2$, $\beta = m^{-2}$, and $L = 2.60$. This solution corresponds to two normal supernumeraries of opposite handedness, with the supernumerary nearest the stump having stump handedness. Solid and dotted contours as in Figure 6. $|\vec{\sigma}|^2/m^2$ has a minimum value of 0 at the center of each region of concentric contours of $|\vec{\sigma}|^2/m^2$, growing to a maximum of $|\vec{\sigma}|^2/m^2 = 1$ at the boundaries. The solid contours have the values 0.1, 0.3 and so on. From J. Totafurno and L.E.H. Trainor (1987) *Journal of Theoretical Biology* **124**, 415–454, Academic Press, with permission

singularities as illustrated in Figure 10. This 2-centered solution represents the production of two supernumeraries of normal structure and therefore is not rotationally symmetric. As is evident from the numerical solution shown in Figure 10, and confirmed by bifurcation analysis, the 2-centered solution is not invariant under reflection in the plane $z = 0$, so that the singularities in the field and their associated supernumeraries are displaced away from $z = 0$. The supernumerary nearest the stump has the handedness of the stump and grafted blastema, while that of the other supernumerary is of opposite handedness. As in the contralateral graft for large L, the continuity of the axial values of the supernumeraries with the neighboring boundary values suggests that the supernumerary nearest the stump consists primarily of stump tissue, whereas that nearest the grafted blastema consists primarily of tissues arising from the grafted blastema. We have found that for the values of α and β considered here, $L_2 = 2.4$.

The 4-centered solution shown in Figure 11 represents two pairs of opposite-handed supernumeraries, each of the supernumeraries being separated

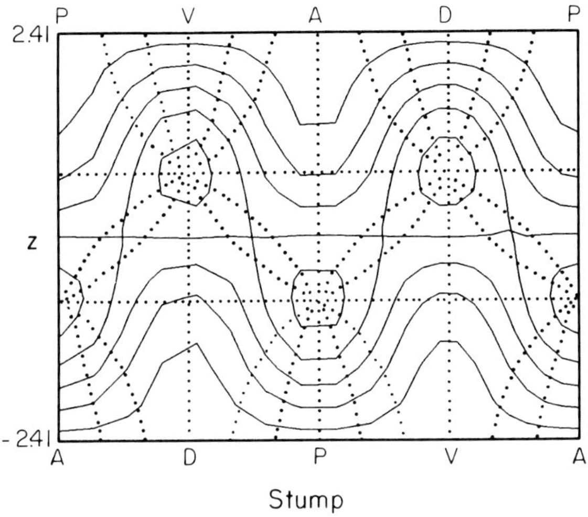

Figure 11: Four-centered solution of the field equations for the 180° ipsilateral graft. Here $\alpha = 2$, $\beta = m^{-2}$ and $L = 4.82$. This solution corresponds to two pairs of opposite-handed normal supernumeraries, with the supernumeraries nearest the stump having stump handedness. Solid and dotted contours as in Figure 6. $|\vec{\sigma}|^2/m^2$ has a minimum value of 0 at the center of each of the regions of concentric contours of $|\vec{\sigma}|^2/m^2$, growing to a maximum of 1 at the boundaries. The solid contours have the values 0.1, 0.3, and so on. From J. Totafurno and L.E.H. Trainor (1987) *Journal of Theoretical Biology* **124**, 415–454, Academic Press, with permission

from one another by 90°. As in the 2-centered solution, the supernumeraries are displaced in the z direction relative to one another, so that now we have two supernumeraries nearest the stump, both having the handedness of the stump and grafted blastema. For the values of α and β considered here, we have found $L_4 = 4.0$.

The rotational symmetry of the model for the ipsilateral graft with APDV inversion ensures that field patterns similar to those of Figures 10 and 11 but translated about the intercalary region by an arbitrary amount are also solutions. This translation of the field pattern occurs in such a way that, as the center of a supernumerary is shifted by an angle ω, the supernumerary itself appears to rotate by an angle ω about its center in a clockwise sense for a right supernumerary and in a counterclockwise sense for a left supernumerary. This is reminiscent of Bateson's mechanical model of supernumerary cockroach limbs in which the supernumeraries rotate about their respective centers as the centers of the supernumeraries are rotated about the primary limb (Bateson 1894, 1992; and see Winfree 1980, pp. 358–359).

We have noted the higher degree of variability in the observed outcomes of ipsilateral grafts as compared to that of contralateral grafts. The multiplicity of steady-state solutions of the model for the ipsilateral graft is in keeping with this observed variability. Which of the solutions is observed in a given experiment depends both upon the length of the intercalary region that forms and upon contingent factors affecting the development of the field in time. Furthermore, because of this multiplicity of steady-state solutions, distinct forms of non-steady-state solutions can be expected depending upon which steady state the field is approaching, thereby contributing to the inherent variability of the results. The various steady-state solutions in the model exist over a range of lengths L of the intercalary region, with qualitatively new forms of solution arising only at the critical lengths for bifurcation. Thus the model provides a relatively size-independent mechanism for the production of supernumeraries, i.e., a similar pattern may develop in a region of tissue with varying sizes, in keeping with the observed behavior in ordinary regeneration (Maden 1981).

Acknowledgments

WB would like to thank Joseph Frankel for sharing, over many years, his knowledge, experience and enthusiasm for *Tetrahymena* and morphogenesis. WB also thanks the Natural Sciences and Engineering Research Council of Canada for continued financial support.

References

Amit, D.J. (1978) *Field Theory, the Renormalization Group, and Critical Phenomena.* Toronto: McGraw-Hill.

Bateson, W. (1894; 1992) *Materials for the Study of Variation, Treated with Especial Regard to Discontinuity in the Origin of Species.* Baltimore: John Hopkins University Press.

Brandts, W.A.M (1993) A field model of left-right asymmetries in the pattern regulation of a cell. *IMA Journal of Mathematics Applied in Medicine & Biology* **10**, 31–50.

Brandts, W.A.M (1995) Relevance of field models to global patterning in ciliates. In D. Beysens, G. Forgacs and F. Gaill, eds., *Interplay of Genetic and Physical Processes in the Development of Biological Form.* Singapore: World Scientific, pp. 240–246.

Brandts, W.A.M, and Trainor, L.E.H (1990a) A non-linear field model of pattern formation: intercalation in morphallactic regulation. *Journal of Theoretical Biology* **146**, 37–56.

Brandts, W.A.M, and Trainor, L.E.H (1990b) A non-linear field model of pattern formation: application to intracellular pattern reversal in *Tetrahymena*. *Journal of Theoretical Biology* **146**, 57–86.

Bryant, S.V., French, V. and Bryant, P.J. (1981) Distal regeneration and symmetry. *Science* **212**, 993–1002.

Bryant, S.V. and Iten, L.E. (1976) Supernumerary limbs in amphibians: experimental production in *Notophthalmus viridescens* and a new interpretation of their formation. *Developmental Biology* **50**, 212–234.

Bryant, S.V. and Iten, L.E. (1977) Intercalary and supernumerary regeneration in regenerating and mature limbs. *Journal of Experimental Zoology* **202**, 1–16.

Bryant, P.J. and Simpson, P. (1984) Intrinsic and extrinsic control of growth in developing organisms. *Quarterly Review of Biology* **59**, 387–415.

Ding, D.-F., Totafurno, J. and Trainor, L.E.H. (1992) Bifurcation analysis of a nonlinear field model of regeneration. *Journal of Mathematical Biology* **30**, 355–377.

French, V., Bryant, P.J. and Bryant, S.V. (1976) Pattern regulation in epimorphic fields. *Science* **193**, 969–981.

Goodwin, B.C. (1976) *Analytical Physiology of Cells and Developing Organisms*. New York: Academic Press.

Goodwin, B.C. and Trainor, L.E.H. (1980) A field description of the clearage process in embryogenesis. *Journal of Theoretical Biology* **85**, 757–770.

Goodwin, B.C. and Trainor, L.E.H. (1983) The ontogeny and phylogeny of the pentadactyl limb. In B.C. Goodwin, N. Holder and C.C. Wylie, eds., *Development and Evolution*. Cambridge: Cambridge University Press, pp. 75–98.

Goodwin, B.C. and Trainor, L.E.H. (1985) Tip and whorl morphogenesis in *Acetabularia* by calcium-regulated strain fields. *Journal of Theoretical Biology* **117**, 79–106.

Goss, R.J. (1969) *Principles of Regeneration*. New York: Academic Press.

Kumazawa, H. (1979) Homopolar grafting in *Blepharisma japonicum. Journal of Experimental Zoology* **207**, 1–16.

Lewis, J. (1981) Simpler rules for epimorphic regulation: the polar coordinate model without polar coordinates. *Journal of Theoretical Biology* **88**, 371–392.

Ma, S.-K. (1976) *Modern Theory of Critical Phenomena.* Don Mills, Ontario: W.A. Benjamin, Inc.

Maden, M. (1980) Intercalary regeneration in the amphibian limb and the rule of distal transformation. *Journal of Embryology and Experimental Morphology* **56**, 201–209.

Maden, M. (1981) Morphallaxis in an epimorphic system: size, growth control and pattern formation during amphibian limb regeneration. *Journal of Embryology and Experimental Morphology* **65** (Supplement), 151–167.

Maden, M. (1982) Supernumerary limbs in amphibians. *American Zoologist* **22**, 131–142.

Maden, M. (1983) A test of the predictions of the boundary model regarding supernumerary limb structure. *Journal of Embryology and Experimental Morphology* **76**, 147–155.

Maden, M. and Mustafa, K. (1982) The structure of 180° supernumerary limbs and a hypothesis on their formation. *Developmental Biology* **93**, 257–265.

Maden, M. and Turner, R.N. (1978) Supernumerary limbs in the axolotl. *Nature* **273**, 232–235.

Mittenthal, J.E. (1981) The rule of normal neighbors: a hypothesis for morphogenetic pattern regulation. *Developmental Biology* **88**, 15–26.

Muneoka, K. and Bryant, S.V. (1984) Cellular contribution to supernumerary limbs in the axolotl, *Ambystoma mexicanum. Developmental Biology* **105**, 166–178.

Nelsen, E.M. and Frankel, J. (1986) Intracellular pattern reversal in *Tetrahymena thermophila.* I. Evidence for reverse intercalation in unbalanced doublets. *Developmental Biology* **114**, 53–71.

Nelsen, E.M., and Frankel, J. (1989) Maintenance and regulation of cellular handedness in *Tetrahymena. Development* **105**, 457–471.

Papageorgiou, S. and Holder, N. (1983) The structure of supernumerary limbs formed after 180° rotation in the newt *Triturus cristatus*. *Journal of Embryology and Experimental Morphology* **74**, 143–158.

Pescitelli, M.J., Jr. and Stocum, D.L. (1980) The origin of skeletal structures during intercalary regeneration of larval *Ambystoma* limbs. *Developmental Biology* **79**, 255–275.

Russell, M.A. (1985) Positional information in insect segments. *Developmental Biology* **108**, 269–283.

Shi, X., Lu, L., Qui, Z., He, W. and Frankel, J. (1991) Microsurgically generated discontinuities provoke heritable changes in cellular handedness of a ciliate, *Stylonychia mytilus*. *Development* **111**, 337–356.

Stock, G.B., Krasner, G.N., Holder, N. and Bryant, S.V. (1980) Frequency of supernumerary limbs following blastemal rotations in the newt. *Journal of Experimental Zoology* **214**, 123–126.

Stocum, D.L. (1982) Determination of axial polarity in the urodele limb regeneration blastema. *Journal of Embryology and Experimental Morphology* **71**, 193–214.

Suhama, M. (1990) The regulation of homopolar doublets to singlets in *Glaucoma*. *Journal of Protozoology* **37**, 344–352.

Tank, P.W. (1978) The occurrence of supernumerary limbs following blastemal transplantation in the regenerating forelimb of the axolotl, *Ambystoma mexicanum*. *Developmental Biology* **62**, 143–161.

Tank, P.W. (1981) Pattern formation following 180° rotation of regenerating blastemas in the axolotl, *Ambystoma mexicanum*. *Journal of Experimental Zoology* **217**, 377–387.

Tevlin, P. and Trainor, L.E.H. (1985) A two vector field model of limb regeneration and transplant phenomena. *Journal of Theoretical Biology* **115**, 495–513.

Totafurno, J. and Trainor, L.E.H. (1987) A non-linear vector field model of supernumerary limb production in salamanders. *Journal of Theoretical Biology* **124**, 415–454.

Turner, R.N. (1981) Probability aspects of supernumerary production in the regenerating limbs of the axolotl, *Ambystoma mex-*

icanum. *Journal of Embryology and Experimental Morphology* **65**, 119–126.

Turing, A.M. (1952) The chemical basis of morphogenesis. *Philosophical Transactions of the Royal Society of London, Series B* **237**, 37–72.

Wallace, H. and Watson, A. (1979) Duplicated axolotl regenerates. *Journal of Embryology and Experimental Morphology* **49**, 243–258.

Winfree, A.T. (1980) *The Geometry of Biological Time*. New York: Springer-Verlag.

Winfree, A.T. (1984) A continuity principle for regeneration. In G.M. Malacinski and S.V. Bryant, eds., *Pattern Formation: A Primer in Developmental Biology*. New York: Macmillan, pp. 103–124.

Symmetry Breaking Bifurcations

John Totafurno and L.E.H. Trainor

1 Introduction

What makes a scientific theory beautiful? Universality, simplicity, parsimony, generality, predictive power, accuracy, depth of insight, analytical power, ability to unify, invariance to contingences, symmetry? Beauty is, of course, in the eye of the beholder, but we expect that many of these characteristics would appear on the lists of most of our readers. With respect to supernumerary production as described in a previous essay (Brandts and Totafurno), one of us (LEHT) was immediately struck by the richness of the symmetries of the problem and by the opportunity this symmetry would provide for developing a theory, or at least a model with some aspect of beauty. In this chapter we would like to comment on the power of symmetry analysis in the development of a field model (Totafurno and Trainor 1987; hereinafter referred to as the TT model) and describe its use in determining the nature of the solutions. Despite the crudeness of the TT field model, it is this reliance upon symmetry that permits the model to work so well in accounting for the positions and structure of the supernumerary limbs which arise in contralateral and ipsilateral grafts in salamanders. Furthermore, as interpreted in the context of the TT model, this interesting biological example provides a rich illustration of the theory of symmetry breaking bifurcations.

In this chapter we attempt to show analytically that the solutions of the TT model representing supernumerary production in the 180° ipsilateral graft arise as symmetry breaking bifurcations. This is achieved using a theorem of Cicogna (1981) and Vanderbauwhede (1982) which is stated at the end of section 4 and applied to the model in section 5. For the reader not interested in ploughing through the more technical parts of sections 4 and 5, let us briefly outline the gist of the argument. The full symmetry group of the model is described in section 5.1. A solution of the model may be found, called the symmetric solution, which is invariant under the action of the full symmetry group of the model and which exists for the entire range of the control parameter L, the length of the intercalary region. Now the nonlinear equations of the TT model may be linearized about this symmetric solution. The theorem of Cicogna (1981) and Vanderbauwhede (1982) requires us to look at the eigenvalue problem for this linearized operator defined by equation (35) and in particular the dimensionality of the eigenspace corresponding to the eigenvalue

0. We show that equation (35) may be reduced to a pair of ordinary differential equations as given by equation (38). Solutions of equation (38) (with $\xi_n = 0$) that are invariant to subgroups of the full symmetry group of the model and in particular to subgroups with only limited rotational symmetry are shown to span only one-dimensional spaces. By the theorem of Cicogna (1981) and Vanderbauwhede (1982) we therefore conclude that these solutions arise as symmetry breaking bifurcations from the symmetric solution. Their form, to lowest order, is given in section 5.5.

2 Fields, Symmetries and the TT Model

The TT field model (see Brandts and Totafurno, this volume) borrows conceptually from the notion of "developmental fields," long recognized by the biological community (see e.g. Wolpert (1969) for a discussion of the field concept in biology), and particularly in this present instance, on the concept of field implied in the polar coordinate model (clock-face model) of French, Bryant and Bryant (1976, hereinafter referred to as FBB). Essentially the TT model puts some dynamical teeth into the phenomenological FBB model by hypothesizing an explicit field which is as smooth as possible given the special circumstances pertaining, and which bridges (in the style of physical field models) over the intercalary region of new tissue generated following transplant in such a manner as to satisfy two different boundary conditions at the edges of the region (corresponding to stump and blastema tissues).

From the standpoint of spatial symmetry, a field model provides a natural arena for its analysis. Let $\sigma(x)$ represent some developmental field defined at positions x over the intercalary region of the regenerating limb. A symmetry, $\tilde{g}$, of this field σ may then be defined as an operator that maps σ into itself, i.e.,

$$\tilde{g}\sigma = \sigma \ . \tag{1}$$

Implicit in (1) is that this relation holds at every point where σ is defined. A symmetry of the field therefore reflects the organization of the field as a whole. In this sense symmetries have a nonlocal character. They say something about the overall character of the field. Rather than focussing on the local details of the field, symmetries concentrate on the interrelationships of the field at different points. It may therefore be expected that the behavior of a model with respect to symmetry is relatively insensitive to many of the details of the model. (In the TT approach the field behavior is determined by a set of dynamical equations.)

Furthermore, the symmetry of a field has a certain discrete nature; the field either possesses the symmetry or it does not. One may therefore expect that despite the detailed changes in the field solution of model equations as the model parameters are changed, the symmetries of the solution are relatively insensitive to the variation of these parameters. In fact symmetry change in the solution might be expected to occur only under special circumstances. This situation will be discussed further below.

Let us apply these notions to supernumerary production. In particular consider the case of the ipsilateral graft with 180° rotation of the blastema (see Brandts and Totafurno, this volume). This graft involves the apposition, at any point about the stump, of tissue on the stump with blastema tissue derived from the opposite side of the limb. This suggests that points about the intercalary region are equivalent with respect to mismatch of field values at the edges of the intercalary region. No special positions are evidently defined by the boundaries at which to expect supernumeraries in the ipsilateral graft. We refer to this symmetry of the ipsilateral region of the ipsilateral graft as "rotational symmetry." Before explicitly defining this symmetry in the context of the intercalary field, we first consider two crucial aspects of the analysis: the dimensionality of the region over which the field is defined and the dimensionality of the field itself. However, in keeping with the above discussion concerning the robustness of symmetries, we simplify the analytical expression of this symmetry by idealizing the geometry of the intercalary region and the functional form of the boundary field values.

As discussed in Brandts and Totafurno (this volume), experiment indicates that the information necessary for successful grafts is contained in the surface tissues of stump and blastema, rather than in the deep structures such as muscle and bone. Thus one may limit consideration to a field on a two-dimensional surface sheet. This sheet was taken for simplicity to be quasi-cylindrical. The topology of a cylindrical sheet is equivalent to a flat rectangular sheet, corresponding to making a longitudinal cut down one side of the cylinder and flattening it out into a rectangle, with the two opposite sides corresponding to the cylindrical ends where the field values must match the stump and blastema field values, respectively, and with the remaining two edges of the rectangle corresponding to the two sides of the longitudinal cut, and thus required to have identical field values (see, for example, Figure 5 of Brandts and Totafurno, this volume). As to the nature of the developmental field defined on this flattened cut cylindrical surface, Einstein's principle (Iooss and Joseph 1980, p. v) that "everything should be made as simple as possible, but not simpler," was used. A scalar field is too simple because it does not allow unique field values to be specified in a continuous manner over the whole two-dimensional sheet. A vector field is ideally suited to the problem since its components along two orthogonal axes in the rectangle can be taken to represent tissue field values on the dorso-ventral and antero-posterior axes. Moreover, one can use the remaining degree of freedom, the vector magnitude, to represent the proximo-distal distance down the limb from shoulder to tip (here taken to be 0). The presence of a supernumerary limb with normal structure is marked by a complete circle of axial values (anterior to dorsal to posterior to ventral to anterior) surrounding a distal singularity (here 0), in keeping with the FBB identification.

The underlying rotational symmetry of the 180° ipsilateral graft proposed here is incorporated in a simple albeit somewhat artificial manner by assuming that the intercalary region has cylindrical geometry and by assuming that the vectorial field values at the boundaries rotate in a uniform fashion as one passes through the ventral, anterior, dorsal and posterior locations about the limb. The appearance of a supernumerary in this graft is an event in which positions about the intercalary region become inequivalent and complete rotational symmetry of the field is lost. We expect, as discussed above, that the symmetries of the solutions which occur when rotational symmetry is lost are not dependent on the idealized nature of the intercalary region and the boundary field values as chosen here.

In the TT model, rotational symmetry is associated with the invariance of the intercalary field to the combined operations of a rotation of the field *en masse* about the axis of the cylinder (which generally will yield a field that does not match with boundary field values) with a global rotation of the field values (which restores the agreement of the field with the given boundary field values). Let $R(\omega)$ be a rotation of the intercalary cylinder through angle ω about its axis and let $r(\omega)$ be a rotation by an angle ω of the two-dimensional field vector $\vec{\sigma}(\vec{x})$ where $\vec{x}$ is the position vector on the cylinder. Rotational symmetry is then expressed by invariance of the field to the operation $R(\omega) \otimes r(\omega)$ defined as

$$R(\omega) \otimes r(\omega)\vec{\sigma}(\vec{x}) = r(\omega)\vec{\sigma}(R(\omega)\vec{x}) \ . \tag{2}$$

Invariance for arbitrary ω indicates complete rotational symmetry, whereas limited rotational symmetry is indicated by invariance for only certain values of ω.

The development of normal supernumeraries and the concomitant presence in the model of isolated singularities in the field, i.e., by isolated points where $\vec{\sigma} = 0$, are associated with limited rotational symmetry of the field. In the ipsilateral graft, the boundary conditions preserve the rotational symmetry of the field equations. Therefore solutions of the model for the ipsilateral graft that represent the production of supernumeraries are the result of spontaneous symmetry breaking.

3 Supernumerary Production as a Symmetry Breaking Bifurcation

As discussed above, the production of supernumeraries in the ipsilateral graft corresponds to field solutions which lack complete rotational symmetry. The challenge for a model of this process is the generation of such an asymmetrical result from a basically symmetrical situation without imposing the asymmetry in an ad hoc fashion.

To this end, the dynamical equations in the TT model for the field were constructed so as to be invariant under the action of $R(\omega)\otimes r(\omega)$ (the complete symmetry group is described below). Einstein's principle was again used in selecting the explicit form of these equations subject to the constraint that the equations be rotationally symmetric. To insure a smooth field, a Laplacian term was included. To insure robustness to small perturbations, a linear term in the field was included. Finally a cubic term was added. This nonlinear term was included to simulate the observed coupling between the dorsal-ventral and anterior-posterior axes of the field and in order to permit the possibility of bifurcations of the field solution to fields that correspond to the reduced rotational symmetry associated with the formation of supernumerary limbs. [a]

A symmetry breaking bifurcation mechanism for the generation of solutions corresponding to the production of supernumeraries may be motivated by considering the field patterns of the intercalary region of the ipsilateral graft as the angle of rotation θ of the blastema prior to grafting is varied. For small θ, one might expect that the dorsal, ventral, posterior and anterior values of the field are arranged in lines over the intercalary region such that, starting at the stump boundary, they simply twist around the cylindrical intercalary region to match up with the rotated blastema boundary. This "twist solution" is rotationally symmetric in the sense defined above and therefore does not represent the production of supernumeraries.

Experimentally, as the graft angle θ is increased, we know that at some critical angle supernumeraries begin to appear (Maden and Turner 1978, Turner 1981) signaling the breaking of the rotational symmetry. This suggests that the solutions corresponding to supernumeraries arise in the model, at some critical angle θ, as bifurcations from rotationally symmetric solutions. Past the

[a] Intriguingly, even if this cubic term is not included, the resulting linear field equations have solutions for the 180° graft with the same symmetries as those of the nonlinear equations. However, these solutions occur only for discrete values of the control parameter in contrast to the more realistic behavior of the nonlinear equations in which a given solution exists over a range of values of the control parameter. Furthermore, the linear model puts no constraints on the norm of the bifurcating solutions in contrast to the nonlinear equation which forces a definite dependence of the norm on the control parameter (see Ding, Totafurno and Trainor 1992).

critical angle the twist solutions presumably become unstable while solutions representing supernumerary production stabilize.

Although the motivation for viewing the production of ipsilateral supernumeraries as a symmetry breaking bifurcation came from thinking of θ as a "control parameter," it has turned out to be more tractable in solving the model equations to set $\theta = 180°$ and look for bifurcations as the length L of the intercalary region is varied. In any case, this corresponds in a natural way to the generation of new intercalary tissue by the organism as it bridges between stump and blastema. As anticipated, as L is increased beyond a critical value, both pairs and eventually quadruples of supernumeraries are predicted to arise in the model via symmetry breaking bifurcations.

4 Theory of Symmetry Breaking Bifurcations

Preliminary to our discussion we review briefly symmetry breaking in a well-known physical model. Consider the second order ferromagnetic phase transition that occurs as the temperature is lowered through the Curie point, T_c. Above T_c, the total magnetization M is 0 with atomic spins randomly oriented; below T_c a net total magnetization $M \neq 0$ along some axis is exhibited breaking the original symmetry. For the magnetized state below T_c we have, to lowest order in M, that $T_c - T = \alpha M^2$, where $\alpha > 0$. The equation of state near T_c may then be written as $\alpha M^3 - M(T_c - T) = 0$. At $T = T_c$, two branches of solutions of the equation of state bifurcate from one another, with one of the branches corresponding to a broken symmetry. Technically, at the bifurcation point $M = 0$, $T = T_c$ the implicit function theorem does not hold so that M cannot be expressed as an invertible function of $T_c - T$ at $T = T_c$.

The theory of symmetry breaking bifurcations is a generalization of the notions of the preceding paragraph. The terminology and techniques are those of function spaces and functional analysis in which Hilbert's realization that functions may usefully be viewed as vectors is fully exploited (see Weyl (1944) for a treatise on Hilbert's work). Furthermore the bifurcation of one solution branch from another permits the possibility of a kind of perturbation theory in which the unknown solution of one branch is determined in terms of the known solution of the other branch in a region of parameter space "near" the bifurcation point. Even though the solutions so generated may have a limited range of validity, the symmetries they possess may in fact characterize the solution branch more generally.

Let Y be a real Banach space (i.e. a complete, normed, real-valued function space; see this essay's Appendix and Griffel 1981) and $\mathbf{R}$ be the space of real numbers. Consider a nonlinear equation of the form

$$F(y, \lambda) = 0 \tag{3}$$

where $F\colon Y \times \mathbf{R} \to Y$, is a nonlinear analytic map, with $y \in Y$, and where $\lambda \in \mathbf{R}$ is a real control parameter. Let there be a trivial solution $y = 0$ (akin to the $M = 0$ solution of the equation of state discussed above) satisfying $F(0, \lambda) = 0$ for all $\lambda \in \mathbf{R}$. Furthermore, suppose for some $\lambda_0 \in \mathbf{R}$ that the Fréchet derivative of F with respect to y,

$$d_y F(0, \lambda_0) = \mathcal{L}_0 \ , \tag{4}$$

is a non-invertible Fredholm linear operator with zero index. Using the Liapunov-Schmidt method, equation (3) can be reduced to an equivalent finite set of scalar equations in a finite number of scalar unknowns in a neighborhood of $y = 0$ and $\lambda = \lambda_0$ (Vanderbauwhede 1982). The method therefore yields the so-called bifurcation equation. To implement the method, we first introduce several definitions (see Appendix). Let

$$V = \ker\, \mathcal{L}_0 \ , \quad W = \text{range}\ \mathcal{L}_0 \tag{5}$$

be complementary subspaces in Y. Also assume that P and Q are projection operators that map Y onto V and W, respectively, i.e., for any $y \in Y$,

$$v = Py \ , \tag{6a}$$

and

$$w = Qy \ . \tag{6b}$$

Finally let

$$\mu = \lambda - \lambda_0 \ . \tag{7}$$

One can then obtain an invertible equation

$$QF(v + w, \lambda_o + \mu) = 0 \tag{8}$$

having a unique solution

$$w = w(v, \mu) \ , \tag{9}$$

where w is of order $o(\| v \|, \mu)$. It is the existence of a nontrivial kernel V of the linearized operator at some λ_0 that is a necessary condition for bifurcation. The bifurcation equation can be expressed, using equation (9), in the form

$$\Phi(v, \mu) = PF(v + w(v, \mu), \lambda_0 + \mu) = 0 \ . \tag{10}$$

If the map F is covariant with respect to a compact topological group G, then w and Φ will also be covariant with respect to G, i.e., for every group operator $g \in G$,

$$gw(v,\mu) = w(gv,\mu) \tag{11a}$$

and

$$g\Phi(v,\mu) = \Phi(gv,\mu) \ . \tag{11b}$$

Let $\tilde{G}$ be a subgroup of G. $\tilde{G}$ is called the isotropy subgroup of G at a point $v \in V$ if for every $\tilde{g} \in \tilde{G}$ we have $\tilde{g}v = v$. We define the fixed-point subspace of an isotropy subgroup $\tilde{G}$, $V^{\tilde{G}}$, to be the set of all points $v \in V$ that have the isotropy subgroup $\tilde{G}$, i.e.

$$V^{\tilde{G}} \equiv \{v \in V | \tilde{g}v = v, \forall \tilde{g} \in \tilde{G}\} \ . \tag{12}$$

The basic questions in symmetry-breaking bifurcation theory are the following:

1. Are there selection rules for the isotropy subgroup of the bifurcating solution and what are they?

2. Is there a method for guaranteeing the existence of a bifurcating solution having a given isotropy subgroup?

Cicogna (1981) and Vanderbauwhede (1982) have given partial answers to these questions. They showed that in a neighbourhood of $y = 0$, $\mu = 0$, there exists a unique bifurcating solution branch $\tilde{v}$ to $\Phi = 0$, or equivalently $\tilde{y} = \tilde{v} + \tilde{w}(\tilde{v},\mu)$ to $F = 0$, having the isotropy subgroup $\tilde{G}$ provided that:

1. the trivial solution goes unstable at the critical point in a non-degenerate fashion,

and

2. the fixed point subspace $V^{\tilde{G}}$ (see equation (12)) of $\tilde{G}$ has dimension 1, i.e. dim $V^{\tilde{G}} = 1$.

The isotropy subgroup $\tilde{G}$ describes the residual symmetry of $\tilde{y}$ in this case. Furthermore, to lowest order in μ, the bifurcating solution has the form

$$\tilde{y} = \eta(\mu)\tilde{v} \ , \tag{13}$$

where η is a real function of μ, and $\tilde{v} \in V^{\tilde{G}}$.

5 Application of the Theory of Symmetry Breaking Bifurcations to the TT Model

In order to apply the theory of symmetry breaking bifurcations to the TT model we consider the ipsilateral graft with $\theta = 180^\circ$ and take as the control parameter the length L of the intercalary region. We briefly summarize here the maximum symmetry group of the problem and the isotropy subgroups relevant to various bifurcating solution branches and the use of the theorem of Cicogna (1981) and Vanderbauwhede (1982) to show the solutions with these symmetries arise via bifurcations. Further details may be found in Ding et. al. (1992).

5.1 *Symmetry Groups of the* 180° *Ipsilateral Graft*

We have already described the rotational symmetry of the 180° ipsilateral graft and introduced the operator $R(\omega) \otimes r(\omega)$ (see equation (2)). This operator is a symmetry of the ipsilateral graft in that both the field equations and the boundary field values of the TT model are invariant under its action for arbitrary ω. Thus if $\vec{\sigma}$ is a solution of the TT model, then $R(\omega) \otimes r(\omega)\vec{\sigma}$ is also a solution for arbitrary ω.

The remaining elements of the maximum symmetry group G of the 180° ipsilateral graft are similar in form to $R(\omega) \otimes r(\omega)$ and may be obtained as follows. The field equations of the TT model possess the symmetries of the field energy given by

$$|\nabla u|^2 + |\nabla v|^2 - \alpha(u^2 + v^2) + \frac{\beta}{2}(u^2 + v^2)^2 \tag{14}$$

where the field $\vec{\sigma} = (u, v)$, α and β are positive constants, and ∇ is the gradient operator for the cylindrical surface (see Brandts and Totafurno, this volume). For convenience let us introduce cylindrical coordinates (z, ϕ) where the z-axis corresponds to the axis of the cylindrical intercalary region with the plane $z = 0$ halfway between the proximal and distal boundaries. The field energy is invariant to the full symmetry group of the cylinder which is conventionally written (Landau and Lifshitz 1965)

$$D_{\infty h} = \{E, R(\omega), \Sigma_v(\nu) \cdot \Sigma_h, \Sigma_h, R(\omega) \cdot \Sigma_h, \Sigma_v(\nu)\}\ , \tag{15}$$

where E is the identity, $R(\omega)$ is a rotation of the cylinder through angle ω about the z-axis, Σ_h is a reflection in the horizontal plane $z = 0$, and $\Sigma_v(\nu)$ is a reflection in a vertical plane containing the z-axis and oriented at an azimuthal angle $\phi = \nu$. (The vertical planes parametrized by ν and $\nu + \pi$ are identical.) The field energy is also invariant to the global action on the field vector $\vec{\sigma}$ of the group

$$C_{\infty v} = \{e, r(\omega), \sigma_v(\nu)\} \ , \tag{16}$$

where $e, r(\omega)$ and $\sigma_v(\nu)$ have a similar action to their uppercase counterparts except that they are defined to operate on the two-dimensional field vectors $\vec{\sigma}$. Specifically, e is the identity, $r(\omega)$ is a rotation of $\vec{\sigma}$ by an angle ω, and $\sigma_v(\nu)$ is a reflection of $\vec{\sigma}$ in the line making an angle ν with the vector $\vec{\sigma} = (1,0)$. In terms of the elements of these two groups we have found that G is given by

$$\begin{aligned} G = \ & \{E \otimes e, R(\omega) \otimes r(\omega), \Sigma_v(\nu) \cdot \Sigma_h \otimes \sigma_v(\nu - \pi/2), \\ & \Sigma_h \otimes r(\pi), R(\omega) \cdot \Sigma_h \otimes r(\omega + \pi), \Sigma_v(\nu) \otimes \sigma_v(\nu)\} \ , \end{aligned} \tag{17}$$

where, as for $R(\omega) \otimes r(\omega)$, the product $A \otimes B$ is defined by

$$A \otimes B\vec{\sigma}(\vec{x}) = B\vec{\sigma}(A\vec{x}) \ , \tag{18}$$

where $\vec{\sigma}(\vec{x})$ corresponds to the field vectors as a function of position $\vec{x}$ over the intercalary region, A is an element of $D_{\infty h}$, and B is the corresponding element of $C_{\infty v}$.

Since G is isomorphic to $D_{\infty h}$, the group of spatial transformation $D_{\infty h}$ characterizes G. We therefore refer to the field $\vec{\sigma}$ which is invariant under for example $R(\omega) \otimes r(\omega)$ as being rotationally invariant for some ω.

5.2 Isotropy Subgroups

In what follows we focus our attention on the 2–centered, 4–centered and twist solutions of the TT model. The 2– and 4–centered solutions possess 2 and 4 isolated singular points, respectively. In contrast, the twist solution has no such singular points; in the twist solution lines of constant direction of the field $\vec{\sigma}$ twist around the intercalary region joining identical values of $\vec{\sigma}$ on the stump and blastema boundaries. Inspection of these solutions of the TT model shows that their associated isotropy subgroups $\tilde{G}$ are subgroups of the maximum symmetry group G given by equation (17) that are isomorphic to the subgroups of $D_{\infty h}$ known as D_{1d}, D_{2d}, and D_{∞}, respectively. We denote these subgroups of G by $\tilde{D}_{1d}$, $\tilde{D}_{2d}$ and $\tilde{D}_{\infty}$.

In order to obtain linear representation of these isotropy subgroups, we may express the operations used to define G in matrix form as follows. Operating on spatial coordinates (x, y, z) one has

$$E = \begin{pmatrix} 1 & & \\ & 1 & \\ & & 1 \end{pmatrix} , \tag{19}$$

$$R(\omega) = \begin{pmatrix} \cos\omega & -\sin\omega & 0 \\ \sin\omega & \cos\omega & 0 \\ 0 & 0 & 1 \end{pmatrix} , \tag{20}$$

$$\Sigma_v(\nu) = \begin{pmatrix} \cos 2\nu & \sin 2\nu & 0 \\ \sin 2\nu & -\cos 2\nu & 0 \\ 0 & 0 & 1 \end{pmatrix} , \tag{21}$$

and

$$\Sigma_h = \begin{pmatrix} 1 & 0 & 0 \\ 0 & 1 & 0 \\ 0 & 0 & -1 \end{pmatrix} , \tag{22}$$

for rotation angle ω and reflection plane with angle ν. Operating on the components of the 2–dimensional vector $\vec{\sigma}$ one similarly has

$$e = \begin{pmatrix} 1 & \\ & 1 \end{pmatrix} , \tag{23}$$

$$r(\omega) = \begin{pmatrix} \cos\omega & -\sin\omega \\ \sin\omega & \cos\omega \end{pmatrix} , \tag{24}$$

and

$$\sigma_v(\nu) = \begin{pmatrix} \cos 2\nu & \sin 2\nu \\ \sin 2\nu & -\cos 2\nu \end{pmatrix} . \tag{25}$$

In terms of these matrices, a representation of $\tilde{D}_{1d}$ is generated by the set

$$\{E \otimes e, \Sigma_v(\pi/2) \otimes \sigma_v(\pi/2), \Sigma_v(\pi) \cdot \Sigma_h \otimes \sigma_v(\pi/2)\} . \tag{26}$$

$\tilde{D}_{1d}$ is the isotropy subgroup for which the spatial part is D_{1d} consisting of rotations by 2π about the z-axis, reflections in a single vertical reflection plane (here with orientation $\phi = \pi/2$), and rotations by π about a single axis in the horizontal plane $z = 0$ that is perpendicular to the vertical reflection plane. Similarly a representation of $\tilde{D}_{2d}$ is generated by the set

$$\{E \otimes e, R(\pi) \otimes r(\pi), \Sigma_v(0) \otimes \sigma_v(0), \Sigma_v(\pi/2) \otimes \sigma_v(\pi/2), \\ \Sigma_v(\pi/4) \cdot \Sigma_h \otimes \sigma_v(-\pi/4), \Sigma_v(3\pi/4) \cdot \Sigma_h \otimes \sigma_v(\pi/4)\} \; . \tag{27}$$

The isotropy subgroup $\tilde{D}_{2d}$ has spatial part D_{2d} consisting of rotations by π about the z-axis, reflections in two orthogonal vertical planes (here with orientation $\phi = 0$ and $\pi/2$), and rotations by π about 2 orthogonal axes in the horizontal plane $z = 0$ making an angle of $\pi/4$ with the vertical reflection planes. Finally, a representation of $\tilde{D}_\infty$ is generated by the set

$$\{E \otimes e, R(\omega) \otimes r(\omega), \Sigma_v(\nu) \cdot \Sigma_h \otimes \sigma_v(\nu - \pi/2)\} \; . \tag{28}$$

Here the isotropy subgroup of the twist solution, $\tilde{D}_\infty$, has spatial part D_∞ consisting of all the transformations generated by rotations by an arbitrary amount about the z-axis, and rotations by π about any axis in the horizontal plane $z = 0$.

5.3 Eigenspace of the Linearized Operator $\mathcal{L}_0$

Let the bifurcating solution be written

$$\vec{\sigma} = \vec{\sigma}_s + \vec{y} \tag{29}$$

where $\vec{\sigma}_s$ is a solution of the TT model that is invariant under the maximum symmetry group G given by equation (17) and where $\vec{y}$ is taken to be a small perturbation to $\vec{\sigma}_s$. We refer to $\vec{\sigma}_s$ as the symmetric solution and find that it has the form, using a 2-component vector matrix notation,

$$\vec{\sigma}_s(z, \phi) = g_0(z) \begin{pmatrix} \cos\phi \\ \\ \sin\phi \end{pmatrix} \; , \tag{30}$$

where (z, ϕ) are cylindrical coordinates for the intercalary region, and where $g_0(z)$ is an odd function of z that is dependent upon the control parameter L. (See section 6(d) of Ding et al. (1992) for an analytic expression for $g_0(z)$.)

The dynamical equations of the TT model may, in the steady-state limit, be written in the form of equation (3) as

$$F(\vec{y}, L) = 0 \tag{31}$$

where the nonlinear operator F is given by

$$F(\vec{y}, L) = \nabla^2\vec{y}+\alpha\vec{y}-\beta(|\vec{\sigma}_s|^2\vec{y}+2\vec{\sigma}_s(\vec{\sigma}_s\cdot\vec{y}))-\beta[|\vec{y}|^2\vec{\sigma}_s+2(\vec{\sigma}_s\cdot\vec{y})\vec{y}+|\vec{y}|^2\vec{y}] \ , \tag{32}$$

where we have explicitly indicated that $\vec{y}$ is a vector and that the control parameter is L. At the stump and blastema boundaries (here taken to be $z = \pm L/2$), $\vec{y} \equiv 0$ since the inhomogeneous boundary conditions of the 180° ipsilateral graft are satisfied by $\vec{\sigma}_s$. Furthermore, it is assumed that $\vec{y}$ is single-valued so that

$$\vec{y}(z, \phi) = \vec{y}(z, \phi + 2\pi) \ . \tag{33}$$

Now the linearized operator $\mathcal{L}_o$ of equation (4) may be seen from equation (32) to be given by

$$\mathcal{L}_o = \nabla^2 + \alpha - \beta|\vec{\sigma}_s|^2 - 2\beta\vec{\sigma}_s(\vec{\sigma}_s\cdot\ ) \ . \tag{34}$$

Solutions $\vec{y}$ of the eigenvalue problem

$$\mathcal{L}_0\vec{y} = \xi\vec{y} \tag{35}$$

where ξ is a real eigenvalue, define the eigenspace of $\mathcal{L}_0$. The subspace V of equation (5) corresponds to those $\vec{y}$ for which $\xi = 0$. Furthermore, for the nonlinear differential equations of TT that are first order in time, the eigenvalue spectrum of the operator $\mathcal{L}_o$ for a given value of the control parameter L determines the stability of the symmetric solution. Perturbations $\vec{y}$ having the form of an eigenfunction of $\mathcal{L}_o$ will either be damped out or grow in time depending upon whether the corresponding eigenvalue ξ is either negative or positive, respectively. Stability of the symmetric solution for a given value of L requires that the eigenvalue spectrum of $\mathcal{L}_o$ be bounded from above and negative. For the trivial solution (i.e. the symmetric solution) to go unstable at the critical point, L_c, in a non-degenerate fashion as required by the theorem of Cicogna and Vanderbauwhede requires that the maximum eigenvalue of $\mathcal{L}_o$ change sign from negative to positive as L is varied through the critical point.

To solve equation (35), we make the transformation of variables

$$\vec{y} = r(\phi)\vec{w} \ , \tag{36}$$

where $r(\phi)$ is given by equation (24), and expand $\vec{w}$ in a Fourier series in ϕ,

$$\vec{w}(z, \phi) = \Sigma_{n=-\infty}^{\infty}\vec{C}_n(z)e^{in\phi} \ , \tag{37}$$

where

$$\vec{C}_n = \begin{bmatrix} C_{n1} \\ C_{n2} \end{bmatrix} .$$

Note that because $\vec{w}(z,\phi)$ is a real vector field, the $\vec{C}_N$ functions must satisfy the reality conditions

$$\vec{C}_n^* = \vec{C}_{-n} ,$$

where $\vec{C}_n^*$ is the complex conjugate of $\vec{C}_n$. Substitution of equations (36) and (37) into equation (35) reduces the linear eigenvalue problem to pairs of ordinary differential equations

$$\frac{d^2}{dz^2}\vec{C}_n(z) + A_n\vec{C}_n(z) = \xi_n\vec{C}_n(z) \tag{38}$$

for the $\vec{C}_n(z)$ with homogeneous boundary conditions

$$\vec{C}_n(z = \pm L/2) = 0, \tag{39}$$

where the dependence of the eigenvalue ξ on n is explicitly indicated, and where

$$A_n = \begin{bmatrix} a - n^2 - 3\beta g_0^2(z) & -2in \\ 2in & a - n^2 - \beta g_0^2(z) \end{bmatrix} . \tag{40}$$

It turns out (Ding et. al. 1992) that the set of eigenfunctions with some given isotropy subgroup $\tilde{G}$ is accompanied by a minimum n, which we denote $\tilde{n}$, and a corresponding maximum eigenvalue $\xi_{\tilde{n}0}$ where the 0 indicates that $\vec{C}_{\tilde{n}}$ has no zeroes in the open internal $(-L/2, L/2)$. Determination of the eigenvectors $\vec{y}$ that span the fixed-point subspace $V^{\tilde{G}}$ is now reduced to the determination of the eigenfunctions of equation (38) corresponding to an eigenvalue of 0 for n equal to $\tilde{n}$ (*i. e.* $\xi_{\tilde{n}0} = 0$). (We assume here no accidental degeneracy which would result in two or more different values of n for a given isotropy subgroup having the same value of $\xi_{\tilde{n}0}$.) Nontrivial $V^{\tilde{G}}$ occurs for the critical value of the control parameter L_c at which the symmetric solution becomes unstable to perturbations with symmetry $\tilde{G}$. Moreover it turns out that since at the critical point the eigenfunction $\vec{C}_{\tilde{n}}$ has no zeros in the open interval $(-L/2, L/2)$, bifurcation occurs to solutions with the lowest gradients and hence maximum smoothness possible subject to the constraints imposed by the given isotropy subgroup.

5.4 The Fixed-Point Subspace $V^{\tilde{G}}$

Let the real and imaginary parts of $\vec{C}_n$ be denoted by $\vec{f}_n$ and $\vec{h}_n$, respectively so that

$$C_{nj}(z) = f_{nj}(z) + ih_{nj}(z) , \tag{41}$$

where $j = 1, 2$. The requirement that a solution of $\vec{y}$ of equation (35) belongs to the fixed-point subspace $V^{\tilde{G}}$ imposes constraints on the components of $f_{nj}(z)$ and $h_{nj}(z)$. In particular, we now consider the individual cases where $\tilde{G}$ is equal to $\tilde{D}_{1d}$, $\tilde{D}_{2d}$, and $\tilde{D}_{\infty}$ defined by equations (26), (27), and (28), respectively.

For $\tilde{G} = \tilde{D}_{1d}$ we find that the minimum allowed n, which we denote $\tilde{n}$, equals 1 and that $f_{\tilde{n}1}$ and $h_{\tilde{n}2}$ both vanish, while both $f_{\tilde{n}2}$ and $h_{\tilde{n}1}$ are even functions of z. Thus at the critical point, denoted L_2, for bifurcation to 2-centered solutions with isotropy subgroup $\tilde{D}_{1d}$, equation (38) reduces to

$$\frac{d^2 h_{11}}{dz^2}(z) + (a - 1 - 3\beta g_0^2(z))h_{11}(z) - 2f_{12}(z) = 0 \tag{42}$$

and

$$\frac{d^2 f_{12}}{dz^2}(z) + (a - 1 - \beta g_0^2(z))f_{12}(z) - 2h_{11}(z) = 0 \ ,$$

with boundary conditions

$$h_{11}(\pm L/2) = f_{12}(\pm L/2) = 0 \ .$$

Numerical solution of equation (42) shows that $f_{12}(z)$ and $h_{11}(z)$ are even functions so that $\vec{C}_1(z) = \vec{C}_1(-z)$ and

$$\dim V^{\tilde{D}_{1d}} = 1 \ . \tag{43}$$

Furthermore it is found that as L increases through L_2 from below, ξ_{10} changes from negative to positive. The symmetric solution therefore becomes unstable to perturbations with the symmetry $\tilde{D}_{1d}$ in a non-degenerate fashion.

A similar analysis for $\tilde{G} = \tilde{D}_{2d}$ yields $\tilde{n} = 2$, and both f_{22} and h_{21} vanish, while both f_{21} and h_{22} are even functions of z. Thus at the critical point, L_4, for bifurcation to the 4-centered solution, equation (38) reduces to

$$\frac{d^2 f_{21}}{dz^2}(z) + (a - 4 - 3\beta g_0^2(z))f_{21}(z) + 4h_{22}(z) = 0 \tag{44}$$

and

$$\frac{d^2 h_{22}}{dz^2}(z) + (a - 4 - \beta g_0^2(z))h_{22}(z) + 4f_{21}(z) = 0 \ ,$$

with boundary conditions

$$f_{21}(\pm L/2) = h_{22}(\pm L/2) = 0 \ .$$

Again numerical solution of equation (44) shows that $\vec{C}_2(z) = \vec{C}_2(-z)$ and hence

$$\dim V^{\tilde{D}_{2d}} = 1 \ , \tag{45}$$

and that the symmetric solution becomes unstable to perturbations with the symmetry $\tilde{D}_{2d}$ in a non-degenerate fashion.

Finally for $\tilde{G} = \tilde{D}_\infty$, it is required that $\tilde{n} = 0$, and that h_{01} and h_{02} both vanish, while both f_{01} and f_{02} are even functions of z. For the special case of $\tilde{n} = 0$, equation (38) yields two uncoupled equations for the components of $\vec{C}_0$. By the Sturm comparison theorem (see Ahmad and Lazer 1978), at a given L the maximum eigenvalue corresponding to the second component is strictly greater than that of the first. Hence at the critical point L_T, for bifurcation to the twist solution, equation (38) reduces to

$$\frac{d^2 f_{02}}{dz^2}(z) + (a - \beta g_0^2(z)) f_{02}(z) = 0 \ , \tag{46}$$

with boundary conditions

$$f_{02}(\pm L/2) = 0 \ .$$

Numerical solution of equation (46) shows that $f_{02}(z)$ is indeed an even function so that $\vec{C}_0(z) = \vec{C}_0(-z)$ and

$$\dim V^{\tilde{D}_\infty} = 1 \ . \tag{47}$$

It is thus found that also in this case the symmetric solution becomes unstable to perturbations with the symmetry $\tilde{D}_\infty$ in a non-degenerate fashion.

5.5 *The Bifurcating Solutions*

In subsection 5.4 we have shown that both requirements of the theorem of Cicogna and Vanderbauwhede listed in the previous section are satisfied for bifurcation from the symmetric solution $\vec{\sigma}_s(z,\phi)$ to solutions having isotropy subgroups $\tilde{D}_{1d}$, $\tilde{D}_{2d}$ and $\tilde{D}_\infty$, with critical values of the control parameter L_2, L_4 and L_T, respectively. We therefore conclude that the 2-centered, 4-centered and twist solutions arise in the TT model via symmetry breaking bifurcation from the symmetric solution.

Furthermore the following form for the bifurcating solutions is obtained. For the 2-centered solution

$$\begin{aligned} \vec{\sigma}(z,\phi) &= \vec{\sigma}_s(z,\phi) + \eta(L) r(\phi)[\vec{C}_1(z)e^{i\phi} + \vec{C}_1^*(z)e^{-i\phi}] \\ &= \vec{\sigma}_s(z,\phi) + 2\eta(L) \begin{pmatrix} \cos\phi & -\sin\phi \\ \sin\phi & \cos\phi \end{pmatrix} \cdot \begin{pmatrix} -h_{11}(z)\sin\phi \\ f_{12}(z)\cos\phi \end{pmatrix} , \end{aligned}$$

where $\vec{\sigma}_s$ is given by equation (30), f_{12} and h_{11} are given by equation (42), and where the function $\eta(L) = 0$ for $L \leq L_2$, whereas, $\eta(L) \neq 0$ for $L > L_2$.

For the 4-centered solution

$$\begin{aligned}\vec{\sigma}(z,\phi) &= \vec{\sigma}_s(z,\phi)+\eta(L)r(\phi)[\vec{C}_2(z)e^{2i\phi}+\vec{C}_2^*(z)e^{-2i\phi}] \\ &= \vec{\sigma}_s(z,\phi)+2\eta(L)\begin{pmatrix}\cos\phi & -\sin\phi \\ \sin\phi & \cos\phi\end{pmatrix}\cdot\begin{pmatrix}f_{21}(z)\cos 2\phi \\ -h_{22}(z)\sin 2\phi\end{pmatrix},\end{aligned}$$

where f_{21} and h_{22} are given by equation (44), and where $\eta(L)=0$ for $L\le L_4$, whereas, $\eta(L)\neq 0$ for $L>L_4$. Finally the twist solution has the form

$$\vec{\sigma}(z,\phi) = \vec{\sigma}_s(z,\phi)+\eta(L)r(\phi)\vec{C}_0(z) \tag{48}$$

$$= \vec{\sigma}_s(z,\phi)+\eta(L)f_{02}(z)\begin{pmatrix}-\sin\phi \\ \cos\phi\end{pmatrix}, \tag{49}$$

where f_{02} is given by equation (46), and where, $\eta(L)=0$ for $L\le L_T$, whereas $\eta(L)\neq 0$ for $L>L_T$. For each of these solution, the theorem of Cicogna and Vanderbauwhede guarantees the existence of the real functions $\eta(L)$ defined for values of L just beyond the respective bifurcation points L_2, L_4 and L_T.

The bifurcation of these solutions from the symmetric solution is shown graphically in Figure 1. Here we plot the field energy given by equation (14) per unit area of the intercalary region as a function of the length of the intercalary region L for the different solutions of the TT equations. As may be seen from this figure, the symmetric solution acts as a kind of backbone from which bifurcate the branches of the field energy for each of the bifurcating solutions. The apparent bifurcation points for each of the numerically obtained branches are consistent with the values of $L_2=2.41$, $L_4=4.04$ and $L_T=3.49$ obtained here using the bifurcation analysis (recall that the radius of the intercalary region R has been taken to be 1).

6 Discussion

Two remarkable observations stand out in the analysis we have summarized here. The first is that a relatively simple field model works so well in accounting for the positions and the characters of the supernumerary limbs which arise in contralateral and ipsilateral grafts in amphibians. The second is how powerfully the theory of symmetry breaking bifurcations is illustrated in this interesting example from biology. We comment below on these two aspects separately.

Crucial to the analysis was the experimental indication that the information necessary for successful grafts is contained in the surface tissues of stump

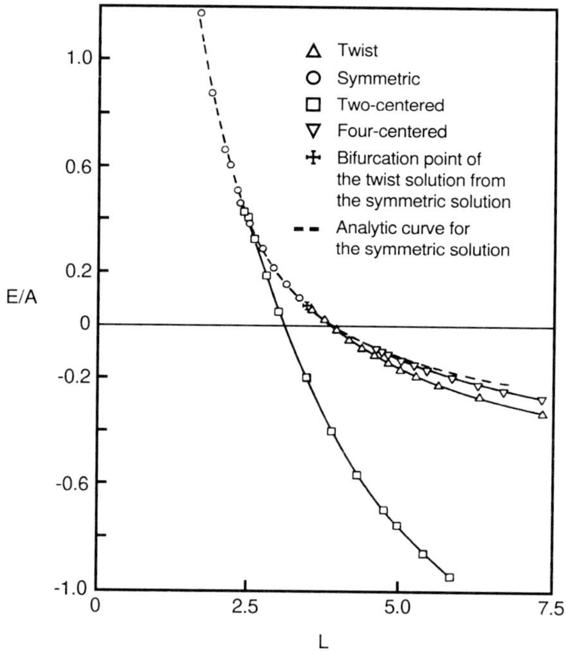

Figure 1: The field energy per unit area, E/A (where the area $A = 2\pi L$), as a function of L for the symmetric, 2-centered, twist, and 4-centered solutions of the 180° ipsilateral graft. Here $\alpha = 2$ and $\beta = m^{-2}$, where m is the norm of the field at the boundaries of the intercalary region. The points indicated (except for the point at which the twist solution bifuractes from the symmetric) were obtained numerically. The solid lines merely join these points. In contrast, the dashed line is determined from an analytic expression. Slightly modified from J. Totafurno and L.E.H. Trainor (1987) *Journal of Theoretical Biology* **124**, 415–454, Academic Press, with permission.

and blastema, rather than in the deep structures such as muscle and bone. This meant that one could limit consideration to a field on a two-dimensional surface sheet. The sheet was taken for simplicity to be quasi-cylindrical. The topology of a cylindrical sheet is a rectangle, corresponding to making a longitudinal cut down one side of the cylinder and flattening it out into a rectangle, with two opposite ends corresponding to the cylindrical ends where the field values must match the stump and blastema field values, respectively, with the remaining two sides of the rectangle corresponding to the two sides of the longitudinal cut, and thus required to have identical field values (see Figure 2).

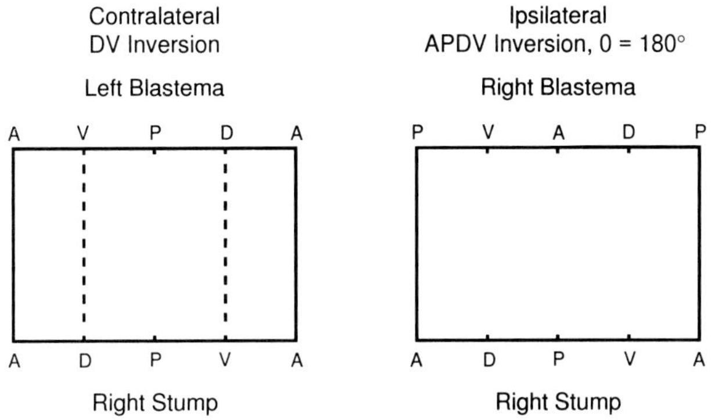

Figure 2: The intercalary region of both a contralateral graft with DV inversion and of an ipsilateral graft with APDV inversion showing the axial values of the stump and blastema boundaries (A = anterior, P = posterior, D = dorsal, V = ventral). In the contralateral case, two points of maximum difference in the axial values of the juxtaposed stump and blastema occur (marked by the dashed lines), whereas in the ipsilateral case, no distinctive points about the intercalary region are defined by the boundaries.

Finally a choice must be made of the nature of the developmental field defined on this flattened out cylindrical surface. Einstein's principle (Iooss and Joseph 1980, p. v) that "everything should be as simple as possible, but not simpler," was used. A scalar field is too simple because it does not allow unique field values over the whole two-dimensional sheet. A vector field is ideally suited to the problem since its components along two orthogonal axes in the rectangle can be taken to represent values on the ventral-dorsal and anterior-posterior axes of the intercalary region. Moreover, one can use the freedom remaining of vector magnitude to represent the proximal-distal distance down the limb from shoulder to tip (here taken as varying from one to zero). The theory was completed by borrowing an apparently successful hypothesis from FBB, namely that limb outgrowth (distal transformation) is presumed to occur wherever a complete circle of azimuthal values (dorsal to anterior to ventral to posterior to dorsal) surrounds a distal singularity (here zero).

Having defined the nature of the morphogenetic field, it was then necessary to come up with a dynamics for the field to obey. Again Einstein's principle was used. A Laplacian was included to ensure a smooth field, a linear term was included to ensure robustness to small perturbations, and a nonlinear term added which could simulate the observed coupling between the dorsal-ventral and the anterior-posterior fields, and which ensured that symmetry bifurcations could occur, presumably corresponding to the formation of supernumerary limbs.

Now the remarkable thing is that this theory worked, and in detail beyond the optimistic expectations of its authors. Perhaps the most remarkable feature of the theory was the complete correspondence between the appearances and locations of theoretical supers and the experimental observations, on the one hand, and with the mathematically predicted symmetry reductions of the bifurcated solutions on the other. The original symmetry is that of a cylinder, essentially, a symmetry broken by the appearance of singularities, but in a manner precisely predicted for the broken symmetry subgroups permitted by the mathematical theory.

The success of this model and the bifurcation analysis pertaining to it, indicates that the concept of a developmental field in biology is at least useful from a descriptive point of view, and suggests that it may even have a substantial physical basis (yet to be identified). Aside from its possible significance in developmental biology per se, the field approach provides another interesting example of a physical manifestation of symmetry breaking in bifurcation theory.

References

Ahmad, S. and Lazer, A.C. (1978) An N-dimensional extension of the Sturm separation and comparison theory to a class of non-selfadjoint systems. *SIAM Journal on Mathematical Analysis* **9**, 1137–1150.

Cicogna, G. (1981) Symmetry breakdown from bifurcation. *Lettere al Nuovo Cimento* **31**, 600–602.

Ding, D.-F., Totafurno, J. and Trainor, L.E.H. (1992) Bifurcation analysis of a nonlinear field model of regeneration. *Journal of Mathematical Biology* **30**, 355–377.

French, V., Bryant, P.J. and Bryant, S.V. (1976) Pattern regulation in epimorphic fields. *Science* **193**, 969–981.

Griffel, D.H. (1981) *Applied Functional Analysis.* Chichester, England: Ellis Horwood.

Iooss, G. and Joseph, D.D. (1980) *Elementary Stability and Bifurcation Theory.* New York: Springer-Verlag. This quotation, attributed to Albert Einstein, appears on p. v.

Landau, L.D. and Lifshitz, E.M. (1965) *Quantum Mechanics.* Third edition. Translated by J.B. Sykes and J.S. Bell. Oxford: Pergamon Press.

Maden, M. and Turner, R.N. (1978) Supernumerary limbs in the axolotl. *Nature* **273**, 232–235.

Totafurno, J. and Trainor, L.E.H. (1987) A non-linear vector field model of supernumerary limb production in salamanders. *Journal of Theoretical Biology* **124**, 415–454.

Turner, R.N. (1981) Probability aspects of supernumerary production in the regenerating limbs of the axolotl, *Ambystoma mexicanum*. *Journal of Embryology and Experimental Morphology* **65**, 119–126.

Vanderbauwhede, A. (1982) *Local Bifurcation and Symmetry*. Research Notes in Mathematics, Volume 75. London: Pitman.

Weyl, H. (1944) David Hilbert and his mathematical work. *Bulletin of the American Mathematical Society* **50**, 612–654.

Wolpert, L. (1969) Positional information and the spatial pattern of cellular differentiation. *Journal of Theoretical Biology* **25**, 1–47.

Appendix

1. Banach space: a normed vector space in which every Cauchy sequence is convergent. For a normed vector space Y, a Cauchy sequence is a sequence $\{y_n\}$ where $y_n \in Y$ such that for any $\xi > 0$, there is a number N such that $||y_n - y_m|| < \xi$ for $n, m > N$.

2. Fréchet derivative: Consider a mapping $F\colon Y \to Y$, where Y is a normed vector space, that is defined on a neighborhood of $y_0 \in Y$.

   The Fréchet derivative of F at y_0, $d_y F(y_0){:}Y \to Y$ is a continuous linear operator for which

   $$F(y + y_0) = F(y) + d_y F(y_0)h + o(h)$$

   as $h \to 0$ where $h \in Y$ and where $o(h)$ stands for any mapping $\phi(h){:}Y \to Y$ such that $||\phi(h)||/||h|| \to 0$ as $h \to 0$.

3. Range and Kernel: Consider a linear operator $\mathcal{L}{:}X \to Y$ defined on some subspace X of a vector space Y.

   The range of $\mathcal{L}$, denoted range $\mathcal{L}$, is the set

   $$\{y = \mathcal{L}x | x \in X\}$$

   The kernel of $\mathcal{L}$, ker $\mathcal{L}$, is the set

   $$\{x | \mathcal{L}x = 0, x \in X\}$$

4. Fredholm linear operator of zero index: Consider a linear map $\mathcal{L}{:}X \to Y$ where X and Y are Banach spaces and $\mathcal{L}$ is bounded, i.e. $||\mathcal{L}x|| < m||x||$ for some number m for all $x \in X$. $\mathcal{L}$ is said to be a Fredholm linear operator if

   (i) dim ker $\mathcal{L} < \infty$;

   (ii) range $\mathcal{L}$ is closed, i.e. range $\mathcal{L}$ contains all its limit points; and

   (iii) codim range $\mathcal{L} < \infty$.

   It is a Fredholm operator of zero index if dim ker $\mathcal{L}$ = codim range $\mathcal{L}$.

Surface Shape and Form in Morphogen

Trevor N. Hart and L.E.H. Trainor

1 Introduction

One of the most fascinating and complex phenomena in nature is the development of a multicellular adult organism from the fertilized egg. This phenomenon involves two basic interrelated processes: the shaping and movement of cells and tissues relative to each other (morphogenesis), and the specialization of these tissues to their specific function in the mature organism (differentiation). Each of these processes involves very complex interactions at both the intra- and intercellular levels. Within the cell, gene expression is carefully controlled so that enzymes and structural proteins are made available to the cell only when they are needed. Between cells, communication (mediated by diffusing substances, tight cell junctions, or other means) insures that cells cooperate to form a consistent developmental pattern, and that cell phenotype is consistent within a given region of tissue. While modern experimental biology, using the techniques of molecular and cell biology, embryology and anatomy (Berrill and Karp 1976, Brachet and Alexandre 1986, and Bard 1990 for history and introductions), has done much to outline some of the detailed mechanisms in development, we are still a long way from unraveling the detailed processes that control development.

One fruitful approach to a qualitative understanding of development has been the idea of a developmental "pattern" or "field." A pattern is a dynamical entity, defined throughout some region of the developing system, that in some way characterizes and determines the local pathway of development. A developmental pattern may or may not be some physically measurable quantity: It may be something as simple as the local concentration of a diffusible substance, as in the reaction-diffusion models of Turing and others (Turing 1952, Meinhardt 1982); a diffusible substance linked to a strain field (Goodwin and Trainor 1985); or it may be something as abstract as "positional information," as in the models of Wolpert (1971) and Totafurno and Trainor (1987). What is important is not so much which physical components of the cell define the field, but rather the behavior of the field. By applying simple models for developmental fields, much of the basic behavior of how regions of tissue become determined can be explained.

The field approach to development has focused largely on the problem of the determination and differentiation of tissues (see also the discussions in the essay by Brandts and Totafurno). An equally important aspect of development is the characteristic changes in shape that occur in the embryo over time. Can field theory be applied to understand this aspect of developmental change? There a several ways in which the field approach is quite amenable to this problem. First, shape is relatively easy to define within a field description, since we readily keep track of the positions and movements of each cell or region of the embryo. Second, since changes in tissue position and shape must be generated by internal forces, an approach which treats the embryo as a deformable medium can be applied to the problem. The principal questions then become: How do we describe the dynamics of shape, how do underlying changes in "positional information" generate shape, and how does shape influence the formation of new developmental patterns?

This essay presents a field approach to the problem of describing shape in morphogenesis. In this approach, the developing system (the interest here is primarily in the early embryo) is regarded as a thermodynamic system, with a local thermodynamic state defined at each position in the organism. A local concept of shape is used, so that the most important aspects of shape are characterised as a field that varies continuously throughout the organism and without reference to an external coordinate system. The constraints on possible shapes and the generation and stability of new shape are formulated using a model of the free energy of a shape based local shape descriptors. The interaction of other patterns with shape parameters can be expressed as free energy terms which include terms associated with these patterns and the shape terms together. The free energy minimum of the system thus combines effects of the pattern and shape together; in this way the properties of the pattern affect the shape and vice versa. This gives a consistent framework for formulating the interaction between the shape and other developmental patterns.

The remainder of our essay is divided into three parts. Section 2 will present the basic theory of the present approach to the problem of shape in morphogenesis. Section 3 presents an application of this theory to understanding the mechanism responsible for gastrulation, a key developmental event in the early embryo. Section 4 will address the future of the problem of shape in development and will present a possible definition of "form" in development based on the present theory.

2 Morphogenesis, Surfaces and Fields

The study of shape, in three dimensions, can essentially be characterized as the study of surfaces embedded in Euclidean space. The reason for this is that a three-dimensional object can, in many cases of interest, be characterized by the surface which forms its boundary with the rest of the world. A surface, or collection of surfaces, helps define what is inside of the object and what is outside. Hence, the understanding of surfaces is central to our understanding of shape. In biology, one is often interested in objects which are in fact cellular sheets, know as *epithelia*, formed from cells which adhere to each other in layers one or two cells in thickness. In development, these sheets play a very important role in the formation of the embryo, forming the primary tissue layers that are the basis for all later tissue and organ formation. It is the shape of these layers that are our primary interest.

The basic tool for our field approach to surfaces in morphogenesis is the theory of differential geometry of surfaces. This theory seeks to understand the three-dimensional nature of a surface in terms of a description that is intrinsic to the surface itself. The surface description begins with a parameterization in terms of a mapping from the plane to the surface, so that each point (locally, at least) has a corresponding point on the plane. Then, at each point, the local shape is characterized by two types of quantities: the metric, which characterizes the infinitesimal distances of arcs along the surface, and the curvature, which characterizes the magnitude and direction of the local curvature of the surface. These quantities depend on the choice of coordinate system, but are covariant: There is a consistent way to transform them from one coordinate system to another. Actually, they transform as second-rank tensors.

Since our interest is in the process of shape change, we need to be able compare the geometry at different points in time. To do this consistently, the parameterization of the surface is required to be fixed for all time. The state at a fixed but arbitrary time is chosen as the "reference state" for the system, and the shapes at other times are compared to each other through this reference state. In essence, the parameterization identifies each point on the surface uniquely and it is through this parameterization that other properties defined at that point can be identified.

To formulate a field theory from this description of a developing surface, we need to list the constraints acting on the field variables. The first is a set of compatibility conditions, which are related to the embedding of the surface in three-dimensional space, and which take the form of a set of differential equations relating the metric and curvature tensors. The details of these conditions need not concern us. Next are constraints associated with boundary conditions on the surface. Finally, the principal conditions will come from the requirement that the total free energy for the system be a local minimum. From this we obtain conditions not only concerning the equilibrium of the surface shape, but also from other field variables that may be influencing the development of the system.

We now give a more detailed treatment of surface geometry and present the full development of surface field theory. Due to restrictions of space, we cannot give full technical detail to either the surface geometry or to surface field theory; hence, many facts will be stated with little justification or proof. The interested reader may consult various references (Stoker 1969, do Carmo 1976, Hart 1989).

The most basic description of a surface specifies the three-dimensional position of each point of the surface. Given a surface parameterization $u = (u^1, u^2)$, the position with respect to a fixed external coordinate system of each point on the surface will be denoted here by $\vec{R}(u)$.

Although this is not our surface description of choice, it will be useful to refer to since we will ultimately have to solve for $\vec{R}(u)$ in order to determine the surface shape. At a given point on the surface we can define the first derivatives of $\vec{R}$ with respect to the coordinate parameters:

$$\vec{R}_\alpha(u) = \frac{\partial \vec{R}}{\partial u^\alpha}(u), \quad \alpha = 1, 2. \tag{1}$$

These pair of vectors define the tangent plane (e.g. Stoker 1969, do Carmo 1976). The unit normal $\hat{n}(u)$ to this plane is the normalized cross product of these two vectors, and is defined as long as the coordinate system is nonsingular. The first geometrical quantity to define is the metric, a second-rank tensor:

$$g_{\alpha\beta}(u) = \vec{R}_\alpha(u) \cdot \vec{R}_\beta(u). \tag{2}$$

The metric tensor characterizes the local area and local arc lengths on the surface. Given a function $f(u)$ defined on the surface, we can specify the invariant integral over a region S of the surface as

$$I(f, S) = \int_S f(u)\sqrt{g}\, d^2u, \tag{3}$$

where $g = g_{22}g_{11} - g_{21}g_{12}$ is the determinant of the metric tensor and $d^2u = du^1\, du^2$ is the product of the surface parameter differentials. The value of $I(f, S)$ is, as the name implies, independent of the choice of surface parameterization: It is a property intrinsic to the surface. As is customary in tensor algebra, the inverse of the metric tensor is defined here by raised indices:

$$g^{\alpha\beta} = (g)^{-1}_{\alpha\beta}. \tag{4}$$

The two forms of the metric tensor are used to raise and lower indices on tensors defined on the surface, i.e. to convert tensors between their covariant, contravariant and mixed forms.

The other geometrical quantity we will need is the curvature tensor, defined in terms of the second derivatives of $\vec{R}(u)$ as

$$b_{\alpha\beta} = \frac{\partial^2 \vec{R}}{\partial u^\alpha \partial u^\beta} \cdot \hat{n}. \tag{5}$$

The significance of this tensor is that the principal curvatures of the surface at a given point are the inverse of the eigenvalues of this tensor. Thus, the curvature tensor characterizes the degree, and the direction, to which the surface locally bends away from the tangent plane. The metric and curvature tensors are not completely independent. They must satisfy a set of differential equations, known as the compatability conditions or the Mainardi-Codazzi equations (do Carmo 1976). In addition, the metric and curvature uniquely define the surface. Thus, they represent a complete set of geometrical descriptors from which we can construct a surface field theory.

We cannot directly use the metric and curvature tensors as our field variables because we need to identify surface points over time during the evolution of the surface. To explain this, we need to understand that all equilibrium and dynamical field properties are determined by the free energy, which is a function of our surface field variables and is a scalar. In particular, it needs to be invariant with respect to changes in the choice of surface parameterization. However, we can see that a change in the choice of parameterization is equivalent to an actual physical change to the surface, where the parameterization remains the same but the surface is stretched and compressed within the same surface profile. As we have noted, the free energy of the surface must change. Thus, the free energy must be a function of some other parameters in addition to the metric and curvature of the surface.

The way out of this problem is to define the free energy not in terms of the metric and curvature directly, but in terms of *changes* in the metric and curvature. Consider the surface as it changes over time. Select a reference state of the surface to be its state at a fixed time $t = t_0$, and denote the corresponding metric and curvature tensors as $\tilde{g}_{\alpha\beta}$ and $\tilde{b}_{\alpha\beta}$. Then, with the parameterization fixed, define the metric and curvature at an arbitary time $t = t_1$ in the usual way. Now we have a clear way to distinguish between simple changes of coordinate system, which should leave the free energy unchanged, and actual physical changes to the surface, which will in general result in changes to the free energy. Our choice of field variables will be a combination of the metric and curvature tensors at $t=t_0$ and $t=t_1$. We define the *stretching deformation tensor*,

$$s^\alpha{}_\beta = \tilde{g}^{\alpha\lambda} g_{\lambda\beta}, \tag{6}$$

and the *bending deformation tensor*,

$$q^{\alpha}{}_{\beta} = \tilde{g}^{\alpha\lambda}(\tilde{b}_{\lambda\beta} - b_{\lambda\beta}). \tag{7}$$

Here and elsewhere, we make use of the summation convention over repeated indices. Note that t_1 is regarded as a variable, while t_0 is fixed; so that the geometric tensors with tildes are constant, while the ones without tildes are dynamically variable. The whole function of the tilde surface is to define the surface parameterization. These quantities are directly related to the deformation tensors used in continuum mechanics. In fact, our approach to gastrulation is a manifestation of the theory of thin elastic shells.

Having defined a set of field variables to characterize surface shape, we now need to specify the conditions which determine both equilibrium states and dynamical behavior. The key to this is to determine an expression for the local free energy at an arbitrary point on the surface, and then use the principles of nonequilibrium thermodynamics. It should be kept in mind that while the developing system is a nonequilibrium system, thermodynamically speaking, it is quite reasonable to use standard nonequilibrium theory in which we can view the system as being in local thermodynamic equilibrium. In this view, we may define, at each point in the body, the usual thermodyanmic quantities, such as entropy, temperature, pressure, free energy, etc. The principal thermodynamic variables will be the free energy density, the deformation tensors, and "unspecified" variables controlled by cellular processes.

The local thermodynamic quantity of greatest interest is the free energy density, since the minima of the free energy characterize equilibrium states of the system. The free energy density is not an independent quantity, but can be expressed in terms of a set of thermodynamic variables. Using our surface field description, we will express the free energy density at a surface point u in terms of the stretching and bending deformation tensors at that point, as well as other field variables in the system (denoted collectively as $\{\phi_i\}$),

$$F = F(\{s^{\alpha}{}_{\beta}\}, \{q^{\alpha}{}_{\beta}\}, \{\phi_i\}). \tag{8}$$

This dependence may include surface derivatives of the field variables as well as the actual values at u. The braces $\{\}$ indicate this functional dependence. The total free energy may then be defined by integrating over the whole surface,

$$\mathcal{F} = \int F(\{s^{\alpha}{}_{\beta}\}, \{q^{\alpha}{}_{\beta}\}, \{\phi_i\})\sqrt{g}\, d^2u. \tag{9}$$

The condition that a given state is an equilibrium is that

$$\delta\mathcal{F} = \delta\left(\int d^2u\sqrt{g}F\right) = 0 \tag{10}$$

for all virtual displacements of the system variables (and possibly their derivatives).

To develop a model of surface shape, we assume that the free energy density can be expressed as a sum of three terms: one that is dependent on the surface-shape field only (i.e. on $s^{\alpha}{}_{\beta}$ and $q^{\alpha}{}_{\beta}$), another involving all other variables but independent of shape variables, and an "interaction term," involving all variables but conforming to a particular form. The free energy density is thus taken to be

$$F = F_{shape}(s^{\alpha}{}_{\beta}, q^{\alpha}{}_{\beta}) + F_{other}(\{\phi_i\}) + F_{int}(s^{\alpha}{}_{\beta}, q^{\alpha}{}_{\beta}, \{\phi_i\}). \tag{11}$$

In order for this expression to be useful, we need to make some assumptions about the form of each term. Since the present discussion is concerned fundamentally with the properties of shape, we will assume that all other field variables which might influence shape are in fact constant locally. These other field variables will in essence behave as parameters in the theory. This implies that F_{other} is a constant, which we can take to be zero (since a constant can be absorbed into one of the other terms). For the other terms, it is desirable to use the simplest possible form that might be useful. Since the free energy density transforms as a scalar under change of surface coordinates, its dependence on $s^{\alpha}{}_{\beta}$ and $q^{\alpha}{}_{\beta}$ must be in terms of scalar invariants that can be formed from these tensors. We will take the simplest of these forms, the invariants $s^{\alpha}{}_{\alpha}$ and $q^{\alpha}{}_{\alpha}$ that can be formed by simply contracting the tenors. The exact form we take, which will be further explained below, is

$$F_{shape} = \frac{A}{2}(s^{\alpha}{}_{\alpha} - 4) + \frac{B}{2}(q^{\alpha}{}_{\alpha})^2 \tag{12}$$

where A and B arc positive constants. Finally, the form for the interaction free energy is basically the same as F_{shape}, except that the shape terms are one order lower and the multiplying parameters are functions of the other field variables:

$$F_{int} = -T(\phi_i) - W(\phi_i)q^{\alpha}{}_{\alpha}. \tag{13}$$

Since we are not explicitly dealing with ϕ_i, we can simply take T and W as functions of the surface coordinates alone.

To understand the form of the shape term, F_{shape}, assume that the interaction term is zero. In the absence of other effects (as implied by the interaction), the surface will remain in a given equilibrium state. By convention this state will be defined as the *reference surface*, characterized by $s^{\alpha}{}_{\beta} = \delta^{\alpha}{}_{\beta}$ and $q^{\alpha}{}_{\beta} = 0$. The free energy is then given by

$$\mathcal{F} = \int d^2u\sqrt{g}\left[\frac{A}{2}(s^{\lambda}{}_{\lambda} - 4) + \frac{B}{2}(q^{\lambda}{}_{\lambda})^2\right] \tag{14}$$

and the required conditions are that $\delta\mathcal{F} = 0$ and $\delta^2\mathcal{F} > 0$ for all variations in the stretching and bending deformation tensors, $s^{\alpha}{}_{\beta}$ and $q^{\alpha}{}_{\beta}$ respectively. At first appearance, it would seem that neither condition could be satisified, due to the fact that apparently only the linear term of $s^{\alpha}{}_{\beta}$ occurs in the integral of equation (14). Normally, the condition of a stable equilibrium requires that the linear term vanish and the quadratic term be positive definite. Both of these conditions are satisfied for the bending-deformation term $q^{\lambda}{}_{\lambda}$ in equation (14). However, what is not directly apparent from this equation is that the $\sqrt{g}$ term is dependent on the stretching deformation tensor. To understand this dependence, recall that the latter is defined by

$$s^{\alpha}{}_{\beta} = \tilde{g}^{\alpha\lambda} g_{\lambda\beta}. \tag{15}$$

Since $g = \det(g_{\alpha\beta})$ and the determinant has the multiplicative property, $\sqrt{g}$ can be rewritten as

$$\sqrt{g} = (\tilde{g})^{-\frac{1}{2}} (\det(s^{\alpha}{}_{\beta}))^{\frac{1}{2}}. \tag{16}$$

To understand the effect this dependence has on the the free energy, consider the expansion in terms of $s^{\alpha}{}_{\beta}$ about the reference state, $s^{\alpha}{}_{\beta} = \delta^{\alpha}{}_{\beta}$:

$$\sqrt{g} \approx \sqrt{\tilde{g}} \left[1 + \frac{1}{2}(s^{\alpha}{}_{\alpha} - 2) + \cdots \right] . \tag{17}$$

Inserting this expression into equation (14) and rewriting the second term, we get

$$\mathcal{F} \approx \int d^2u \sqrt{\tilde{g}} \left[1 + \frac{1}{2}(s^{\lambda}{}_{\lambda} - 2) \right] \left[-A + \frac{A}{2}(s^{\lambda}{}_{\lambda} - 2) + \frac{B}{2}(q^{\lambda}{}_{\lambda})^2 \right] . \tag{18}$$

On expanding the integrand, we find the first-order terms in the deformation tensors vanish, fulfilling the first equilibrium condition for the reference state. Similarly, equation (17) can be expanded to second order; an analogous calculation shows that $\delta^2\mathcal{F} > 0$ holds if A and B are positive.

Now we are prepared to deal with the interaction terms in the free energy density. Unlike the shape terms, which might actually appear without other terms, the interaction terms in the free energy density can really only produce a meaningful model when other terms are present. Nonetheless, it is convenient to talk about the total free energy contribution of these terms separately. This contribution can be written as

$$\mathcal{F}_{int} = -\int d^2u \sqrt{g} \left[T + W(q^{\lambda}{}_{\lambda}) \right] . \tag{19}$$

The form is similar to the shape contribution to the free energy, except that the terms are one lower in order. And, as in this case, the shape dependence of the area term, $\sqrt{g}$, plays an important role. Consider a shape that is near the reference state, so that the expansion (17) is valid. Then equation (19) can be written as

$$\mathcal{F}_{int} \approx -\int d^2u\sqrt{\tilde{g}}\left[1+\frac{1}{2}(s^{\lambda}{}_{\lambda}-2)\right]\left[T+W(q^{\lambda}{}_{\lambda})\right]. \qquad (20)$$

On expanding this expression, we find that the integral contains primarily first-order terms in the deformation tensors. Thus, if the functions T and W are nonzero, the effect on these terms when combined with the shape terms is to force the system out of the reference state. The reference state cannot be an equilibrium because the condition $\delta\mathcal{F} = 0$ is no longer satisfied.

These terms have a simple and important physical interpretation. Consider equation (19) with $W = 0$ and imagine there is some small change to the surface. Then the change in the interaction part of the free energy would be

$$\delta\mathcal{F} = -\int d^2u\delta(\sqrt{g})T. \qquad (21)$$

Since $d^2u\,\delta(\sqrt{g})$ is the change in the local surface area, this expression is an integral form of the thermodynamic relation

$$\delta F = \tau\delta A, \qquad (22)$$

where τ is the surface tension and A is the surface area. Thus, T is a local thermodynamic quantity that is analogous to surface tension, the fundamental difference being the presence of the minus sign in front of the first expression. The interpretation of this is that positive surface tension acts to contract the area of a surface, whereas a positive value of T acts to expand the area of the surface. Thus, T represents a force (or lateral pressure) acting within the surface caused by fields other than those depending directly on surface shape (i.e. $s^{\alpha}{}_{\beta}$ and $q^{\alpha}{}_{\beta}$). We will refer to T as the *stretching interaction function.*

To understand W, the second term in equation (19), recall that the bending deformation tensor, $q^{\alpha}{}_{\beta}$ (see equations (7) and (5)), measures the change of surface curvature between the current surface shape and the reference shape. It does not measure the curvature in an absolute sense, but rather relative to the direction of the surface normal. That is, if curvature increases in the direction of surface normal, the overall curvature change is positive, whereas if the curvature increases in the opposite direction, it is negative. Consider equation (19) with $W > 0$. The net effect of this term is to cause the surface to curve in the direction of the surface normal, since this increases the curvature and hence decreases the free energy (recall the minus sign in front of equation (19)). Thus, we can interpret W as a force that acts on the surface, bending it preferentially in a particular direction. We refer to W as the *bending interaction function.*

The complete model for the free energy of the system is

$$\mathcal{F} = \int d^2u\sqrt{g}\left[\frac{A}{2}(s^\alpha{}_\alpha - 4) + \frac{B}{2}(q^\alpha{}_\alpha)^2 - T - Wq^\alpha{}_\alpha\right]. \qquad (23)$$

The effect of the first two terms is to maintain the system at the equilibrium reference state, while the effect of the latter two is to drive the system towards a new state. The balance of these effects will be for the system to find a free energy minimum at a new surface shape. In the next section we will apply this model to the phenomenon of gastrulation.

3 The Application of Surface Field Theory to Gastrulation

Gastrulation is an important and dramatic process in the development of the embryo (Berrill and Karp 1976, Bard 1990). It is the first stage at which the embryo develops into a distinctly nonuniform shape, and initiates the separation of the primary germ-cell layers. Before gastrulation, all epithelial cells comprise a single layer, basically spherical in shape, which surrounds an interior filled with fluid (e.g., as in ascidians) or fluid and mesenchyme cells (e.g., as in amphibians). After gastrulation, the embryo becomes double-layered, with one layer on the inside and the other on the outside, and the interaction of these two layers becomes fundamental to further development. The onset of gastrulation is marked by a thickening of the cell layer at the *vegetal pole.* The layer then folds inward, beginning at one point and subsequently spreading to a significant portion of the layer, until part of the layer is folded completely inside the remaining layer (see Figure 1). This localized infolding is a special case of a general process found in embryonic tissues known as *invagination.*

This section is devoted to applying the techniques of the surface field theory of the previous section to the problem of gastrulation. The primary question of interest will be how the underlying forces in the surface, represented by the interaction funtions T and W of equation (23), are related to the shape produced. These functions will be expressed directly in terms of surface parameterization, using a convenient functional form. Having selected these functions, which act in essence as "fix sources," the resulting shape is then calculated numerically. Since the exact nature of the forces which are responsible for gastrulation forces remains experimentally undetermined, these simulations

Figure 1: Surface cross sections illustrating gastrulation in the sea urchin. Intended to illustrate basic features only.

serve as a means of comparing several possible mechanisms. In addition, they will serve to illustrate the theory presented in the previous section.

There are a number of experimental facts which we can set down regarding invagination of the gastrula (Ettensohn 1985). First, the forces which cause the deformation are not the result of external pressure on the layer, but instead are generated within the layer, either by cells pushing against each other or by expansion/contraction of individual cells. This conclusion comes from studies in which the invaginating region of tissue is excised and isolated from the rest of the embryo, and it is found that a considerable amount of tissue deformation occurs. Second, forces act to deform the shape of the individual cells rather than to cause cells to shift in position relative to one another. Thus the cell layer behaves more like a solid than a fluid. This makes it more likely that the theory of section 2 will be valid, since it is more related to the theory of solids than of fluids.

Several mechanisms have been proposed to explain the origin of the forces responsible for gastrulation and other tissue invaginations. The *differential adhesion model*, proposed by Gustafson, Wolpert and others (Gustafson and Wolpert 1962, 1967; Nardi and Kafatos 1976; Mittenthal and Mazo 1983), suggests that forces could be generated by changes in the adhesive properties that exist between cells in the sheet. Changes in adhesion in one region of the tissue with respect to another would produce localized forces, which in turn tend to lower (increase) the energy in the sheet through an increase (decrease) in their area of contact. The *contractile element model*, introduced by Burnside (Burnside 1971; Odell, Oster, Alberch and Burnside 1981), proposes that forces are generated by the contraction of filaments found on the outer or apical surface of the sheet and a preferred direction of folding exists because the contraction occurs along only one side. The contraction would presumably originate in the cytoskeleton and the natural agent of contraction would be actin-myosin filaments (the basic agents of muscle contraction), which are often found there.

Each of these models implies a distinct way in which forces act on the tissue sheet. In the differential adhesion model, compression or expansion forces act uniformly across the layer thickness, at least according to most interpretations of the model. According to the discussion at the end of section 2, this implies that the bending interaction term W must be zero. The stretching term T in this equation alone represents the force generated by this mechanism. In the contractile element model on the other hand, forces are produced mainly on the apical side and therefore do not act uniformly across the surface. Therefore, both T and W terms are nonzero in the interaction free energy. In fact (Hart 1989, p. 63) W is related to T through

$$T = \overline{T} \quad \text{and} \quad W = h\overline{T}/2, \tag{24}$$

where $\overline{T}$ is an arbitrary function to be specified and h is the surface thickness. By contrast, the differential adhesion model has these terms given by

$$T = \overline{T} \quad \text{and} \quad W = 0. \tag{25}$$

Thus, the distinct nature of these models is reflected directly by the free energy approach of section 2. As for the time dependence of these functions, it is convenient and reasonable to assume that their change over time is much slower than the time for the system to achieve equilibrium. The system is regarded as changing in a quasi-static manner.

In order to directly apply this approach, an assumption must be made regarding the spatial dependence of T and W. Since tissues in the blastula are to some extent determined, it seems plausible that the forces generating invagination might be localized in the vegetal region. The *threshold model* (Hart and Trainor 1989) assumes the sheet may be roughly divided into two regions: an "active" region which defines the location of the the initial invagination and an "inactive" region which is, effectively, the rest of the sheet. The interaction function has a maximum value in the active region and is approximately zero in the inactive region. Its value decreases rapidly to zero at the interface. For the model of gastrulation, the embryo before deformation (i.e. the blastula) will be taken as a spherical shell and the active region a symmetric cap centered at the vegetal pole. Adopting a spherical-polar coordinate system for the sphere, a general function which satisfies these requirements is

$$f(\theta) = 1/[1 + \exp(-\omega(\cos(\theta) - \cos(\theta_0)))]. \tag{26}$$

The function $\overline{T}$ in equations (24) and (25) is defined by multiplying this function by an appropriate constant. One question to answer is whether T should be positive or negative in the active region. This can be determined by observing that the layer becomes compressed there, as indicated by the thickening of the layer. Thus T must be negative (see discussion, end of section 2). The function $f(\theta)$ is a form of the logistic function, which is often used to model cooperative phenomena. The parameter θ_0 represents the "cutoff" position, which determines the size of the active region, while the parameter ω determines the effective width of the transition between the active and inactive regions.

Given a form for the free energy density, we must find a shape which minimizes the total free energy $\mathcal{F}$ in equation (9). Even for the simplest of surfaces, this is impossible to do analytically, and thus a numerical approximation must be employed. The numerical calculations presented here employ a finite difference-successive approximation scheme to minimize equation (9). The shape of the surface was found by employing a steepest descent method (Press, Flannery, Teukolsky and Vetterling 1988). In these calculations, the surface was assumed to be a surface of revolution. The surface was forced to have a polar axis. This assumption is possible because the free energy density has the same symmetry.

Simulations were performed for three different parameter values for the threshold function (26). These functions are shown in Figure 2, with the function $-\overline{T}$ (refer to equations (24) and (25)) in arbitrary units as a function of the coordinate θ. Curve "a." represents a tension which has a very high value at $\theta = 0$ and falls off rapidly as a function of θ, while curves "b." and "c." represent tensions which have a lower value at $\theta = 0$ but decrease more slowly.

Numerical solutions consistent with the contractile element model, with interaction functions given by equation (24), will be presented first. Typical solutions are shown in Figure 3 for the $\overline{T}$ distributions given in Figure 2. Each of Figures 3A, B and C correspond respectively to the three $\overline{T}$ distributions given by the curves labelled a., b., c. in Figure 2. Comparison of the Figures 3A and 3B to early gastrulas shows good qualitative agreement and therefore the threshold model represents a distribution of body stress that could be approximately correct. These solutions correspond to the limiting values of T_0 for which the numerical scheme, as implemented, converges to an acceptible solution.

Figure 3C suggestes a possible explanation for exogastrulation, an abnormal developmental event characterized by an outfolding of vegetal plate. In this case the $\overline{T}$ is similar to the one for Figure 3B except that it decreases more slowly to zero as θ increases. According to these simulations, an outfolding could occur even though the forces in the layer are pushing it inward. If

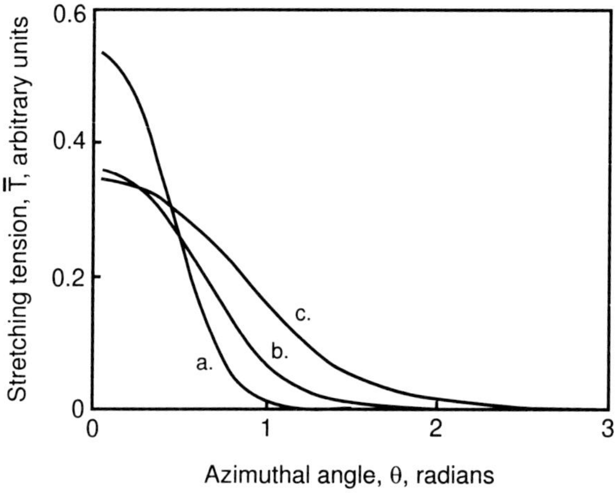

Figure 2: Plot of $-\overline{T}$ versus θ for the tensions used in the numerical solutions. θ is given in radians. Parameter values are: a. $T_0 = -1.0$, $\theta_0 = 0.18$, $\omega = 10.$; b. $T_0 = -0.70$, $\theta_0 = 0.18$, $\omega = 5.0$; c. $T_0 = -0.58$, $\theta_0 = 0.52$, $\omega = 3.0$.

the active region is too wide and if $\overline{T}$ is too flat, the sheet will "pucker in" on the sides, pushing the topmost part of the vegetal pole outwards. Once this has happened, this topmost part will continue to be pushed outwards, and the path to becoming an exogastrula is irreversibly taken. One consequence of this explanation is that the fundamental way forces are generated is the same in both the gastrula and the exogastrula, and the changes necessary to produce the abnormal state might be quite subtle in nature.

A numerical solution consistent with the differential adhesion model, with interaction functions given by equation (25), is shown in Figure 3D. This numerical solution corresponds to $-\overline{T}(\theta)$ given by the curve a. in Figure 2. In spite of the large contraction which occurs at the vegetal pole, the resulting shape is simply flattened and has no infolded dimple. While the absence of the dimple does not prove that a body tension without a moment cannot produce a gastrula, it suggests that, if it could, it would demand a tension very much larger than required by a body tension with a moment. Comparison with our corresponding findings in the case of contractile elements (Figure 3A) shows that the contractile element model is a far more attractive mechanism to explain gastrulation.

An important aspect of invagination that any theory of gastrulation must account for is the thickening of the layer *prior* to the formation of a visible folding (placode formation). The contractile element model, although certainly

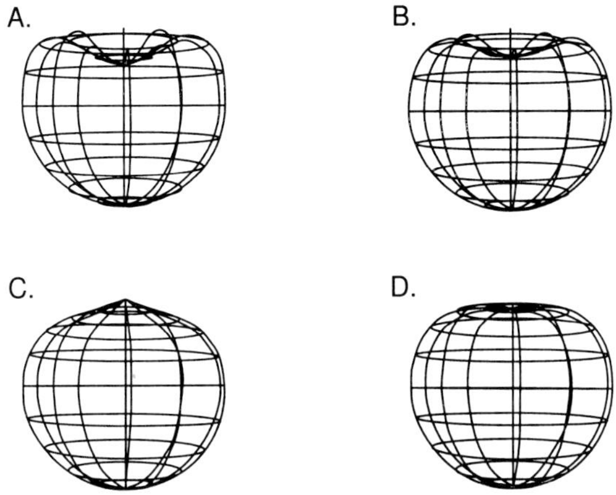

Figure 3: A., B. The three-dimensional surfaces corresponding to the equilibrium deformations of the tension curves a. and b. of Figure 2, with body stresses corresponding to the contractile element hypothesis. C. The three-dimensional surfaces corresponding to the equilibrium deformations of the tension curve c. of Figure 2, with body stresses corresponding to the contractile element hypothesis. D. The deformed surface corresponding to the tension curve a. of Figure 2, with the body stresses corresponding to the differential adhesion hypothesis.

supported by the simulations presented here, is unable to account for this aspect of the phenomenon (Ettensohn 1985). It is not that our simulations do not imply a change in layer thickness; in fact, the simulations shown in Figures 3A and 3B (those consistent with the contractile element model) have thickness changes of over 50%. This calculation is based on corresponding changes in surface area and the assumption that the material is incompressible (Hart 1989). However, the problem is the timing, since the simulations don't explain why this thickening happens prior to the surface folding inward. It is interesting to compare the surface thickness found in these simulations with those of Figure 3D (which is consistent with the differential adhesion model). Here, the surface thickness increases almost 80%, although no infolding occurs. Thus, this simulation, in essence, would explain placode formation but not gastrulation!

These observations suggest a possible model for tissue invagination which is capable of explaining both placode formation and infolding. First, we emphasize that the simulations are *consistent* with the aforementioned models, but could be consistent with other models as well. The simulations shown in Figure 3A–C result from a free energy density based on the interaction terms of equation (24), while those behind Figure 3D result from the interaction terms of equation (25). The former is consistent with a mechanism that generates forces preferentially on one side of the layer, while the latter is consistent with one that generates forces uniformly across the layer thickness. Now imagine there is some subsequent mechanism which begins by generating forces uniformly. The result is that the surface becomes compressed, the placode forms, but the surface remains unfolded. Now, if there were some subsequent mechanism by which the inner (basal) surface would deteriorate and lose its ability to sustain the forces of compression, then the layer would begin to fold inward. This stage would be analogous to the contractile element model, in terms of the free energy analysis presented here.

This mechanism is thus comprised of two stages: the first, in which forces act uniformly across the surface to form the placode, and the second, in which the integrity of the inner part of the surface breaks down and the surface infolds. Is such a mechanism consistent with the known biology of the epithelial layer? We know that the epithelial layers are not symmetric, so it is entirely possible that the stress limits on the inner and outer parts of the layer are different. One significant difference is the presence of the extracellular matrix (EM) on the apical (outer) surface of the layer. The EM could have significant ability to resist deformation and there is already evidence it plays an important role in gastrulation in sea urchins (Lane, Koehl, Wilt and Keller 1993). A possible scenario would involve the placode forming by a general compression due to adhesive forces generated uniformly across the cell sheet. During this process, extracellular matrix would quite likely continue to be deposited on the apical surface. Invagination could then be initiated by a relaxation in the adhesive forces. In consequence, the surface would tend to expand back to its original shape. However, the newly deposited EM would resist the expansion, resulting in a net inward bending force on the layer and the subsequent invagination. This is certainly not the only possible mechanism which might generate such a two-stage sequence, but it seems consistent with current experimental understanding.

The numerical results presented here certainly are not able to prove that one mechanism or another is responsible for invagination. However, they are useful for identifying general features that are helpful in comparing mechanisms. For example, the results clearly suggest that the way forces are generated across the thickness of the surface is an important question any mechanism must address. They also suggest that bending terms are necessary for invagination to occur. The results show that predicting surface shape is not a trivial process. The exogastrula solution in Figure 3C is a totally unexpected consequence of the model. This clearly shows that analytical and numerical models have a key role to play in understanding which particular shapes any particular mechanism is able to produce.

4 Toward a Theory of Form

The work presented in this essay is a first step toward a theory of shape in morphogenesis. It gives us a simple framework, which has at its foundation physical theory, to develop models of morphogenesis for specific systems. This framework is based on field theoretic concepts, in which shape is dynamically related to other morphogenic agents, expressed as fields, through a free energy function and possibly subject to other constraints (such as conservation laws). Although mathematical solutions to these models may be very complex, it should be possible to predict stable states of the shape-morphogen system numerically, thereby giving rise to predictions which can be verified by experiment. An example of such a solution, in which the internal "morphogen" field is constant, was given in section 3.

It is clear, however, that a shape description by itself has some major shortcomings. While the model is capable of giving very precise predictions, it is impossible to expect it will exactly correspond to experiment, since there is so much variation, due to embryo size, genetic factors, external conditions, and so on, even within a given species. This limitation is not confined to the present model, but applies to any model of shape, as such. Thus we can hope only for a qualitative agreement. However, certainly one of the chief objectives of theoretical work in this area is to identify common features in the development of a wide range of organisms and attempt to describe them with some type of unified theory. We might call such an approach a theory of *form*.

How might we precisely define a concept of form? There is certainly no agreed upon definition currently in the field. D'Arcy Thompson (1961) defined form as "a diagram of forces," meaning a collection of underlying causes which give rise to an outer shape. In this definition, two shapes which are generated from the same underlying physical principles, in which the constituent components of the physical principle bear the same relationships to each other, would be described as being of the same form. Thompson's work largely involved the study of shapes of shells and horns using growth models, where the underlying principles are fairly simple. There is little doubt that the task will be much harder for the invaginating embryo mainly because there is no obvious single growth principle with could describe the process. In any event, a precise definition of the equivalence of "physical principles" will have to be supplied, and this would be one of the challanges in taking this approach to a theory of form.

C.H. Waddington (1966) has presented a slightly different, although not inconsistent, view of form in biology. In this view, form is thought of as a generalization of shape, in which shapes that are related in certain ways may be thought of being equivalent, or of having the same "form." For example, we have can readily tell on inspection that two gastrulas from totally different species (in most cases) have the same form, because the basic characteristics are the same: it is double-layered, formed from a roughly spherical shape, with a large region pushed inside the rest to form a pocket. We might even make this geometrically precise. It has spherical topology and has a large simply connected region of positive mean curvature and a large simply connected region of negative mean curvature (mean curvature is defined as $K = (K_1 + K_2)/2$ where K_1 and K_2 are the principal curvatures; see also the essay by Cummings, this volume). This type of classification of shapes would lie somewhere between topology and geometry; topology regards the blastula and gastrula as equivalent, while geometry would regard gastrulas as distinct if their curvatures were different at some point.

These two views of form can be brought together by using the processes involved in creating a shape as the means to defining its form. This approach makes sense because we are principally interested in understanding how a shape is created. Let us formulate this approach within the context of the theory presented in section 2, with the physical principle defining a "form" being the gobal minimum of the free energy function. We need not be concerned with the exact form of the local free energy density, only that it is a function of the stretching and bending invariants, as well as possibly other local fields. It need not be homogeneous over the body; hence the free energy densities used in section 3 would certainly qualify. Within the context of this theory, shapes that could be generated by the theory, using different sets of parameters for the model, will be comparable in a direct way using the model. In Thompson's definition, they were generated by the same underlying cause. Clearly, however, this is not what we want, since the model presented in section 3 is able to produce both the gastrula and exogastrula using different sets of parameters for the model. Thus, we need a more restrictive way of associating shapes with the same form.

One possible solution is to try to formulate the idea of form in a nonbiological context, and see what can be transfered over to the biological one. One place where an idea like form occurs is in the physical theory of states of matter. A pure substance such as water occurs in three different states (solid, liquid and gas states), which might be regarded as "forms" in the sense that each has different outward physical characteristics. More to the point, we identify a whole region of phase space, encompassing a range of physical parameters (density, temperature, pressure) with the same state. The change from one state to another is characterized by a phase transition, involving discontinuty in the derivatives of the free energy. By analogy, we could, within the context of the free energy model of section 2, define two shapes to belong to the same form if we could continuously alter the parameters of the model in such a way that the total free energy of the system, and all of its derivatives, are continuous. Conversely, if a discontinuity in the free energy occurs, then the shapes would be of different form. It should be emphasized here that free energy means the total free energy of the system, which is the integral over the whole body of the free energy density.

There are a number of nice features in defining form in this way. First, form is defined not in terms of shape only, but in terms of shape *and* underlying properties (the free energy model). This frees us from the temptation of comparing shapes that are produced in very different ways, such as a embryo and a snail shell. At the same time, if we do want to compare shapes from different systems, then we must find a common model which would apply to each one. Thus, it gives us a means to attempt such a comparison. For example, we may wish to compare the processes of gastrulation and neurulation and ask whether the "forms" generated are, in this sense, the same. We would have to build a common free energy model, and then agree on some way of mapping shapes to each other. The correspondence of "transition points" (free energy discontinuties) would then tell us how these process are the same and how they are different.

Another nice feature is that transition points are likely to be very important in determining developmental pathways. An example is the gastrulation model of section 3. As the active region of stress is widened, the resulting shape goes from the gastrula to the exogastrula (compare Figures 3A and 3C, respectively). There is almost certainly a discontinuity in the free energy going between these two shapes, and by our definition they would represent different forms. If the shape is near the transition point, then a small perturbation could either send it down the normal (gastrula) or abnormal (exogastrula) pathways.

We would intuitively expect the gastrula and exogastrula to be different forms, since they do represent different developmental pathways. What about the gastrula and the blastula (the latter being, in our model, the undeformed sphere)? Does the model have a transition point in going between these two shapes? At this point, we cannot be certain about the answer to this question. It is possible that, at the point in time where the vegetal pole begins to fold inwards, a transition point must be involved. On the other hand, it is possible too that the free energy could change completely smoothly. It is clear, however, that for our definition of form to work, we will want a transition point, since development is fundamentally the process of *appearance* of form. The blastula and gastrula must be of different form. If there is no transition point between the shapes, then we must be prepared to either look for a different model, or re-examine this definition of form.

The approach to form presented here is of course preliminary in nature. Work will need to be done to determine whether or not classifying shapes in terms of "transition points" in the free energy is meaningful for morphogenetic processes. However, our approach offers a way that to answer some basic questions about morphogenesis without resorting to increasingly complex numerical models to generate shape. By using simple, analytical models which have equivalent transition points, and hence equivalent forms, we could have a theory for which results would be relatively tractable and for which hypotheses would be easy to test. Such a program would seem to be a step forward in achieving an eventual theory of morphogenesis.

5 Concluding Remarks

While shape is a fundamentally important aspect of morphogenesis, it has remained the most difficult to describe and treat in theoretical models. The work presented here represents an attempt to address this problem within the context of developmental field theory, based on the underlying physics and biology of the system as currently understood. The goal is not simply to produce numerical simulations of shape, but also to develop a general framework in which models of shape can be compared. This framework leads directly to a notion of "biological form."

References

Bard, J. (1990) *Morphogenesis: The Cellular and Molecular Processes of Developmental Anatomy*. New York: Cambridge University Press.

Berrill, N.J. and Karp, G. (1976) *Development.* New York: McGraw-Hill.

Brachet, J. and Alexandre, H. (1986) *Introduction to Molecular Embryology*. Berlin: Springer-Verlag.

Brière, C. and Goodwin B.C. (1988) Geometry and dynamics of tip morphogenesis in *Acetabularia. Journal of Theoretical Biology* **131**, 461–475.

Burnside, B. (1971) Microtubules and microfilaments in newt neurulation. *Developmental Biology* **26**, 416–441.

do Carmo, M.P. (1976) *Differential Geometry of Curves and Surfaces.* Englewood Cliffs, NJ: Prentice-Hall.

Ettensohn, C.A. (1985) Mechanisms of epithelial invagination. *Quarterly Review of Biology* **60**, 289–307.

Goodwin, B.C. (1976) *Analytical Physiology of Cells and Developing Organisms.* New York: Academic Press.

Goodwin, B.C and Trainor, L.E.H. (1985) Tip and whorl morphogenesis in *Acetabularia* by calcium-regulated strain fields. *Journal of Theoretical Biology* **117**, 79–106.

Gustafson, T. and Wolpert, L. (1962) Cellular mechanisms in the morphogenesis of the sea urchin larva. Change in shape of the cell sheets. *Experimental Cell Research* **27**, 260–279.

Gustafson, T. and Wolpert, L. (1967) Cellular movement and contact in sea urchin morphogenesis. *Biological Reviews of the Cambridge Philosophical Society* **42**, 442–498.

Hart, T.N. (1989) *Viscoelastic Models of Morphogenesis.* PhD Thesis, Department of Physics, University of Toronto, Canada.

Hart, T.N. and Trainor, L.E.H. (1989) Geometrical aspects of surface morphogenesis. *Journal of Theoretical Biology* **138**, 271–296.

Lane, M.C., Koehl, M.A.R., Wilt, F. and Keller, R. (1993) A role for regulated secretion of apical extracellular matrix during epithelial invagination in the sea urchin. *Development* **117**, 1049–1060.

Meinhardt, H. (1982) *Models of Biological Pattern Formation.* London: Academic Press.

Mittenthal, J.E. and Mazo, R.M. (1983) A model for shape generation by strain and cell-cell adhesion in the epithelium of an arthropod leg segment. *Journal of Theoretical Biology* **100**, 443–483.

Nardi, J.B. and Kafatos, F.C. (1976) The differential adhesion model: gradient of a non-diffusible cell surface parameter. *Journal of Embryology and Experimental Morphology* **36**, 489–512.

Odell, G.M., Oster, G., Alberch, P. and Burnside, B. (1981) The mechanical basis of morphogenesis. *Developmental Biology* **85**, 446–462.

Press, W.H., Flannery, B.P., Teukolsky, S.A. and Vetterling, W.T. (1988) *Numerical Recipes in C.* New York: Cambridge University Press, p. 318.

Stoker, J.J. (1969) *Differential Geometry.* New York: Wiley-Interscience.

Thompson, D'A.W. (1961) *On Growth and Form.* Abridged edition. Cambridge: Cambridge University Press.

Totafurno, J. and Trainor, L.E.H. (1987) A non-linear vector field model of supernumerary limb production in salamanders. *Journal of Theoretical Biology* **124**, 415–454.

Turing, A. (1952) The chemical basis of morphogenesis. *Philosophical Transactions of the Royal Society of London, Series B* **237**, 37–72.

Waddington, C.H. (1966) *Principles of Development and Differentiation.* New York: Macmillan.

Wolpert, L. (1971) Positional information and pattern formation. *Current Topics in Developmental Biology* **6**, 183–224.

III

Development:

Principles of Self-Organization

Generic Dynamics of Morphogenesis

Brian Goodwin

1 Introduction

Development, the process whereby an adult organism with all its intricacies of form is generated from simple beginnings such as a fertilized egg, is a very complex process. In a fruit fly, a frog or a water lily thousands of genes are involved, their products interacting in patterns that change in time and space within the developing embryo, somehow involved in generating the structures that emerge reliably generation after generation — limbs, eyes, guts, gonads; roots, shoots, leaves and flowers. From a molecular perspective the task of understanding such overwhelming complexity is very daunting. Does the genetic program that defines the sequence of gene activities during development need to be described in detail before we can understand the principles of morphogenesis, how the form of an organism is generated? If physics and chemistry had proceeded this way, from molecules to macroscopic behavior, it is unlikely that the gas laws, the properties of liquids as expressed in hydrodynamics, or even the laws of motion would ever have been discovered. Biology may be fundamentally different from physics and chemistry in this respect, lacking macroscopic regularities. However, there is persistent evidence that general developmental principles are at work across the diversity of species, and recent models suggest that morphogenesis may be considerably simpler than the molecular level of analysis suggests.

2 Developmental Fields

Because organisms have spatial order, any mathematical model of their form is a field theory of some kind. D'Arcy Thompson (1917) described the shapes of adult organisms and their parts in terms of mathematical functions expressing structures of minimal energy or area, either static or dynamic. His recognition of the similarity of form generated by a drop of fusel oil falling through paraffin and the shape of a jelly-fish is a classic instance (Figure 1), suggesting that organisms generate their complexity through repeated iterations of basic pattern generators. Another type of generator appears to be operating in spiral structures such as gastropod shells or horns. Thompson described the invariant growth rules that result in such forms, and how variation in a few basic parameters could generate the diversity observed. Raup and Michelson

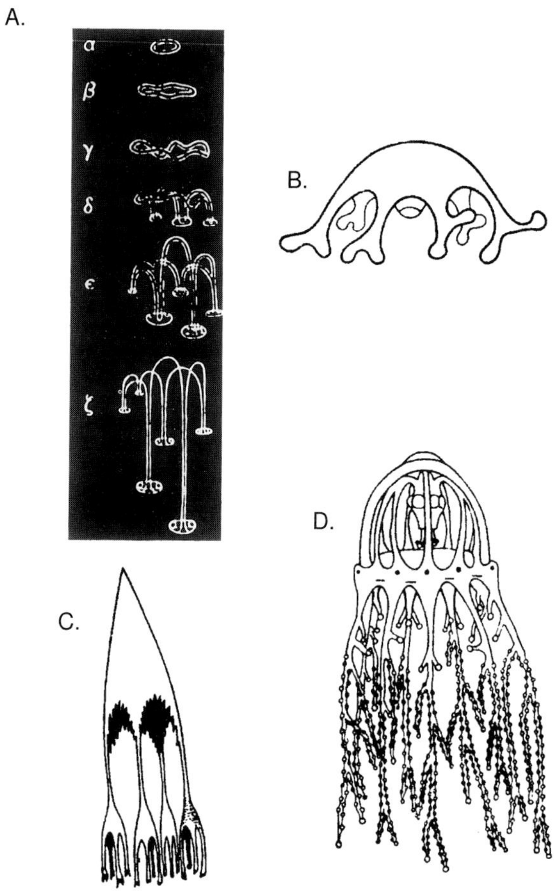

Figure 1: A. Patterns of ink falling through water. B. Medusoid of the species *Cladonema* C. Pattern of fusel oil falling through paraffin. D. Medusoid of the species *Syncoryme*. From D'Arcy W. Thompson (1961; reissued 1966) *On Growth and Form*. Abridged edition edited by J.T. Bonner. Copyright © Cambridge University Press. Reprinted with permission of Cambridge University Press.

(1965) used this approach to describe the range of possible spiral forms of gastropod shells, and to identify which of these possibilities have been realized by different species, both living and fossil. Organisms do seem to be using rules to generate morphology.

Just how general these rules may be comes from studies showing that organisms as different as cockroaches and amphibians reveal similar principles in the morphogenetic fields involved in generating their limbs. Nature is often the first to perform experiments by producing aberrant forms or monsters that are later studied in laboratories. Among these are individuals with three limbs where one should be, resulting in an eight-legged cockroach or a six-legged salamander. Whenever this occurs the multiple structure has an invariant property: There is always a plane of mirror reflection between adjacent limbs. This regularity of pattern was described by William Bateson (1894, 1992) on the basis of spontaneous teratologies, so it is known as Bateson's Rule. Furthermore, if the multiple limb is on the right side of the body, the structure consists of two right and one left limbs; whereas if it is on the left, it consists of two left and one right.

An explanation of this rule was provided by French, Bryant and Bryant (1976) by means of a model that used polar coordinates to describe the morphogenetic field of the limb. Experiments had shown that multiple limbs arise from disturbances to normal limb development, such as removing a limb (or a limb blastema) and rejoining it to the stump after rotation. Two basic rules of the model were the key to the result: (1) After any disturbance, the morphogenetic field obeys a principle of maximum "smoothness" within its boundaries — that is, adjacent field values should be such that there are no sudden jumps from one value to another. If these values are specified by concentrations of "morphogens," then this smoothness principle is most simply understood as a result of diffusion. (2) A limb forms whenever a complete circle of angular coordinates is produced within the field boundaries. It then emerged that whenever the smoothing operation resulted in the production of extra limbs, they always occurred in opposite pairs. This is the only way the boundary conditions of the whole field could be satisfied. The result is a little like spin conservation in particle physics: Extra particles can be produced in an interaction so long as their spins cancel. Effectively an extra right and left limb cancel one another out; and smoothness also requires that the structures are arranged so that there are planes of mirror symmetry between them.The principles involved can be formulated in topological terms, using the notion of an index of a closed curve in a field (Glass 1977). If we assign to a left limb field the index +1, using the convention that the angular coordinate runs clockwise through one full cycle and so will generate a limb, then any extra limbs in this

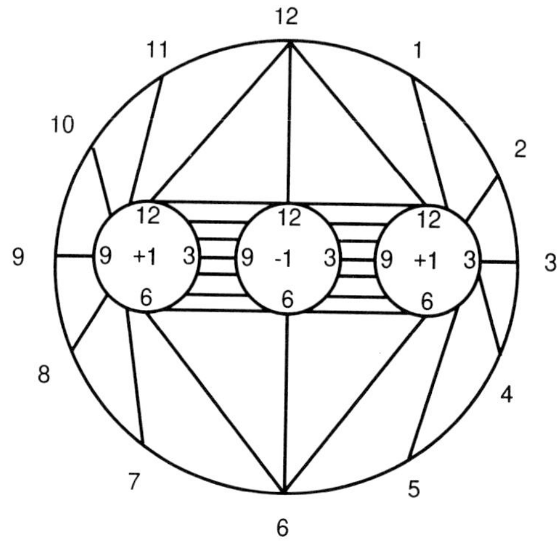

Figure 2: The polar coordinate description of a limb with two extra supernumeraries. The large outer circle represents the proximate boundary (shoulder) of a limb with circumferential or radial values running from 1 to 12, with the convention that the counterclockwise direction represents a left limb and the circle is given the index +1. The small circle within the boundary, also with index +1, represents a more distal position towards the end of the limb. Two supernumerary limbs are represented by the other two circles. In order for the field lines of the morphogenetic field to be coherent, the field values and the indices of the supernumeraries must be as shown, with planes of mirror symmetry conforming to Bateson's Rule, resulting in a left and a right supernumerary. The left (+1) and the right (−1) supernumeraries could in principle cancel one another out by fusion of field lines, leaving a normal structure. This is what happens in *Tetrahymena* when the mirror-symmetric OAs annihilate one another during regulation of the cell back to normal (Frankel 1989).

field must occur in +1 and −1 pairs to preserve the index of the whole field (Figure 2). Another treatment of this problem uses a vector field model of the limb blastema, with an energy function whose minima correspond to the observed results of grafting and regeneration experiments (Totafurno and Trainor 1987).This shows the continuing utility of the approach pioneered by D'Arcy Thompson and the value of physical principles in formulating morphogenetic problems.

Since these studies, another case has been reported that extends the principle in dramatic fashion. Frankel and his colleagues, studying the structures produced on the unicellular protozoon *Tetrahymena*, observed a remarkable phenomenon. The oral apparatus (OA) is a structure used to gather food particles as the cell glides through the water. It is asymmetric, so it can be defined as right or left-handed according to convention. A normal *Tetrahymena* with a single OA (say a right) is like the right side of an organism with a right limb. If another right OA is now introduced into this field, there is a conflict in the field values that can only be resolved by canceling one of the two right OAs. One way of doing this is to produce a left OA, and sure enough, this was observed experimentally (Frankel and Nelsen 1986). Furthermore, the additional structure arises between the two existing ones, resulting in the same mirror reflections that are observed in limbs. Two of these mirror-imaged OAs then move together and annihilate one another, the cell returning to its normal form. An analytical treatment of this problem, again based on a vector field model to capture the handedness of the circumferential field of the cell and of the OAs located in it, describes the phenomena elegantly in terms of energy minimisation (Brandts and Trainor 1990a,b).

Yet another example of mirror-reflected patterns has been explained by basic topological properties of morphogenetic fields. There are four categories of segmentation mutant phenotypes in *Drosophila*, each including cases of mirror reflection of the head-to-tail axis over characteristic distances which are also the wavelengths of gene product distributions for the class. The longest of these are maternal mutants such as *bicaudal* or *dicephalic*, in which nearly half the body is mirrored, resulting in a double-tailed or double-headed monster (Nüsslein-Volhard 1977). The gene products are distributed in gradients extending over roughly half the embryo (Driever and Nüsslein-Volhard 1988). Gap gene products occur in bands extending over 3–4 segments, with mirror-symmetric mutant phenotypes over these distances. The other two categories are called pair-rule and segment polarity, the former including cases of mirror reflection over two segments while the latter have reflections over single segment wavelenths. Different mutants within each class are phase-shifted with respect to the centers of reflection. There is clearly something generic in these patterns. They have been explained by Goodwin and Kauffman (1993) by means of a mapping from genes to pattern that uses ratios of gene product to define a slope or an angle in a polar coordinate map, at the center of which is a phaseless singularity. Mutations that alter gene products such that the ratio map crosses the singularity result in mirror-symmetric patterns because now the map crosses the same ratios twice, in opposite directions: Its index is 0 rather than 1 for the normal map (Figure 3). This "color wheel model" explains the phenomena well, and shows yet another use of an essentially topological argument to understand a property typical of a class of mutants. There seem to be universal principles at work across a great range of phyla, from unicellulars through arthropods to vertebrates, suggesting that we are dealing with some generic dynamic properties of morphogenesis. But how is this to be understood at the molecular level?

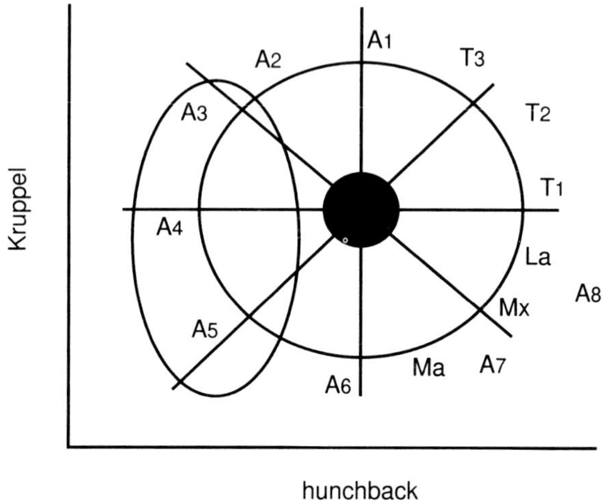

Figure 3: The mapping from two gap gene product distributions (*Kruppel* and *hunchback*) to segments in *Drosophila*, showing the normal map (full circle) and the result of a mutation with loss of *hunchback* activity (ellipse). Because the latter cuts the spokes of the color wheel that defines the mapping twice, once in each direction as one travels in counterclockwise direction around the curve, it defines a mirror-symmetric map and the curve has index 0 as compared with +1 for the normal map. The ellipse in two dimensions describes only the mirror-symmetric part of the mutant phenotype. Other segments are coded by ratio combinations with other gap genes, such as *knirps* and the full mapping has as many dimensions as there are segmentation genes. See text and Goodwin and Kauffman (1993) for details.

A smoothing principle, resulting from diffusion of substances or some other spatial averaging process, is one aspect of morphogenetic field activity; but there must also be a dynamic initiator of pattern, since organisms get more complex as they develop from egg to adult. It was A.M. Turing (1952) who first showed what type of process this might be. He examined the properties of interacting nonlinear chemical reactions whose products diffuse at different rates (different diffusion coefficients) and demonstrated that spatial patterns of these products could arise spontaneously. These reaction-diffusion systems have the property of symmetry breaking from uniform to nonuniform spatial structure, concentrations of the different chemicals developing into standing waves whose wavelength and phase relations depend upon the chemical interactions, parameter values, and boundary conditions of the field. Now called excitable media, these systems have been extensively studied mathematically and by computer simulation (Jahnke, Henze and Winfree 1988), and their spatial patterns have been related to biological forms such as gastropod spirals and insect segments (Meinhardt 1982, 1986) and to animal pigmentation patterns and limb development (Murray 1989). There are also candidates for the molecular species involved in generating these patterns, such as gene products in *Drosophila* segmentation, retinoic acid in limbs (Maden 1982; Tickle, Alberts, Wolpert and Lee 1982; Eichele and Thaller 1987), and something as simple as calcium in unicellular algae (Goodwin, Skelton and Kirk-Bell 1983; Goodwin, Brière and O'Shea 1987; Harrison and Hillier 1985; Harrison, Graham and Lakowski 1988).

However, a problem encountered in many models of pattern formation is stability, or robustness. The patterns generated are often sensitive to initial and boundary conditions. On the other hand, the processes involved in morphogenesis are very robust, as they must be to generate the same form generation after generation. Genetic studies have convinced many investigators that morphogenetic fields as global ordering processes that organise spatial patterns during development play a minor role in embryogenesis, with most of the detailed form of an organism being determined by localised gene activity and interactions. But no one has shown how to derive actual three-dimensional structure from distributed patterns of gene products, and there are cogent arguments that simply knowing when and where different molecules are made in a developing embryo will never be enough to explain morphogenesis. Molecular composition is not, in general, sufficient to define form, a principle well understood in physics and chemistry (Goodwin 1985). It is necessary also to specify spatial order — i.e., the nature of the field to which the molecules contribute by interaction. To resolve this impasse, it is useful to examine in some detail a relatively new category of morphogenetic model that includes aspects of previous ones but embeds them in a context of mechanical forces that generate three dimensional structures simulating and explaining basic aspects of actual organismic form.

3 Morphogenesis

It is a good strategy to study a system that is basically simple, but is just complex enough to present a nontrivial example of the problem under investigation. The organism whose life cycle is shown in Figure 4 is one that has served developmental biologists well since it was first cultured in the laboratory by Hammerling (1931). It is the giant unicellular alga, *Acetabularia acetabulum*, whose habitat is the shallow waters around the Mediterranean. The life cycle can be taken to start with the conjugation of two flagellated, haploid isogametes (no males or females in this species), which form a roughly spherical, diploid zygote about 50 microns (μm) in diameter. The spherical

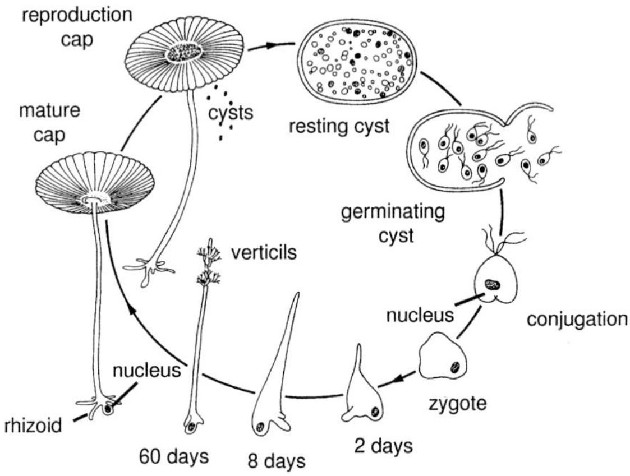

Figure 4: The life cycle of *Acetabularia acetabulum*.

symmetry is broken by the formation of a stalk that grows upwards and a branching root-like structure, the rhizoid, that anchors the alga to rocks on the sea floor. The single nucleus remains in a branch of the rhizoid while the stalk grows and, after several weeks, starts to produce delicate branching elements (bracts) in rings called verticils or whorls. These can be seen in more detail in Figure 5, which shows an alga at 10–12 weeks that has just produced a little structure at the center of the last whorl — a cap primordium. This grows into the mature cap, about 0.5–1.0 cm in diameter, while the stalk at this stage is between 3 and 5 cm long. To the uninitiated eyes, the cap looks multicellular, but the whole organism is a single giant cell that has undergone a fairly complex pattern of morphogenesis.

After several months the alga goes into reproductive mode, the single nucleus dividing many times to produce thousands of haploid nuclei that stream into the cap and differentiate into flagellated gametes within cysts that are released as the cap wall dissolves away under the action of enzymes. Little hatches in the cysts then open to release the gametes, and the life cycle starts anew. How are we to understand this life cycle, which is one of the simpler kind? In studying this in the laboratory, it is more convenient to work with regenerating algae. If the cap is cut off, it is regenerated following exactly the same morphogenetic sequence as during normal development: tip growth, whorl formation, cap formation. The process is highly reproducible, allowing for systematic experimental investigation.

Figure 5: Formation of a cap primordium, after three whorls.

4 Modeling Morphogenesis

Our research depended upon the work of many others (cf. Brachet and Bonnotto 1970) but our particular interest in morphogenesis led us to study those factors and processes that were particularly involved in growth and form. It emerged that calcium is particularly important (Goodwin and Pateromichelakis 1979, Goodwin et al. 1983, Harrison and Hillier 1985) and this led us to a study of the dynamic properties of the cytoskeleton, the network of filaments in the cytoplasm whose mechanical state is modulated by calcium. Calcium exerts its influence by a variety of means: It activates enzymes such as gelsolin that cuts microfilaments; it is involved in initiating contraction of actomyosin filaments; and it influences the polymerization of microtubules and microfilaments. These properties of the cytoskeleton and the effects of calcium derive largely from work on organisms other than *Acetabularia* whose cytoskeletal network may be simpler, lacking microtubules and actomyosin even though the alga can make both tubulin and a myosin epitope which are involved in nuclear division and cyst morphogenesis (Menzel, Jonitz and Elsner-Menzel 1992). The effects of calcium on the cytoskeleton in plants is described by Williamson (1984), Shelanski (1989), and by Menzel and Elsner-Menzel (1989). The way in which calcium affects the mechanical state of the cytoplasm (measured by its elastic modulus) is shown qualitatively in Figure 6.

The cytoplasm forms a thin shell just inside the cell wall whose shape changes during regeneration in the manner shown schematically in Figure 7. Inside the shell of cytoplasm is a liquid-filled cavity, the vacuole, within which

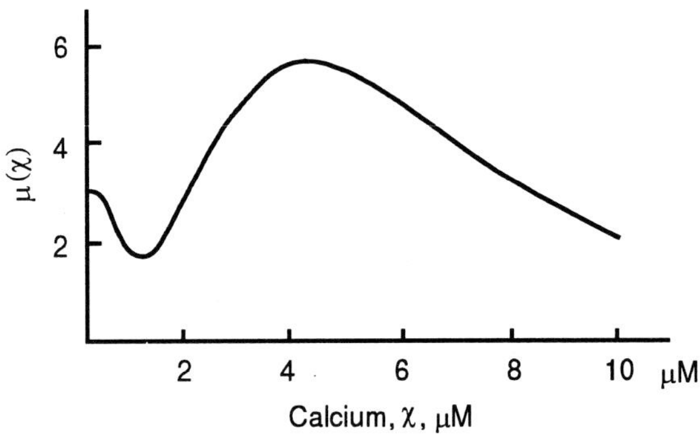

Figure 6: Variation of elastic modulus with free Ca^{2+} concentration (micromolar). Qualitative relations only.

salts and small molecules accumulate, producing an osmotic pressure on the cell wall that keeps the cell in a state of turgor, and gives it a flexible rigidity. Our model consisted of three components only: wall, cytoplasm, and vacuole. The most important of these is undoubtedly the cytoplasm, and it was necessary to derive equations that describe its mechanical properties, the way in which calcium is regulated in the cell, and how calcium influences the state of the cytoplasm. The derivation, carried out by Goodwin and Trainor (1985), followed quite closely an earlier model of cytoskeletal-calcium dynamics described by Oster and Odell (1984). This is based upon the theory of viscoelastic media in which strain (degree of stretching or compression of the cytoplasm) is the primary variable, the theory being adapted to the properties of the cytoplasm, particularly its dependence on calcium. Equations for calcium concentration, which is regulated at resting levels of fractions of a micromole in all eukaryotic cells, describe the combined effects of calcium-binding proteins and membrane pumps, as presented in Hart, Trainor and Goodwin (1989) and Brière and Goodwin (1990). These equations are mutually coupled so that not only do changes in free calcium concentration affect the strain in the cytoplasm, but cytoplasmic strain influences the concentration of free calcium. As a result, the model has the same properties as Turing systems: Spatial patterns of strain and calcium concentration can arise spontaneously as a result of bifurcation and symmetry breaking in the dynamics of the system. In consequence, within a region of cytoplasm such as the regenerating tip it is possible for changes of state and shape to occur starting from initially uniform conditions. Our equations are not the same as Turing's (1952), since no chemical reactions are involved, but they belong to the same class of partial differential equations

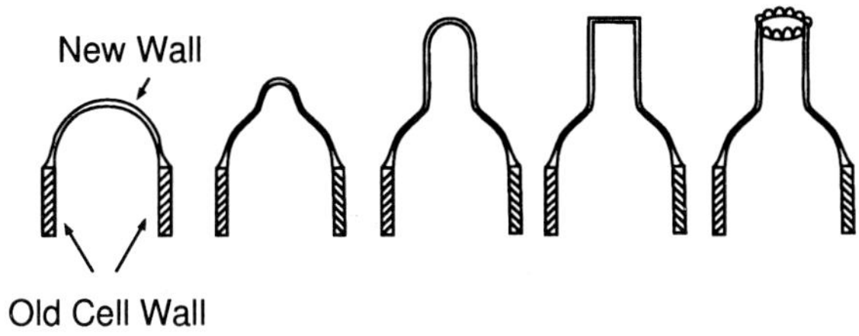

Figure 7: Representation of the morphogenetic changes taking place during regeneration of *Acetabularia*. Proceeding left to right, the five stages schematically illustrated are hemispherical dome, tip formation, growth, tip flattening, and whorl formation.

describing excitable media.

The basic reason for the capacity of the model to bifurcate and generate spatial patterns comes from the interaction between calcium and the cytoskeleton. Wherever free calcium happens to increase in concentration from resting levels, the cytoplasm softens because of the effect of calcium on cytoplasmic elastic modulus (Figure 6). The cytoplasm is closely apposed to the cell wall, which is under pressure from the vacuole, so the cytoplasm is always under tension (being stretched). Wherever it softens, it will stretch more (positive strain), which results in release of more calcium from the bound or sequestered condition, as assumed in the model. So there is a positive feedback loop that results in a local runaway increase in cytoplasmic strain and free calcium concentration, initiating a spatial nonuniformity of state. This increase is then opposed by the reversal of the elastic modulus-calcium relationship (Figure 6) when free calcium increases beyond about 5 μM, and by diffusion of calcium, so the pattern gets stabilized. Conversely, wherever free calcium fluctuates below resting levels there is reduced cytoplasmic strain accompanied by more binding and sequestration of free calcium. These are the dynamic properties that underlie spontaneous pattern formation in the model, as in other models of this type (e.g. Murray 1989).

In order for the shape of the cell to change, it is necessary for the pattern of cell well synthesis to change in response to patterns in the cytoplasm. A growth algorithm was defined in accordance with known properties of wall growth in plants based upon strain (Cleland 1971; Green, Erickson and Buggy 1971). Coupling of wall to cytoplasmic state was assumed to result from wall softening (decrease in wall elastic modulus) as a function of underlying cytoplasmic strain via a mechanism such as a strain-activated membrane pumps. The exact nature of this coupling is not yet experimentally established.

Biologically speaking, our model is extremely simple: an elastic wall capa-

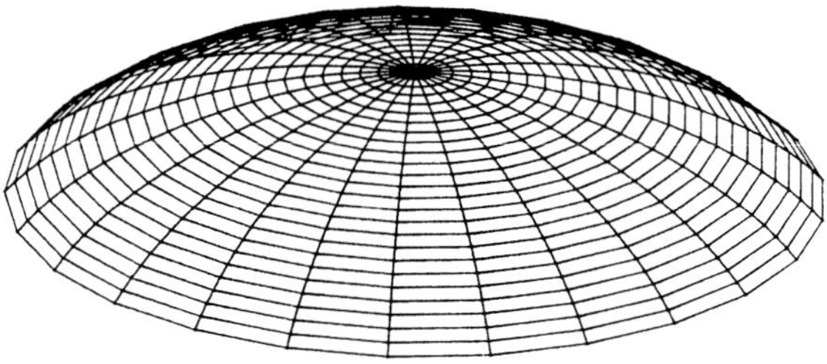

Figure 8: Computer simulation of regeneration in which the cytoplasm and the cell wall are described as shells made up of finite elements which obey equations describing their dynamics as mechanochemical or elastic media, respectively.

ble of growth, an excitable cytoplasm whose state influences that of the wall, and a vacuole that exerts constant pressure on the wall. Mathematically, the model is complex and nonlinear. To observe its behavior in three dimensions it was necessary to carry out a finite element simulation. This was done in collaboration with Christian Brière (Brière and Goodwin 1988). The regeneration domain was described by a set of elements (Figure 8), each of which obeyed the equations of the model. These collectively defined the geometry of the domain in accordance with intrinsic properties such as elastic modulus, and pressure such as that exerted by the vacuole. The boundaries of the regeneration domain were fixed, as they are by the old cell wall in a regenerating alga, and the initial conditions were specified with all the variables starting from spatially uniform patterns. Parameters were chosen so that the cytoplasm was in the bifurcation range where spatial patterns could arise spontaneously, and with values that resulted in wavelengths of variables that were smaller than the domain over which morphogenesis was to occur (the regeneration domain of Figure 8). This was simply to ensure that interesting patterns could arise. The model was then allowed to do its own thing, to follow whatever morphogenetic trajectories were available to it.

5 Morphogenetic Dynamics

The first pattern that developed in the model was the spontaneous formation of gradients in free calcium concentration and in cytoplasmic strain with maxima at the tip of the dome of Figure 8. As a result of this, the wall softened at the top and underwent an elastic deformation due to vacuolar pressure (Figure 9). This was followed by growth of the wall and an extension of the tip. After a period of growth another change occurred. Instead of gradients in cytoplasmic

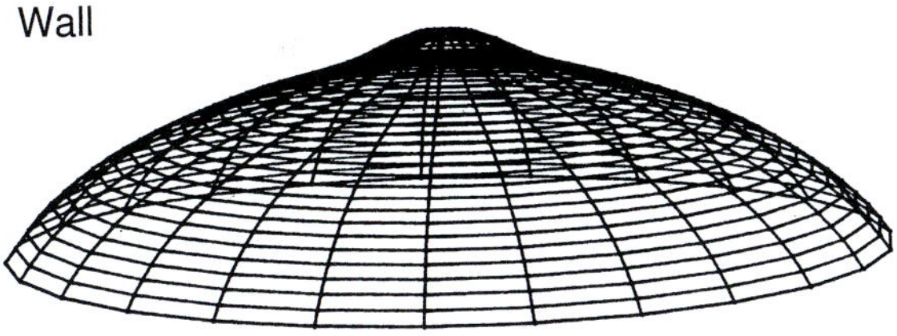

Figure 9: Tip initiation in the model.

variables with maxima at the tip, an annulus of elevated free calcium and strain occurred a short distance back from the tip. This is shown in Figure 10 for calcium, while the actual geometry of the regenerate is shown in the accompanying finite-element simulations showing wall and cytoplasmic shape. This result explained something we had frequently observed in regenerating algae but had never understood. Just before a whorl is initiated, the normally conical tip flattens, as shown in Figure 7. The model suggested that this was due to the transition from calcium and strain gradients with maxima at the tip, to an annulus, maximum wall softening and therefore maximum wall curvature then occuring on this annulus resulting in the flattened tip.

The next stage should be whorl formation. Could the calcium annulus break symmetry and generate a whorl prepattern? The result is shown in Figure 11: Under perturbation, the annulus bifurcates into a series of peaks which have the symmetry of a whorl. These changes of calcium pattern, from gradient to annulus to a ring of peaks accompanying the morphogenetic sequence from tip to whorl formation, have been observed experimentally by Harrison et al. (1988) using the calcium probe chlorotetracycline, which measures membrane-bound calcium. It is to be expected that free calcium will show similar changes, but this remains to be demonstrated.

The model gave further results. As growth continued and the tip continued to extend, the calcium annulus became unstable and collapsed back to a gradient, with a conical tip forming. After a further period of growth an annulus formed again, with the possibility of bifurcating to a whorl prepattern. This sequence is reminiscent of the intermittent production of whorls that occur during algal growth. We were unable to carry out a full simulation of whorl formation because the finite element model is not sufficiently robust to allow stable growth on the fine scale of individual bracts of a whorl. This requires a substantial extension and refinement of the modeling program that has not yet been achieved. Also, we have not yet obtained a structure that looks like a cap. What we do get is an enlarged terminal structure that grows laterally

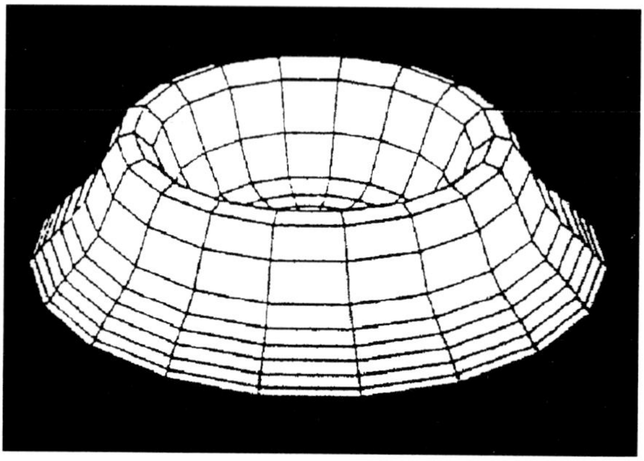

Figure 10: Tip flattening and calcium annulus formation.

but is not nearly as flat as a cap. It appears that more anisotropy is required in the model so that strain levels increase around the rim of the cap more than those elsewhere, a type of hoop strain that has been proposed by Green (1987, 1989) for leaf morphogenesis in higher plants (e.g. Figure 12).

6 Morphogenetic Attractors

Despite these limitations of the model, the results obtained are of considerable interest. The sequence of tip formation, growth, tip flattening, whorl initiation, intermittent repeat of this process as growth continues, and terminal cap-like structure is a result simply of the cycle of cytoplasmic dynamics causing a change of shape which then acts back on dynamics. This defines a moving-boundary process, which is the category to which morphogenesis belongs. Our results suggest that what we are seeing is the expression of a dynamic attractor in morphogenetic space, a robust pattern-generator that arises naturally and inevitably in a system organized as are the interacting shells of wall and cytoplasm, the latter having excitable dynamics.

What is the role of genes in this process? They specify parameter values within a range that allows a morphogenetic trajectory to unfold to a viable form. We do not yet know how large this domain is in parameter space. But the ease with which we found suitable parameter values suggest that it is extensive — a large basin of attraction seems to exist. We do know that we can change most of the parameters without significantly affecting behavior, so the morphogenetic sequence is quite robust. Our conjecture is that it is

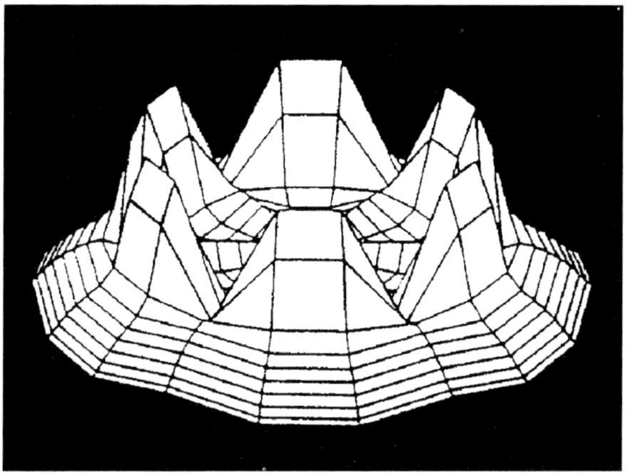

Figure 11: Bifurcation of the calcium annulus to a whorl prepattern.

generic in this space — a structurally stable trajectory. If this is the case, then there should be many species of algae in the taxon containing *Acetabularia* that all share the same basic morphology. And this is indeed the case. The *Dasycladales* are an ancient group of giant unicellular green algae, most of which are known from fossil remains of calcified cell walls that extend back at least to the Cambrian era, 570 million years ago. These, together with the twenty or so living species, are described in a recent monograph by Berger and Kaever (1992). All species have a rhizoid and an axis with bracts, most of them produced in whorls. In the majority of species the bracts are the gametophores (where the cysts and gametes form) and no caps are produced, so *Acetabularia* belongs in this respect to a minority class. However, it continues to produce whorls despite their sterility and apparent lack of function, being transient structures that are shed. We know also that the algae can grow perfectly well without whorls, whose production can be prevented by simply reducing the calcium concentration in the seawater (Goodwin et al. 1983), and these whorl-less stalks can then make caps and produce cysts simply by restoring calcium to normal levels. So why does *Acetabularia* produce whorls of sterile bracts?

An obvious answer is that these structures result from the generic dynamics of morphogenesis in this group of organisms, all of which share the same basic organization. They are produced not because they are useful but because they are natural. This is the way we explain forms in nature — the shape of a diamond crystal, vortices in liquids, elliptical trajectories of planetary movement about the sun. These are the structurally stable forms generated by

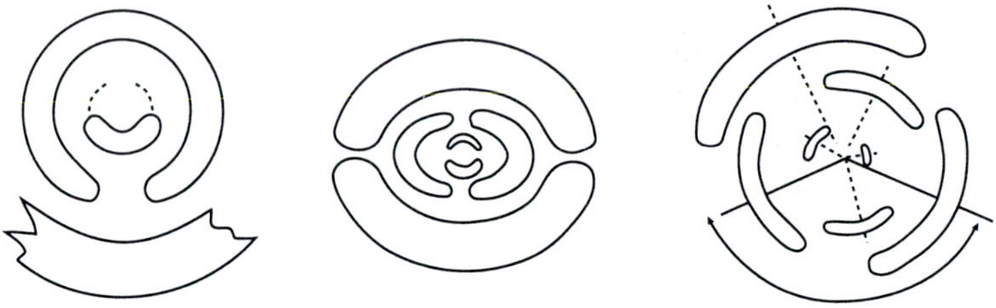

Figure 12: The three basic patterns of leaf phyllotaxis.

particular dynamic processes. In biology the notions of function or adaptation or natural selection are often used to explain organismic form, but these ideas address only the question why a structure persists. They fail to explain why the structure exists — i.e., what kind of dynamic process makes the form possible as one of its stable solutions. This gets at the generative roots of the problem, as described also by Newman and Comper (1990). Seventy-nine years ago D'Arcy Thompson made precisely the same point in the introduction to his book *On Growth and Form* (1917, p. 23): "So long and so far as 'fortuitous variation' and the 'survival of the fittest' remain ingrained as fundamental and satisfactory hypotheses in the philosophy of biology, so long will these satisfactory and specious causes tend to stay severe and diligent enquiry ... to the great arrest of future discovery".

Is there a genetic program that directs the morphogenetic sequence in *Acetabularia*? That genes are involved in the process there is no doubt — they set the values of the parameters. How many genes are involved we do not know, but certainly every one of the 26 parameters must involve the contributions of many genes. For example, the effective diffusion constant of free calcium in the cytoplasm, one of the sensitive parameters that affects the dynamics, is dependent upon the concentration and affinity constants of any protein that binds calcium such as calmodulin, calcitonin, and any analogue of troponin involved in the cytoskeleton. Concentration regulation of proteins involves other genes, so there could be as many as ten genes involved in specifying this one parameter. The restoring force of the cytoplasm or the elastic modulus of the cell wall depend on dozens of gene products that contribute to the mechanical properties of these cellular components. We could probably account for several hundred genes in our model, all acting polygenically on the parameters. Not one of these do we need to know in detail, since we are looking for generic properties — classes of process, types of form that can be generated.

Of course, the more detailed information we have, the more accurate the model. But the clue to robust morphogenesis may be in the coupling between different components of the overall process. An excitable cytoplasm generates an initial gradient that is stabilized by plastic deformations of the tip, whose growth then makes possible the appearance of the next mode, an annulus. The geometry then changes again with tip flattening, creating the conditions for a third bifurcation to the pre-whorl pattern, which is stabilized by further changes of shape as the bract primordia grow into a whorl. Thus a hierarchical cascade of sequential symmetry breaking occurs and complex form arises from a dynamically stable trajectory. This is reminiscent of D'Arcy Thompson's medusa (Figure 1). The principles here can be generalised to a conjecture that morphogenesis in viable species is an intrinsically robust process that arises from the coupling between different dynamic modes in a hierarchically organised system (Goodwin 1993; Goodwin, Kauffman and Murray 1993, Goodwin 1994). A number of different candidates for this hierarchical, coupled system are emerging (e.g., Hunding, Kauffman and Goodwin 1990; Lacalli 1990; Mittenthal, Baskin and Reincke 1992). Out of this and other work on these complex systems may come the resolution of the tension between dynamic models and genetic programs that combines both in a new category of morphogenetic field that can be used as the basis for a classification of the generic forms available to life. This would open the door to a rational taxonomy of living forms (Ho 1990) and provide a context for understanding how evolution has explored morphospace (Gould 1991).

References

Bateson, W. (1894; 1992) *Materials for the Study of Variation, Treated with Especial Regard to Discontinuity in the Origin of Species.* Baltimore: Johns Hopkins University Press.

Berger, S. and Kaever, M.J. (1992) *Dasycladales: An Illustrated Monograph of a Fascinating Algal Order.* Stuttgart: Thieme.

Brachet, J. and Bonnotto, S (1970) *Biology of Acetabularia.* London: Academic Press.

Brandts, W.A.M. and Trainor, L.E.H. (1990a) A non-linear field model of pattern formation: intercalation in morphallactic regulation. *Journal of Theoretical Biology* **146**, 37–56.

Brandts, W.A.M. and Trainor, L.E.H. (1990b) A non-linear field model of pattern formation: application to intracellular pattern reversal in *Tetrahymena*. *Journal of Theoretical Biology* **146**, 57–86.

Brière, C. and Goodwin, B.C. (1988) Geometry and dynamics of tip morphogenesis in *Acetabularia*. *Journal of Theoretical Biology* **131**, 461–475.

Brière, C. and Goodwin, B.C. (1990) Effects of calcium input/output on the stability of a system for calcium-regulated viscoelastic strain fields. *Journal of Mathematical Biology* **28**, 585–593.

Cleland, R. (1971) Cell wall extension. *Annual Review of Plant Physiology*. **22**, 197–222.

Driever, W. and Nüsslein-Volhard, C. (1988) A gradient of *bicoid* protein in *Drosophila* embryos. *Cell* **54**, 83–93.

Eichele, G. and Thaller, C. (1987) Characterisation of concentration gradients of a morphogenetically active retinoic acid in the chick limb bud. *Journal of Cell Biology* **105**, 1917–1923.

Frankel, J. (1989) *Pattern Formation: Ciliate Studies and Models*. New York: Oxford University Press.

Frankel, J. and Nelsen, E.M.(1986) Intracellular pattern reversal in *Tetrahymena thermophila*. II. Transient expression of a *janus* phenocopy in balanced doublets. *Developmental Biology* **114**, 72–86.

French, V., Bryant, P.J. and Bryant, S.V. (1976) Pattern regulation in epimorphic fields. *Science* **193**, 969–981.

Glass, L. (1977) Patterns of supernumerary limb regeneration. *Science* **198**, 321–322.

Goodwin, B.C.(1985) What are the causes of morphogenesis? *BioEssays* **3**, 32–35.

Goodwin, B.C. (1993) Development as a robust natural process. In W.D. Stein and F.J. Varela, eds., *Thinking About Biology: An Invitation to Current Theoretical Biology*. Santa Fe Institute Studies in the Sciences of Complexity. Lectures, Volume III. Reading, MA: Addison-Wesley, pp. 123–148.

Goodwin, B.C. (1994) *How the Leopard Changed Its Spots*. London: Weidenfeld and Nicolson.

Goodwin, B.C., Brière, C. and O'Shea, P.S. (1987) Mechanisms underlying the formation of spatial structure in cells. In R.K. Poole and A.P.J. Trinci, eds., *Spatial Organization in Eukaryotic Microbes*. Oxford: IRL Press, pp. 1–9.

Goodwin, B.C. and Kauffman, S.A. (1993) Mapping gene activities into morphological patterns in *Drosophila*. In H.G. Othmer, P.K. Maini and J.D. Murray, eds., *Experimental and Theoretical Advances in Biological Pattern Formation.* NATO Advanced Science Institute Series, Volume 259. New York: Plenum Press, pp. 87-99.

Goodwin, B.C., Kauffman, S.A. and Murray, J.D. (1993) Is morphogenesis an intrinsically robust process? *Journal of Theoretical Biology* **163**, 135–144.

Goodwin, B.C. and Pateromichelakis, S. (1979) The role of electrical fields, ions, and the cortex in the morphogenesis of *Acetabularia. Planta* **145** , 427–435.

Goodwin, B. C., Skelton, J. C. and Kirk-Bell, S. M. (1983) Control of regeneration and morphogenesis by divalent actions in *Acetabularia mediterranea. Planta* **157**, 1–7.

Goodwin, B.C. and Trainor, L.E.H. (1985) Tip and whorl morphogenesis in *Acetabularia* by calcium-regulated strain fields. *Journal of Theoretical Biology* **117**, 79–106.

Gould, S.J.(1991) The diversity of the Burgess Shale arthropod fauna and the limits of cladistic analysis: why we must strive to quantify morphospace. *Paleobiology* **17**, 411-423.

Green P.B. (1987) Inheritance of pattern: analysis from phenotype to gene. *American Zoologist* **27**, 657-673.

Green, P.B. (1989) Shoot morphogenesis, vegetative through floral, from a biophysical perspective. In E. Lord and G. Barrier, eds., *Plant Reproduction: From Floral Induction to Pollination.* American Society of Plant Physiology Symposium Series, Volume 1. New York: American Society of Plant Physiology, pp. 58–75.

Green P.B., Erickson, R.O. and Buggy, J. (1971) Metabolic and physical control of cell elongation rate. In vitro studies in *Nitella. Plant Physiology* **47**, 423–430.

Hammerling, J. (1931) Entwicklung und Formsbildungsvermogen von *Acetabularia mediterranea. Biologische Zeitschrift* **51**, 633–647.

Harrison, L.G., Graham, K.T. and Lakowski, B.C. (1988) Calcium localisation during *Acetabularia* whorl formation: evidence supporting a two-stage hierarchical mechanism. *Development* **104**, 255–262.

Harrison, L.G. and Hillier, N.A. (1985) Quantitative control of *Acetabularia* morphogenesis by extracellular calcium: a test of kinetic theory. *Journal of Theoretical Biology* **114**, 177–19.

Hart, T.N., Trainor, L.E.H. and Goodwin, B.C. (1989) Diffusion effects in calcium-regulated strain fields. *Journal of Theoretical Biology* **136**, 327–336.

Ho, M-W. (1990) An exercise in rational taxonomy. *Journal of Theoretical Biology* **147**, 43–57.

Hunding, A., Kauffman, S.A. and Goodwin, B.C. (1990) *Drosophila* segmentation: supercomputer simulation of prepattern hierarchy. *Journal of Theoretical Biology* **145**, 369–384.

Jahnke, W., Henze, C. and Winfree, A.T. (1988) Chemical vortex dynamics in three-dimensional excitable media. *Nature* **336**, 662–665.

Lacalli, T.C. (1990) Modelling the *Drosophila* pair-rule pattern by reaction-diffusion: gap input and pattern control in a 4-morphogen system. *Journal of Theoretical Biology* **144**, 171–194.

Maden, M. (1982) Vitamin A and pattern formation in the regenerating limb. *Nature* **295**, 672–675.

Meinhardt, H. (1982) *Models of Biological Pattern Formation.* London: Academic Press.

Meinhardt, H. (1986) Hierarchical inductions of cell states — a model for segmentation in *Drosophila. Journal of Cell Science Supplement* **4**, 357–381.

Menzel, D. and Elsner-Menzel, C. (1989) Induction of actin-based contraction in the siphonous green alga *Acetabularia (Chlorophycea)* by locally restricted calcium influx. *Botanica Acta* **102**, 164–171.

Menzel, D. Jonitz, H. and Elsner-Menzel, C. (1992) The cytoskeleton in the life cycle of *Acetabularia* and other related species of dasyclad green algae. In: D. Menzel, ed., *The Cytoskeleton of the Algae.* Boca Raton, FL: CRC Press, pp. 195–217.

Mittenthal, J.E., Baskin, A.B. and Reincke, R.E. (1992) Patterns of structure and their evolution in the organisation of organisms: modules, matching, and compaction. In J.E. Mittenthal and A.B. Baskin, eds., *Principles of Organization in Organisms.* Santa Fe Institute Studies in the Sciences of Complexity, Volume XIII. Reading, MA: Addison-Wesley, pp. 321–332.

Murray, J.D. (1989) *Mathematical Biology.* New York: Springer-Verlag.

Newman, S.A., and Comper, W.D. (1990) 'Generic' physical mechanisms of morphogenesis and pattern formation. *Development* **110**, 1–18.

Nüsslein-Volhard, C. (1977) Genetic analysis of pattern formation in the embryo of *Drosophila melanogaster*. Characterisation of the maternal effect mutant *Bicaudal*. *Roux's Archiv für Entwicklungsmechanik* **183**, 244–268.

Oster, G.F. and Odell, G.M. (1984) The mechanochemistry of cytogels. *Physica D* **12**, 333–350

Raup, D.M. and Michelson, A. (1965) Theoretical morphology of the coiled shell. *Science* **147**, 1294–1295.

Shelanski, M.L. (1989) Intracellular ionic calcium and the cytoskeleton in living cells. *Annals of the New York Academy of Sciences* **56**, 121–136.

Thompson, D'A.W. (1917) *On Growth and Form.* Cambridge: Cambridge University Press. (See also the 1961 Cambridge University Press abridged edition of this classic.)

Tickle, C., Alberts, L., Wolpert, L. and Lee, J. (1982) Local application of retinoic acid to the limb bud mimics the action of the polarizing region. *Nature* **298**, 564–566.

Totafurno, J. and Trainor, L.E.H. (1987) A non-linear vector field model of supernumerary limb production in salamanders. *Journal of Theoretical Biology* **124**, 415–454.

Turing, A. M. (1952) The chemical basis of morphogenesis. *Philosophical Transactions of the Royal Society of London, Series B* **237**, 37–72.

Williamson, R.E. (1984) Calcium and the plant cytoskeleton. *Plant, Cell and Environment* **7**, 431.

Toward a Model of Growth and Form in Living Systems

Fred W. Cummings

1 Introduction

More than forty years have passed since the introduction of the famous "Turing" (1952) model of pattern formation in embryogenesis. Turing's seminal idea has spawned a number of more specialized models and applications which fall under the classification of "reaction-diffusion" systems (Gierer and Meinhardt 1972; Goodwin 1976; Murray 1977; Kauffman, Shymko and Trabert 1978; Meinhardt 1982; Harrison 1987; Nijhout 1992). Given a surface (invariably a topologically flat surface) this model has been capable of mimicking very many of the patterns seen in nature. Thus this mechanism is seen as a "pattern" generator, as contrasted to a shape, or "morph" generator. It is the purpose of the present essay to discuss a model for a candidate for the "morphogen" (or morphogens), and this model will have a somewhat different conceptual basis than the reaction-diffusion models (R-D models).

This present model has the following features: (1) It is simpler to conceptualize, and contains fewer parameters than R-D models. (2) It is more "robust" than the R-D class of models in that the nonlinearities are not viewed as the dominant feature. In the R-D schemes, the nonlinearities often greatly limit the allowed region of parameter space. In the present picture, the nonlinearities are more properly viewed as perturbations which serve to pick out selected modes over other "degenerate" ones. (3) A promising candidate for the "morphogen(s)" is suggested which is directly in the pathway of control of cell division, orientation of the microtubules and microfilaments, and transcription; in general it is proposed that any "morphogen(s)" should be biochemicals directly in the pathway of many aspects of growth control and form. (4) The steady-state patterns obtained from R-D models are also obtained from the present point of view. (5) Contact is made with the energy-minimization patterning mechanism of Steinberg (1963), as well as the important findings of French, Bryant and Bryant (1976), Bryant and Muneoka (1986), and Bryant (1993). The results of French, Bryant and Bryant point to the crucial importance of the epithelial boundary in a regenerating organ in the determination of the regenerated pattern and form, as is shown in the phenomenon of "regeneration/duplication" across species lines (Cummings 1994).

The present model focuses on the determinants of shape, and necessarily involves pattern formation as well; however, it does not rule out the possibil-

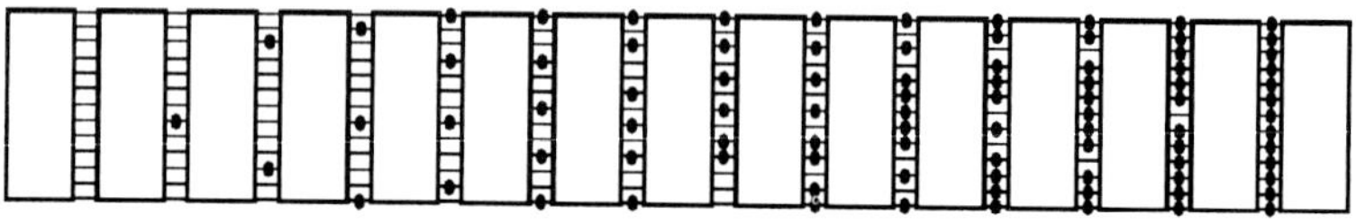

Figure 1: A schematic one-dimensional line of cells connected by two different types of homotypic adhesive molecules. These two types are shown as single lines (R), and single lines with solid dots (B). In the text, these are called "R" and "B" molecules, in order to be as nonspecific as possible. The sum R + B is constant on each cell, and the dynamic is that R and B have an effective repulsive interaction between them; whether this arises from an autocatalytic interaction where a large spatial density of R catalyzes its further production from B (and vice versa), or whether the two types move away from each other in the plane of the membrane, is not specified. The resulting situation shown in the Figure is the result of minimization of energy.

ity that the morphogen envisioned here is acting in concert with patterning mechanisms of some R-D scheme. (We use "morphogen" to distinguish the cell surface molecules we have in mind as giving rise to shape or form from the "patterning" molecules of the R-D schemes, in spite of the use of the term "morphogen" often in the past to describe a pattern-forming mechanism on an already *specified* surface; the Greek root is *morph* = shape). Also, the time dependence of the present model is very different from the usual time dependence of R-D models.

2 A Description of the Model

The essence of the model can most easily be introduced by a one-dimensional schematic. Figure 1 considers a linear array of cells connected by molecules. These molecules come in two types, "red" and "blue," and are perhaps conformational forms of the same molecule. In the more general case of a two-dimensional surface, where the "surface" is viewed as the middle surface through cells of height h, the molecules will connect the cells in these epithelial sheets by their lateral surfaces. The molecules distribute themselves according to the following simple rules.

First, in the absence of interaction between red (R) and blue (B), both R and B will distribute themselves as smoothly as possible over the surface (in the case of the Figure 1, over the length) consistent with any boundary conditions. R and B, if acting independently, have a tendency to avoid their own kind: R avoid R, B avoid B. As the example of Figure 1 illustrates, if R (shown as lines without dots) were held to a high value at the left end then the R would form a linear gradient along the length.

The second rule is that the R and B interact, and this repulsive interaction causes them, in effect, to try to avoid each other; they repel each other, and presence of R tends to cause B to convert to R (or possibly also to migrate in the membrane to regions of lower concentrations of R), and vice versa. The lowest energy configuration, following Steinberg (1963), will be sought such that these two rules are obeyed as best as possible, subject to the constraint below. It is further assumed that each cell in a given field puts out a constant concentration of the sum of R and B. Thus as seen in Figure 1, the total of R (lines without dots) equals the total number of B (lines with dots), and each cell has the same total of R+B. Very complex patterns arise from this simple algorithm on the surfaces of interest, and it yields essentially the same steady-state patterns as the time independent R-D models, although the conceptual basis here is very different.

Part of the present motivation is the striking fact that if one takes a small piece from the length of a hydra (*mediterranea*, say), the small piece will reform itself into a miniature hydra by reforming the head (or foot) at its original end, i.e., it remembers the direction of the original gradient. *Hydra* will accomplish this regeneration down to excised pieces as small as 5% of the original body mass (Bode and Bode 1984). The original density of R is reestablished at the "head" end in the excised piece, as is apparent from the rules above; since energetically speaking, R avoids B, and a given cell has a fixed sum of R+B, then R will collect at one end, and B at the other in the excised piece in the simplest one-dimensional example. It seems simpler to conceptualize this fcat of regeneration (and the regeneration of "high" and "low" positional values) from the present point of view (when this reestablishment of the "high" and "low" points is energetically favored) than from the point of view of R-D schemes with their numerous parameters which need to be (almost numinously) adjusted to accommodate the new situation. The process of size "regulation" is apparent here.

It is envisioned that as "high" and "low" points are established, they become "frozen in" (reminiscent of the hypostome and basal disc of hydra) so that they act as quasi-permanent or permanent boundary regions for the subsequent solution of the equation describing the dynamic. As time progresses, and the area A of the (closed) epithelial sheet increases, further more or less permanent "sources" are established. Growth will occur selectively in those regions which have the highest gradients of morphogen, as regions of R and B "repel" each other.

A heuristic description of the proposed model has been outlined above.

These ideas are easily translated into mathematical form. We do not consider that the particular formulation is important; the ideas sketched above will lead to essentially the same patterns starting from various mathematical formulations; the formulation below is certainly not unique, but perhaps only the simplest which embodies the above rules.

The dynamical behavior will be viewed as imposed partly externally, as an uptake of nutrients from the environment occurs, and gives rise to an increase in the total area A of the (closed) epithelial sheet in time, as cell division and the accompanying growth occurs. Time-dependent changes are also assumed to occur as genes are activated, and these changes are viewed as occurring ultimately via changes in the parameters of the model.

Consider the energy E given by the expression

$$E = \epsilon_0 \int \left[D_1(\nabla\phi_1)^2 + D_2(\nabla\phi_2)^2 + \alpha^2(\phi_1 - c)(\phi_2 - c)\right] dA \tag{1}$$

where ϵ_0 is a constant energy/area. The integral extends over the entire middle surface (Cummings 1994) of total area A; ϕ_1 and ϕ_2 describe the positive densities (number/area) of the two types (R and B molecules in the above description), and D_1, D_2, c, and α^2 are (spatial) constants.

The first two terms of equation (1) describe the tendency of each of ϕ_1 or ϕ_2 alone, in the absence of interaction between them (i.e., $\alpha = 0$), to spread themselves as evenly as possible over the surface, subject to any constraints which may be imposed (such as fixed values of either in a specified region). The minimum energy configuration for each in this simple case would be given by each molecular type independently satisfying the well-known Laplace equation. The third term under the integral tends to a minimum energy when the two types avoid each other; when ϕ_1 is larger than c and ϕ_2 is smaller than c, (and vice versa), the energy is clearly lowered compared to the case of the two densities being large in the same region of space. Not shown for simplicity are terms which lead to higher-order nonlinearities in the equations for the energy minimum condition, although these unspecified terms are to be considered as determinants in the choice of specific spatial modes; they will lift degeneracies, since different modes will have different energies.

The effective "repulsive" interaction between the R and B is expected to be an autocatalytic interaction, in the sense that a high density of R causes the further conversion of type B to type R in this region (or vice versa).

The third rule above imposes the constraint

$$\phi_1 + \phi_2 = 2c \tag{2}$$

which says that the sum of the two densities is a constant at all locations on the middle surface. The constant c is the average concentration of each type integrated over the entire surface; substitution of the constraint of equation (2) into (1) gives

$$E = const. \int \left[(\nabla \Psi)^2 - k^2 \Psi^2 \right] dA \ , \tag{3}$$

$$\Psi \equiv \phi_1 - c = c - \phi_2 \tag{3a}$$

$$k^2 \equiv \alpha^2 / (D_1 + D_2) \ . \tag{3b}$$

(Note that in a more general formulation there may be two different averages, c_1 and c_2, not considered here, but the end result is substantially the same.)

Minimization of the integral of equation (1), subject to the constraint of equation (2), gives in the usual way via the Euler-Lagrange equation (Morse and Feshbach 1961) the well-known Helmholtz equation (again leaving nonlinearities implied rather than expressed)

$$\nabla^2 \Psi + k^2 \Psi^2 = 0 \ . \tag{4}$$

This equation has often been invoked in the past in the context of developmental patterns (Murray 1977, Kauffman et al. 1978, Goodwin and Trainor 1980, Hunding 1983, McNally and Cox 1989, Kauffman 1993). Murray (1981) has noted that the vibrational modes of thin plates correspond generally to the pattern boundaries produced by R-D systems on similar shapes in the steady state; the Helmholtz equation is the time-independent description of vibrational modes on thin plates, and for a given shape, results in allowed "eigen" mode solutions and eigenvalues (Morse and Feshback 1961).

The present model thus envisions that a "pattern" equilibrium is established on a time scale which is shorter than the cell cycle time in a given spatial region. (This could be related to a distribution in cell cycle times (Bryant, Hayamizu and Gardiner 1993).) Thus the time dependence of the present model is very different than that of R-D schemes. Time dependence in the present view has its origin in the (assumed determined externally) change in the total area $A = A(t)$; this area A is contained in the Laplacian operator, or in the "metric" function g (see below). The system is an open system, and there is a given energy uptake (determined by conditions of available nutrients) which is exhibited as cell division and the subsequent cell growth which leads to the increase in the total area A. No commitment is here implied as to the locale of cell division and local growth, as this is an emergent feature of the model.

A second and most important time change is viewed as given by the change of the parameters of the model, and these parameters are viewed as being under genetic control. Put another way, genes set the form of the two functions, Gauss curvature $K(\Psi)$ and the function comprising the third term in equation (1), or the Helmholtz equation (4), this latter being assumed to be in general nonlinear. This will be discussed further in the next section, when the geometry of the surface is introduced, and a model of growth and form is presented.

3 Geometry, Speculation and Genes

In the above formulation, via equation (1), we have envisioned a situation in which each cell is influenced by its neighbors, and each takes on a value of ϕ_1, ϕ_2 or Ψ governed by the condition which gives a minimization of the energy specified by (1). However, we also know that the individual cell has a genetic (homeobox for our purposes here) state, an internal state. This is viewed here as determined by the ϕ_1, ϕ_2, or perhaps by the ratio of the two densities $\phi_{1,2}$. Goodwin and Kauffman have given interesting and persuasive arguments for thinking that the ratio of gene products are key determinants of patterning (Kauffman 1993).

These homeobox genes are the "master" genes of development, special genes which have been remarkably conserved throughout evolution, from plants and nematodes to humans, and are to be distinguished from the more numerous species-specific genes, which are under the control of these master developmental genes. These genes will be seen in the present view as controlling the parameters of the model, or the form of two functions. But first, in order to illustrate which parameters we have in mind, it is necessary to briefly sketch the coupling of the "morphogen" to the surface geometry, and the reverse, which is presently envisioned. The morphogen(s) is viewed as affecting the geometry of the epithelial sheet, via the intermediary of the genome, and in turn, this changed surface sheet will cause an altered morphogen pattern, which will then cause an altered geometry, and so on: Growth and morphology change in time. This general coupling has been discussed more fully in Cummings (1994), to which the reader is referred. The picture of the coupling of morphogen to geometry as given in this reference may be summarized roughly as follows.

The equation for the metric as determined by the Gauss curvature is

$$-\frac{1}{2g}\left(\partial_u^2 + \partial_v^2\right) ln(g) \equiv -\frac{1}{2g}\partial^2 ln(2g) = K \tag{5}$$

and the metric g determines the total area A through the "normalization" condition

$$\int g\, dudv = A \; . \tag{6}$$

Here the coordinates (u, v) lie in the surface and (for simplicity) have been chosen to form an orthogonal coordinate system. The Gauss curvature K, the product of the two principal curvatures at a point, is viewed as determined by the morphogen Ψ, so that $K = K(\Psi)$. In a recent model (Cummings 1994), K depends on three parameters which determine (1) the coupling of K to $\Psi(u, v)$, and (2) three main branches that specify the behavior of the scalar function $K(\Psi)$. When the thickness of the epithelial sheet h is specified, itself also a function of the coordinates (u, v), the "mean curvature" H is also determined, and thus both of the principal curvatures at each point, so that a complete specification of the "sheet" (of thickness h) is given.

The coupling of the morphogen to the geometry K through the metric function $g(u, v)$ is seen by rewriting equation (4) as

$$\left(\partial_u^2 + \partial_v^2\right) \Psi + k^2 g(u, v)\Psi = 0 \ , \tag{7}$$

since the Laplacian operator is related to g by

$$\nabla^2 = \frac{1}{g} \left(\partial_u^2 + \partial_v^2\right) \tag{8}$$

in these convenient conformal coordinates (u, v). Since K depends on $\Psi(u, v)$, we have the evolution: Start with a simple metric g_0, e.g., perhaps a sphere. Compute $\Psi(u, v)$ from equation (7), and $K\left(\Psi(u, v)\right)$ (e.g., from knowledge of the three parameters A, B, C, say, in the model cited). From equations (5) and (6), find the new $g = g_1$, and from g_1, find $\Psi_2(u, v)$, and $K\left(\Psi_2(u, v)\right), \ldots$ and so on. Thus, with a specified $K(\Psi)$ function, the "animal" grows and changes shape. This is a closed system of equations. The different initial conditions, the boundary conditions at the start, determine different organisms; these initial boundary conditions are presumed to be given maternally, or in the original seed. The metric function g is always normalized to A, so that we may form the product $g = A\, G(u, v)$, where G is normed to unity over the surface. Here A is the total area and G is the "shape" function.

It should be noticed that the formulation of the model is independent of any specific coordinate system. This requirement of coordinate independence is a necessary condition which any viable model must satisfy, and is often helpful in guiding one to a correct formulation, or more properly, a key guide in weeding out unsatisfactory ones. The coordinates are freely chosen by the *observer*, while the "field" belongs to the animal; it is important to keep this distinction clearly in mind, that is, of what "belongs" to the observer and what to the animal. Without such a distinction in mind, is easy to consider that a singularity or discontinuity introduced by the choice of coordinate system is something for which one need find a mechanism, or a molecular counterpart, with consequent great loss of time and effort. The R-D theories are generally formulated at the outset so as to satisfy coordinate independence. On the other hand, the famous polar coordinate model (French et al. 1976, see also Gilbert 1991 and Bryant 1993) is an obvious case of a coordinate-dependent model. The discontinuity of the polar coordinate model in the angle coordinate (which runs from 0 to 12) comes immediately to mind. Although the polar coordinate model has enjoyed many successes and has thoroughly influenced the present work, the coordinate-dependent aspect is its weakest point. The effect of fixed boundaries, the need for two morphogens to specify a point on a surface, the shortest intercalation rule and especially its emphasis on local cell-cell interactions have been incorporated into the present model in slightly disguised form. It would seem germane, then, to find a coordinate-free formulation. Perhaps the present formulation can serve as a modest start. (See also Brandts and Totafurno, and Hart and Trainor, this volume.)

In the present picture, the genome (and we refer to the homeobox genes only) has been relegated to the obviously very essential role of determining the function $K(\Psi)$, (viewed here as occurring via control of the parameters which enter its description), and also in determining the function in the second term of the Helmholtz equation; the constants in equation (7) are thus presumably under genetic control, k^2 as well as any constants entering the (here unspecified) nonlinear terms which serve to pick out the desired mode via equation (1) and the condition of energy minimization. The minimum energy will depend on the mode (Kauffman 1993) in the case that nonlinearities are present.

There is the crucial question of *when* and *where* the genome is asked to intervene. The "where" has been addressed through the morphogen function Ψ. Gene products (proteins) activate further genes and have their proper spatial domain of action specified by the morphogen(s); in this sense the morphogen acts as a sort of cofactor, however a cofactor which has a well-specified spatial domain of action. Kauffman (see Kauffman et al. 1978, Kauffman and Ling 1981, Kauffman 1993) has presented a scenario for this sequence of events in the *Drosophila* embryo segmentation patterns and in the partitioning of the wing imaginal disc, based on Waddington's 1940 "trainyard" binary switching idea: At each juncture in time, the genome is asked to make a binary decision, and a given cell records (and stores) the information as e.g. (t = time) t_0: (1); t_1: (1,0); t_2: (1,0,0); t_3: (1,0,0,1); t_4: (1,0,0,1,1); ... which the cell at a particular location records as determined from the morphogen function value relative to a *single* threshold value at the times t_0, t_1, ... This sequence of events occurs as the organism is progressively divided, fractal-like, into smaller and smaller compartments. (In *Drosophila* the first three regions of anterior-posterior gene specification are maternally determined, apparently by diffusion,

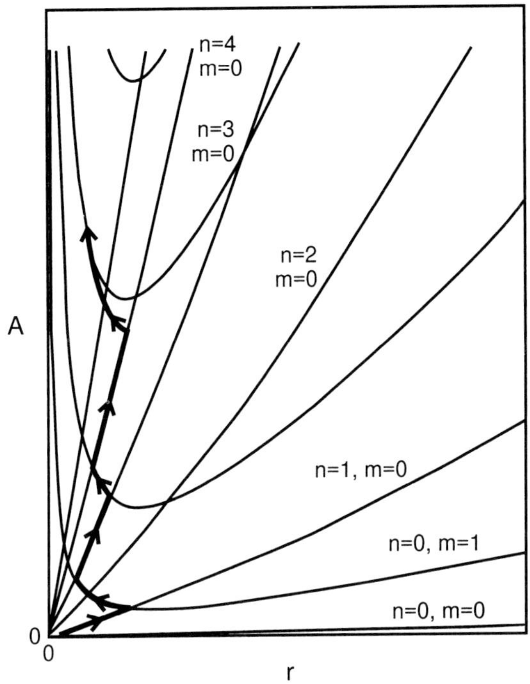

Figure 2: An imaginary "flat worm" is considered. The total area A of the closed "epithelial" (middle) surface of the right-cylindrical "worm" is shown versus the "shape" of the worm $r = a/b$, for the allowed solutions of the model given specified boundary conditions. Here $2\pi a$ is the perimeter of the right cylinder (not necessarily circular), and b is the length along the cylinder (see Figure 2 and the Appendix). The cylinder has been taken as a closed surface, with ends formed of flat areas at each end. As the total area increases, the solution path in the (A, r)-plane can move along one of the curves labelled by the various integers (n, m). As A increases in time, the shape r is seen to change. At each such binary crossing of curves of different (n, m), a decision must be made. A possible sample path is shown by the arrows. It is presumed that the decisions which are made at each binary crossing will be determined by the nonlinearities of the Helmholtz equation, which terms will remove the "degeneracy," but have not been explicitly included in the model.

and a gradient of "transcription factor" first divides the embryo; cellularization does not occur until about the thirteenth cell cycle (Gilbert 1991), so that the present scheme of cell-surface molecules is envisioned as coming into play at that time in this case; in all cases, the initial conditions are "maternally" set, and we imagine that the maternal conditions are not "set" until cycle thirteen in the case of *Drosophila* segmentation patterns).

An important point to recognize is that a continuous change in area A is only possible when accompanied by a change in shape if we are to maintain a solution to the Helmholtz equation as the animal grows. The most discussed situation is that of an "eigenvalue" equation when the shape is kept constant, and the total area is changed. In this case solutions will only exist when the proper "eigenvalues" are attained for certain areas with discrete values. The sphere or the cylinder are typical examples; as the sphere radius R is increased, solutions only exist for certain *discrete* radii R_{lm}, and are the well-known Legendre polynomials (Morse and Feshbach 1961). The cylinder of proportions ("shapes") which have radius/length $= r/l$ fixed also have solutions only for discrete values of the area $A = 2\pi rl$ in the usual way of thinking. Thus as area increases continuously, in order to have a solution exist for all A, it is necessary to have a concomitant continuous shape change. This is easily worked out for the cylinder (see Figure 2 and the Appendix). As the area A increases, the ratio r/l will increase or decrease depending on the initial conditions; eventually a binary decision must be made, when two solution lines in the $(A, s)(sr/l)$ plane corresponding to different integer pairs (n_0, m_0) and (n_1, m_1) intersect. These intersections occur at discrete times and partition the organism into discrete spatial regions, and are thus very suggestive of the points of gene "switching" in Kauffman's picture. The (unspecified by equation (1)) nonlinearities will suffice to "pick out" the path in (area, shape)-space, so that the genes acting before the point of area A where a binary decision is required may be seen as determining the parameters of the nonlinear terms which determine the subsequent mode. In the simple case of the cylinder (where we insist that a cylinder shape be maintained) there is only one parameter defining the "shape," or two parameters in all ($(A, r/l)$ in place of (r, l)). More generally, when we consider other than $K = 0$ situations, we may picture an increasing area A as one of the axes, and the other axes (N in general) defining a sort of "morphospace," where the solution point traverses a minimum energy trajectory as growth occurs.

4 A Promising Molecular Mechanism for the Morphogen

The adhesive molecule protein receptor tyrosine phosphatase is a three-domain, transmembrane receptor-like molecule, and is the counterpart of the well-studied receptor protein tyrosine kinase (RPTK), which is a cell-surface receptor for numerous "growth" factors, and which phosphorylates tyrosine residues in the cytoplasm. The three-domain cell-membrane-spanning protein tyrosine phosphatase ("RPTP") molecule plays the opposite role, and detaches phosphate groups from tyrosine residues on proteins in the cytoplasm (Fischer, Charbonneau and Tonks 1991; Gebbink et al. 1993). It forms homotypic adhesive connections to adjacent cells. Most worthy of note for our purposes here is that RPTP comes in two different, but structurally very similar, types: $RPTP_\alpha$ and $RPTP_\beta$. Further, of most interest is the fact that when insect cells expressing $RPTP_\alpha$ were mixed with cells expressing $RPTP_\beta$ and gently shaken, the two types promptly sorted out into separate spatial regions (see Eijgenraam 1993); there is clearly interaction between the two types. It seems apparent that RPTP plays an equally important role as RPTK in regulating cell growth and differentiation since they act on the same kinds of biochemical switches. While the molecule RPTK is involved in the stimulation of cell division and differentiation by way of various growth factors, at least some of the very diverse extracellular parts of the RPTK molecule, the ligand for the RPTP molecule appears to be of the same type expressed on the surface of the adjacent cell.

The connection between the adhesive properties of RPTP and its enzymatic action within the cell is at present unknown. Although the target proteins within the cell are not all as yet identified, it is clear that the RPTP molecule is in the cell's signaling pathway, and the growth inhibitory effects seem likely. It seems likely that the RPTP will also prove to be a crucial player in the changes in the cytoskeleton (and in our present terms, in control of the form of the Gauss function K). By way of a cascade of intermediate proteins, transcription is also extremely likely to be connected with the same RPTP molecule, and transcription factors an important element in the loop. In view of the theory and experiments of Steinberg (1963) and also the ideas of Edelman (1988) and Singer (1992), an adhesive molecule is already a promising candidate for "morphogen;" however, we wish to emphasize here that we consider "the" morphogen to consist of a complex of elements, of which the adhesive cell-surface molecule is but an integral part of the loop. Further properties of the RPTP molecule, the fact that it comes in two "repelling" forms α and β, (and thus is probable to be a "pattern maker") as well as being involved in the control of cell division, transcription, and control of the cytoskeleton, as well as itself being affected by the gene products in turn, makes it, in our view, a most promising candidate for the role of "morphogen."

S. Byrant et al. (1993) have advanced interesting ideas based on gradients of transcription factors. The present ideas can be viewed as complementary to Bryant's; "the" morphogen is to be interpreted as the "morphogen complex" or "morphogen loop," consisting of all of the elements from cell-surface molecule to gene and return, including transcription factors, these latter being the elements emphasized by Bryant et. al. as playing the role of "morphogen."

It is our view that a "morphogen" should be tied as closely as possible to a "loop" which involves control of cell division adhesion, control of the cytoskeleton, gene transcription and transcription factors, and feedback from the gene to the adhesive cell surface molecule. The present work hopes to point to the ease with which cell surface molecules can specify pattern.

Acknowledgements

This paper is in honor of the many accomplishments of Lynn Trainor in the field of theoretical biology. His numerous seminal contributions have served as an inspiration and a beacon to all. I wish to thank the Aspen Center for Physics for their kind hospitality in Summer 1994.

References

Bode, P.M. and Bode, H.R. (1984) Patterning in Hydra. In G.M. Malacinski and S.V. Bryant, eds., *Pattern Formation: A Primer in Developmental Biology*. New York: Macmillan, pp. 213–241.

Bryant, P.J. (1993) The polar coordinate model goes molecular. *Science* **259**, 471–472.

Bryant, S.V., Hayamizu, T.F. and Gardiner, D.M. (1993) Patterning in limbs: the resolution of positional confrontations. In H.G. Othmer, P.K. Maini and J.D. Murray, eds., *Experimental and Theoretical Advances in Biological Pattern Formation*. NATO Advanced Science Institute Series, Volume 259. New York: Plenum Press, pp. 37–44.

Bryant, S.V. and Muneoka, K. (1986) Views of limb development and regeneration. *Trends in Genetics* **2**, 153–159.

Cummings, F.W. (1994) Aspects of growth and form. *Physica D* **76**, 146–163.

Edelman, G.M. (1988) *Topobiology.* New York: Basic Books.

Eijgenraam, F. (1993) Things start getting sticky for a cell surface enzyme. *Science* **261**, 833.

Fischer, E.H., Charbonneau, H. and Tonks, N.K. (1991) Protein tyrosine phosphatases: a diverse family of intracellular and transmembrane enzymes. *Science* **253**, 401–406.

French, V., Bryant, S.V. and Bryant, P.J. (1976) Pattern regulation in epimorphic fields. *Science* **193**, 969–981.

Gebbink, M.F.G.B., Zondag, G.C.M., Wubbolts, R.W., Beijersbergen, R.L., van Etten, I. and Moolenaar, W.H. (1993) Cell-cell adhesion mediated by a receptor-like protein tyrosine phosphatase. *Journal of Biological Chemistry* **268**, 16101–16104.

Gierer, A. and Meinhardt, H. (1972) A theory of biological pattern formation. *Kybernetic* **12**, 30–39.

Gilbert, S.F. (1991) *Developmental Biology.* Third Edition. Sunderland, MA: Sinauer Associates.

Goodwin, B.C. (1976) *Analytical Physiology of Cells and Developing Organisms.* New York: Academic Press.

Goodwin, B.C. and Kauffman, S.A. (1992) Deletions and mirror symmetries in *Drosophila* segmentation mutants reveal generic properties of epigenetic mappings. In J.E. Mittenthal and A.B. Baskin, eds., *Principles of Organization in Organisms.* Santa Fe Institute Studies in the Sciences of Complexity, Volume XIII. Reading, MA: Addison-Wesley, pp. 269–296.

Goodwin, B.C. and Trainor, L.E.H. (1980) A field description of the cleavage process in embryogenesis. *Journal of Theoretical Biology* **85**, 757–761.

Harrison, L.G. (1987) What is the status of reaction-diffusion theory thirty-four years after Turing? *Journal of Theoretical Biology* **130**, 493–499.

Hunding, A. (1983) Bifuration of non-linear reaction-diffusion systems in prolate spheroids. *Journal of Mathematical Biology* **17**, 223–239.

Kauffman, S.A. (1993) *The Origins of Order: Self-Organization and Selection in Evolution.* New York: Oxford University Press.

Kauffman, S.A. and Ling, E. (1981) Regeneration by complementary wing disc fragments of *Drosophila melanogaster*. *Developmental Biology* **82**, 238–242.

Kauffman, S.A., Shymko, R.M. and Trabert, K. (1978) Control of sequential compartment formation in *Drosophila*. *Science* **199**, 259–270.

McNally, J.G. and Cox, E.C. (1989) Spots and stripes: the patterning spectrum in the cellular slime mold *Polyspondelium*. *Development* **105**, 323–333.

Meinhardt, H. (1982) *Models of Biological Pattern Formation*. New York: Academic Press.

Morse, P.M. and Feshbach, H. (1961) *Methods of Theoretical Physics*. New York: McGraw Hill.

Murray, J.D. (1977) *Lectures on Nonlinear-Differential-Equation Models in Biology*. Oxford: Clarendon Press.

Murray, J.D. (1981) A prepattern mechanism for animal coat markings. *Journal of Theoretical Biology* **88**, 11–19.

Murray, J.D. (1989) *Mathematical Biology*. New York Springer-Verlag.

Nijhout, H.F. (1992) Pattern formation in biological systems. In L. Nadel and D.L. Stein, eds., *1991 Lectures in Complex Systems*. Santa Fe Institute Studies in the Sciences of Complexity. Lectures, Volume IV. Reading, MA: Addison-Wesley, pp. 159–187.

Singer, S.J. (1992) Intercellular communication and cell-cell adhesion. *Science* **225**, 171–177.

Steinberg, M.S. (1963) Reconstruction of tissue by dissociated cells. *Science* **141**, 401–408.

Trainor, L.E.H. and Goodwin, B.C. (1986) Stability analysis on a set of calcium-regulated visco-elastic equations. *Physica D* **21**, 137–141.

Turing, A.M. (1952) The chemical basis of morphogensis. *Philosophical Transactions of the Royal Society of London, Series B* **237**, 37–72.

Appendix. The Cylindrical Flat Worm

In thinking about the question of times $t_0, t_1, \ldots$, of gene activation or switching it may be useful to consider the simplest of situations, the situation of an imaginary "flat worm" in the form of a cylinder closed at both ends. We will impose the following boundary conditions, but the essential points of the discussion will not depend on the particular conditions. Let the "animal" consist of a closed right cylinder with the top closure (taken here as a circle of perimeter $2\pi a$) held at the morphogen value c and the bottom closure of the same form held at the constant value $-c$. Solving equation (7) using a constant metric function g, where $g = a^2$ on the cylinder, we choose the dimensionless coordinates u, v such that u runs the length of the cylinder, from $u_0 = b/2a$ to $-b/2a$, and $0 < v < 2\pi$ is the angular coordinate perpendicular to u. Thus the integral over sides plus ends gives the total area, and

$$\int g\, du dv = A = 2\pi(ab + a^2) \tag{A.1}$$

which equals the area of the (cylinder + circular ends). Here a is the cylinder radius, and b is the length of the cylinder.

The solutions to equation (7) have the form

$$\Psi(u, v) = c\ \sin[(4n+1)\pi au/b]\cos(mv), \quad n, m = 0, 1, 2, 3, \ldots \tag{A.2}$$

and the condition which must be met by the "eigenvalue" $k^2 g$ ($= (ka)^2$ for the section of cylinder) is

$$k^2 g = k^2 a^2 = (4n+1)^2(\pi a/b)^2 + m^2 . \tag{A.3}$$

This may be written (to within a multiplicative constant) as

$$k^2 A = r(r+1)(4n+1)^2 + (1 + 1/r)(m/\pi)^2 , \tag{A.4}$$

where A = total area and $r = a/b$. Thus the eigenvalue condition relates the total area A to the "shape" parameter $r = a/b$. Figure 2 shows the total area A as a function of "shape" r, for various values of the integers n, m.

The first important point to notice from Figure 2 is that *continuous* increase in total area forces a *change in shape* r, for fixed n, m. As long as the shape stays the same, then the area A can only change in "jumps," as the integers n or m change; this is the accustomed way of thinking of eigenvalue problems: In the case of the sphere of radius R, there exist solutions *only* if the sphere has the "correct" radius R; in our cylinder example, if the ratio

$r = a/b$ is fixed, then area can only increase in "quantized" jumps. However as is seen from our simplest example, the area increase causes a continuous change in shape r.

The second point to be made from examination of Figure 2 is that a given trajectory along a line of allowed solutions in the (A, r)-plane requires that a *binary* decision be made at numerous points, and *times*, along the trajectory as $A(t)$ increases. One such possible trajectory has been indicated in Figure 2. At these "decision" points along the trajectory, there can be a change in either integer n or m, and it is such a decision point that we are suggesting as a possible determinant of the times $t_0, t_1, \ldots$ above. The progression through the integers n, m serves as a convenient "counting" mechanism, as spatial compartments and times are marked off in succession.

These binary decisions are expected to be determined via the (unspecified) nonlinear perturbative terms of equation (1) and the resulting terms in equation (7). Energy considerations can be expected to decide in favor of one of the two paths at each time; the trajectories are not distinguished energetically without the nonlinear terms. Such nonlinearities, neglected so far in our discussion, are suddenly seen as crucial elements in the whole scheme. (These terms then add important further counsel beyond Yogi Berra's advice of "if you come to a fork in the road, take it.")

This cylinder "flat worm" example is of course only suggestive of possibilities, and these suggestions are highly speculative. More work is needed simply to attempt to generalize to shapes other than cylinders and cones ($K = 0$) as the "genes" in various areas dictate new shapes via the Gauss function K in the more general case where the K (curvature) is not essentially fixed for all time. And more generally, instead of trajectory A versus r we must think of the system point as wandering in a higher dimensional "morphospace", when $A = A(r_1, r_2, \ldots)$. It would also be essential to try to understand if switching at such "decision" points has a possible or even plausible molecular mechanism.

A brief discussion of an actual possible molecular manifestation of the R and B molecules of the second section is given in section 4.

Living Organization, the Coherence of Organisms and the Morphogenetic Field

Mae-Wan Ho, Yu-Ming Zhou and Julian Haffegee

1 The Problem of Living Organization

The problem of living organization may be stated as follows: How is it that an organism consisting of a multiplicity of tissues and cells and astronomical numbers of molecules of many different kinds can develop and function as a whole? One idea is that it is *coherent*, i.e., it has collective modes of activity that are coupled together and extend over all space-time domains (Ho 1993, Ho 1994, Ho 1995a). This not only accounts for the long-range dynamical order that is so immediately characteristic of living organisms, but also results in a unique energy relationship within the system. In a coherent system, energy from any local source is readily shared throughout the system, and conversely, it can be mobilized to any point from all over the system (Ho 1995a, b; 1996a). That is why the organism can have energy at will, whenever and wherever required, and in a perfectly coordinated way.

The coherence of organisms is a dynamic, nonequilibrium regime maintained by energy flow. To first approximation, it can be described classically as a dissipative structure (Prigogine 1967). A number of us have argued, however, that *quantum* coherence is involved in living organization, and hence a classical description is inadequate. We shall begin by presenting evidence for coherence from a contemporary biochemical perspective, then go on to consider the theoretical arguments and empirical observations in support of quantum coherence in organisms and its implications for morphogenesis.

2 The Coherence of Energy Transfer in Metabolism

2.1 Coupled Energy Transfer

Biochemists have known for a long time that all metabolic reactions are connected in a complex net where the energy-yielding reactions are invariably coupled to the energy-requiring ones, mainly, though not exclusively, through the interconversion of ATP and ADP. The most basic energy transducing reactions — those involved in oxidative and photosynthetic phosphorylation to make ATP — are so well coupled that they are fully reversible and capable of working at close to 100% efficiency. This comes about, in part, through the "solid-state"-like organization of energy-transducing membranes, which provide an organized medium of low dielectric constant, enabling electrons and protons to flow along or across the membrane from one protein complex to another without equilibrating with the bulk phase (Kell and Westerhoff 1985, Ho 1995c).

2.2 Enzymes — Flexi-Molecular Energy Machines

The most important role in energy transduction in living organisms is played by enzymes. Every student in biochemistry knows that enzymes work as catalysts by lowering the activation energy barrier between substrate and product. What is seldom recognized is that metabolic reactions are but the means whereby energy is transformed within living systems as they engage in all the activities that constitute being alive. And the energy is transformed *directly and in discrete quanta* by flexible enzymes working as *molecular energy machines* (Ho 1995d). In other words, the energy is never released or thermalized before it is used, as it would be in a heat engine. Instead, chemical energy in individual molecules bound to individual enzymes is turned into different kinds of work directly by the enzyme. The total amount of work done is just the sum over the number of quanta of energy transduced by an equal number of molecular energy machines. Stored energy is therefore able to go directly from the point at which it is released to where it is utilized with minimum loss or dissipation. The secret of enzyme action is the ability to store the energy transiently in conformational fluctuations involving highly correlated movements of the whole macromolecule, together with the substrate, which enables the chemical reaction to go through its transition state "with the greatest of ease" (Ho 1995d). This has been described as "free-energy complementarity" by physical chemist/enzymologist Rufus Lumry, who contributed a great deal to revolutionize the way we see enzyme action today (Lumry 1994).

2.3 The Solid-State Cell

Another emerging picture which is of great importance to energy transfer is the exquisitely detailed structure of the cell. Many people are already familiar with the typical structure of the cell with its many organelles, internal membranes and cytoskeletal elements. Recently, it is becoming increasingly evident that the so-called "cytosol" is also highly organized, consisting of a "microtrabecular lattice" of actin-like filaments filling up the entire cell and serving as binding sites for most, if not all "cytosolic" enzymes and a large proportion of the metabolites (Clegg 1984). These metabolites are "channeled" or passed directly from one enzyme to the next without equilibrating with the bulk phase.

The cell takes on the characteristics of a special "solid-state" where up to 60% of the cell water may also be structured by the enormous surface area available (Ho 1995c).

The metabolic net, therefore, operates within an intricate, dynamically organized space-time structure that is compartmented and microcompartmented down to the level of single enzymes, so that energy coupling effectively spans the entire range of space-time domains from the macroscopic down to the microscopic and molecular. The organism is thick with activities on every scale (Ho 1994). Energy is ever cascading in both directions from the extremely fast and local through a series of correlated activities in a nested hierarchy of space-time domains to the slow and global. The result is a system kinetically and thermodynamically optimized for energy storage and mobilization, and also maximally communicative and responsive (Ho 1995c).

2.4 Coherence and Metabolic Regulation

Much of the biochemical literature is devoted to metabolic regulation in terms of rate-limiting enzymes, concentrations of key metabolites, and so on. One possibility which has not been considered is the coherence or transparency of the system itself to energy mobilization throughout the range of space-time domains. This will depend on collective activities at every level being correlated simultaneously between levels. In a coherent system, each part of the whole is as much "in control" of all other parts as it is responsive. Such distributed control is already explicitly recognized in the idea of an interconnected metabolic net in *metabolic control analysis* pioneered by geneticist/biochemist Henry Kacser (1987), which analyzes the sensitivity of steady state fluxes and concentrations of metabolites in the system to small perturbations. The results show that "control" is invariably distributed throughout the system, the distribution of control itself changing under different conditions. Thus, global and local are always mutually entangled (Ho 1995c).

Another relevant concept in biochemistry is that the cell or organism is regulated by "energy-charge," reflected in the relative ratios of ATP/ADP/AMP (Atkinson 1977). In a pumped metabolic system, it is to be expected that the degree of coherence will depend, among other things, on the energetic content, or the amount of energy stored and mobilizable, of which energy-charge would be an estimate. As distinct from a system at thermodynamic equilibrium, the temperature of a living system does not uniquely define its energy content. Other factors affecting the degree of coherence may be the extent of coupling between metabolic reactions so that the dissipation of energy is minimized.

Thus, the energy coupling that gives rise to coherence is the direct consequence of the metabolic network operating within the living system, which is dynamically organized down to the detail of individual molecular energy machines. Contemporary developments in biochemistry within the past thirty years lead to a very different view of regulation and the physiology of organisms, and hence of health and well-being. The major shift is from mechanistic reductionism to a holistic, organic perspective which loses none of the analytic precision, and which is only just beginning to be explored (Ho 1993, Ho 1995a, 1996b). Let us now turn to the more theoretical considerations of how coherence may arise.

3 Coherent Excitations and Nonequilibrium Phase Transitions

The quantum physicist Schrödinger (1944) was among the first to point to the difficulty in envisaging how all the molecules of different kinds in even a tiny bacterium could work together in such an orderly, purposeful way. He suggested that the organism must be functioning in collective modes analogous to those responsible for superfluidity in helium or superconductivity in certain metals at temperatures close to absolute zero. That analogy is misleading for the simple reason that the low temperature phase-transition phenomena of superfluidity and superconductivity depend on the minimization of thermal energy in a system at thermodynamic equilibrium. The living organism, in contrast, is an open system maintained far from thermodynamic equilibrium by energy flow.

A more appropriate analogy for how dynamic, long-range order can arise far from thermodynamic equilibrium is the laser. Energy is pumped into a reflecting cavity in which atomic light-emitters are embedded. At low levels of pumping, the atoms emit randomly with respect to one another, thus acting as an ordinary lamp. As the pumping is increased, and more and more energy is reflected or stored within the cavity, the emission of the atoms will be coupled to the radiation field in the cavity, becoming synchronized with it and hence with one another. Thus, at a certain critical power intensity — the *laser threshold* — all the atoms will oscillate in phase and emit together, sending out a giant wave train that is a million times as long as that emitted by individual atoms. In effect, a phase transition has taken place from a regime where individual molecular motions are random with respect to one another, to a regime where all the molecules are moving coherently, hence the dramatic appearance of long-range macroscopic order. In contrast to low temperature superconductivity — an equilibrium phase transition which is unstable to the input of random thermal energy — the laser is a nonequilibrium phase transition *dependent* on the input of energy, and hence stable to perturbation for it can channel random energy into collective modes.

3.1 Coherent Excitations and Biological Rhythms

Beginning in the late 1960s, solid state physicist Herbert Fröhlich developed the first detailed theory of how collective modes of activity, or long-range dynamic order can arise in living systems (Fröhlich 1968, 1980). He pointed out that as all biological molecules are strongly dielectric and are packed rather densely together in the cell, energy from metabolism is not totally dissipated as heat, but could be stored in the form of coupled electric and mechanical vibrations. Under the most general conditions of energy pumping, those vibrations will go into coherent collective modes, which may be a stable or metastable strongly electrically polarized field, or oscillations at different frequencies depending on the molecules or supramolecular structures involved. Fröhlich's ideas have been taken up and extended by others (Duffield 1988, Wu 1994). Duffield (1988), in particular, shows that coherent excitations are an *asymptotically stable global attractor*. In other words, all open systems capable of energy storage will evolve towards coherent excitations when energy is pumped in, and once established, the excitations will be stably maintained.

This important theoretical result says that one does not need special arrangements for coherent excitations to happen. Living systems, and indeed any similar system with the ability to trap and store energy, will spontaneously evolve towards coherent, dynamic order consisting of collective phonon modes or lattice vibrations (electrodynamical waves in solid medium). In the living organism, these collective modes may extend throughout a whole macromolecule as conformational vibrations or an aggregate of macromolecules. They may even involve entire population of cells, such as muscle fibres or neurons in the brain.

The conditions within living systems may be especially favorable for coherent excitations, as the cell is recognized to approach the solid state capable of transmitting and storing phonons, supporting dipole oscillations or other forms of excitations. Collective modes of activity or oscillations are well-known in physiology, and are to be found in the functioning of every organ and tissue. The cellular and tissue oscillations range from split seconds to minutes, and continue through to familiar biological rhythms such as the circadian and circannual. Biological activity is *cyclical* at every level (Ho 1993, Ho 1995a). Many biologists have attempted to identify the "master" clock that sets biological time, or controls particular rhythms. Putative genes have been identified that are supposed to be responsible for how fast or slow the biological clock runs (Goodwin 1995). It may well turn out that the rhythms and oscillations are the range of collective excitations that a coherent, space-time structured system will naturally settle into (Ho 1993), without the need for any particular controlling master-clock or pace-setter gene. The coupled oscillators comprising the whole system are mutually entrained, and mutually communicating via the energy delocalization mechanisms mentioned above (Ho 1995a).

3.2 The Fröhlich State and Sensitivity to Weak Electromagnetic Fields

Beginning in the 1970s in the west (but considerably earlier in the former Soviet Union), an increasing number of reports appeared that suggested living organisms are extremely sensitive to weak electromagnetic fields (Ho, Popp and Warnke 1994). Fröhlich suggested that coherent excitations could be the basis of electromagnetic sensitivity (Fröhlich 1980). As mentioned above, the Fröhlich state is stable to perturbation. So how can it be the basis of sensitivity to weak external electromagnetic fields? Fröhlich's discussion on this point was not very clear. One possibility is "resonance effects" at specific frequencies of the external field. If the frequency is sufficiently close to a particular endogenous coherent excitation regime, it could precipitate the particular phase transition, or it could entrain the regime away from its endogenous frequency. At any rate, the sensitivity may be associated with the instability at phase transition itself, and not with coherent excitation per se. This is consistent with the large body of work in the laboratory of Liburdy, which shows that sensitivity to static magnetic fields as well as microwaves is associated with membrane phase transition temperatures (Liburdy 1994; for more on membrane phase transitions see the chapter by Jones and Tevlin, this volume). It would also explain why biological effects of weak electromagnetic fields are so notoriously difficult to reproduce or pin down. The biological system under investigation must be uniform, and in the identical physiological state at the time of investigation. However, we neither have a theory of the physiological state, nor anything like a theory of the organism. The so-called physiological "state" is actually a vast conglomerate of dynamic processes with an intricate space-time structure (Ho 1993) where innumerable "phases" may coexist, and transitions between regimes may be occurring or not, at any instant.

4 Coherent Photons in Living Organisms

Evidence for coherence in organisms also comes from "biophoton" research carried out over the past 20 years by Popp and his coworkers (Popp, Li and Gu 1992), who found that all organisms emit light spontaneously at very low intensities, which can be detected with a sensitive photon detector. All organisms, tissues or cells can also be stimulated to re-emit at higher intensities by brief exposures to light. The composition of the emitted light covers the entire range of sensitivity of the photon-detector — 200 to 900 nanometers (nm) — and most probably beyond, with approximately equal number of photons over all wavelengths. Popp has referred to that as the "$f(\lambda)$ = constant distribution" (Popp 1986), which is very far away from thermodynamic equilibrium. That means the living system has too much energy, as well as an indeterminate amount of energy compared to a system at thermodynamic equilibrium at the same temperature — typically around 300K.

The same broad band distribution in the emitted light is obtained regardless of whether the system is stimulated with white light, or monochromatic light, suggesting that energy input into any frequency is immediately shared over all frequencies. Indeed, all frequencies seem effectively coupled or linked together so that they may go up and down in concert over a thousand-fold variation in total light emission intensity when the system is perturbed. Furthermore, all parts of the spectrum have identical hyperbolic decay kinetics. It has the general form

$$I = A(t + t_o)^{-1/\delta} \tag{1}$$

where I is the light intensity, A, t_o and δ are constants, and t is the time after light exposure. According to Popp and Li (1992), hyperbolic decay indicates that the light is emitted from a coherent source.

These findings, taken together, suggest that *the organism is a coherent photon store, with equal number of photons distributed over all frequencies which are nonetheless coupled together to a single effective degree of freedom.* That is in line with the biochemical picture and the theoretical considerations of coherent excitations that we have discussed above.

Does the hyperbolic decay kinetics constitute a definitive proof of coherence? To satisfy the proof that hyperbolic decay is *sufficient* for a coherent system, an additional assumption required is that the system is *ergodic*, i.e., it is stationary and probably homogeneous as well, so that a spatial average and a time average of the system will coincide. But as we know, very few real systems of interest, and least of all, living systems, satisfy the condition of ergodicity. This highlights our present ignorance as to the nature of the coherence of organisms. In particular, one would like to know if there are *different* kinds of coherence which characterize an organism as it develops from the initially relatively featureless egg to the highly differentiated and shapely adult.

5 The Coherence of Organisms

The first observations one can make with regard to the developing organism is that there appears to be two approximately "stationary regimes": the oocyte (unfertilized, mature egg), which is relatively undifferentiated in structure and function, and the fully developed, adult organism in quasi-steady state, which is highly differentiated in both structure and function. Organizationally, the oocyte is in a dormant state with perhaps a very simple space-time structure compared to that of the adult, which is very dense and complex. Fertilization triggers the sequence of dramatic events in morphogenesis that transforms the one to the other, beginning with a global pattern determination process that roughs out the body plan of the adult organism. It has been suggested that a *global* dynamic coherence — in the form of the *morphogenetic* field — is instrumental in generating the first patterning during development (Ho 1989). But what distinguishes the coherence of the morphogenetic field from that of the adult organism?

One might be tempted to think of development as a progressive de-coherence and devolution of the globally coherent embryo to locally autonomous regimes. It may be more accurate, however, to regard it as a process of progressive complexification and generation of coherent space-time structure. Differentiated spatial structures, which arise from dynamic patterning processes, in turn circumscribe and define further dynamic time structures. So, it is not so much a loss of coherence as that the coherence becomes increasingly multimodal, expanding in space-time domains in both microscopic and macroscopic directions. The global dynamic coherence at the beginning of morphogenesis is most easily recognized as such on account of its simple space-time structure. The first events — cleavages into blastomeres or multiplication of the nuclei in organisms that go through a syncytial blastoderm stage — are invariably synchronous over the entire embryo, which therefore constitutes a single phase-transition domain. As complexification proceeds with increasing space-time differentiation, the embryo appears more and more asynchronous and heterogeneous. Nevertheless, the differentiated domains remain coupled together as a coherent whole.

While it is intuitively obvious, at least to some of us, that organisms are coherent wholes, it is still far from clear as to what that implies for the organism, nor what its observational consequences are. So, how does one go about investigating the coherence of organisms? A prerequisite is to use techniques and methodologies appropriate to studying living organisms as they are living and developing. Ideally, such techniques should be noninvasive, or minimally invasive, and yet possess sufficient sensitivity to allow the living organism to inform the experimenter of its internal state. But that inevitably makes great demands on the investigator to interpret and to make sense of the observations.

6 The Coherence of the Morphogenetic Field

One of the earliest generalizations to emerge from the study of embryology is the notion of the morphogenetic field which gives a dynamic global order to the embryo at the very outset (Driesch 1891). Driesch's concept of the morphogenetic field was originally inspired by analogy to a magnetic field. Years later, Burr and Northup (1935) detected electric fields around developing and adult organisms from which they postulate the electrodynamical field as the basis of biological organization. Becker (1962) confirmed their findings, demonstrating the existence of a direct current (DC) field on the surface of intact organisms which changes dramatically as the result of both anaesthesia and injury. He also identified DC currents that are specifically associated with limb regeneration in salamanders. Using the more sophisticated vibrating probe technique, Jaffe (1982) was able to detect transembryonic ionic currents in all developing systems examined. These currents typically appear before any overt sign of development, suggesting that they may play a role in organizing the major body axes and patterns in the embryo (Nucitelli 1988).

Thus, the available evidence indicates, at least, that electrodynamical processes are active during the period when the basic body plan is laid down in early embryogenesis. We began a series of studies, using increasingly noninvasive techniques to find out if the "morphogenetic field" responsible for determining body pattern is indeed a coherent electrodynamical field (Ho 1989), and if so, how it might be involved in generating body pattern.

7 Weak Magnetic Fields Perturb Body Pattern

Very young fruit fly embryos, briefly exposed to weak static magnetic fields at a time when pattern-determining processes are known to be active, develop a special kind of global body-pattern transformation in the larvae emerging 24 hours later. These have never been described among the tens of thousands of genetic mutants that have been isolated, nor among the numerous abnormalities resulting from other perturbations (Ho et al. 1992a, Ho et al. 1994). They involve a transformation of the normal segmental body plan into helical configurations spanning a variable number of consecutive segments up to the entire body. That kind of pattern would be expected if the external magnetic field were to interact with molecules that are sensitive to magnetic orientation and which are being phase-ordered over the whole embryo. Alternatively, the external magnetic field might be interacting with moving charges correlated over the entire embryo, that are somehow involved in determining body pattern.

However, the energy of the external field is several orders of magnitude below thermal threshold, so that it could have no effect if individual molecules are moving randomly with respect to one another. On the other hand, if a large number of molecules — of the order of Avogadro's number — are moving coherently, or undergoing phase-ordering simultaneously, then the effect of the external field would be summed over all of the molecules to give the global transformations that were observed. The effect is most likely operating during some nonequilibrium phase transition when, as mentioned above, the sensitivity to external electric or magnetic fields is greatly increased. Thus, the sensitivity of pattern formation to weak magnetic fields suggests that coherent electrodynamical processes are involved in determining body pattern.

8 Cooperative Emission of Coherent Light

Further evidence for coherence comes from the remarkable phenomenon of superdelayed luminescence exhibited by these same embryos (Figure 1). This occurs in synchronously developing populations in which intense, often prolonged and multiple flashes of light are re-emitted with delay times of minutes to hours after a single brief light exposure (Ho, Xu, Ross and Saunders 1992b; Ho and Popp 1993). The phenomenon occurs only in synchronous populations that are stimulated with light during the first 40 minutes of development. The superdelayed flashes themselves do not obviously correlate with specific embryonic events; they give information on the physical state of the embryos at the time of light stimulation rather than at the time when the flashes are taking place.

Superdelayed re-emission takes several forms. It may appear as a single sharp burst of light (< 20 seconds in duration) or a succession of sharp bursts separated by long periods of background emission. It may also come in a single prolonged flash or a series of them, each sustained for tens of minutes. For the multiple prolonged flashes, the total luminescence produced — in excess of the background emission — shows significant first- and second-order dependence on the number of embryos present. All these observations suggest that superdelayed luminescence results from cooperative interactions among embryos within the entire population, so that most, if not all, of the embryos are emitting light simultaneously.

What is the source of the light emitted? Light is generally emitted from an excited atom or molecule when an electron in the outermost shell, having been

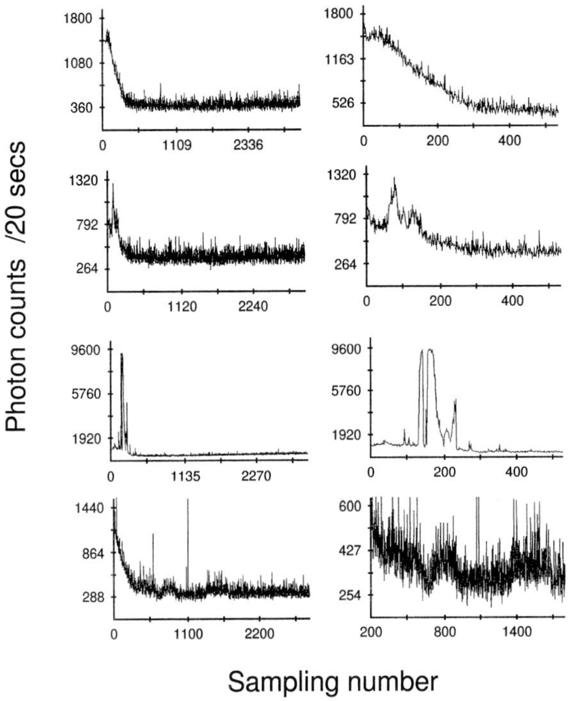

Figure 1: Superdelayed luminescence in *Drosophila*. Continuous recordings of aggregated photon count per 20 seconds from synchronously developing embryos. Top trace, controls, not exposed to light. The other traces are of embryos previously exposed to white light for 1 minute and exhibiting different forms of superdelayed luminescence. Traces on the right are expanded versions of those on the left.

promoted to an excited energy level by, say, a collision with another molecule or absorption of energy by other means, falls back into a lower energy level. Light emission does not always occur, however. The excited electron can often start to move, thus becoming an electric current, or it can be involved in a chemical reaction. The electron can also relax back to the ground state nonradiatively, that is, instead of emitting photons, it can give off energy as phonons, or as heat.

In a solid-state system, which the living cell approximates, outer electrons are neither localized to single atoms or molecules, nor do they have single energy levels. Instead, continuous and delocalized bands of frequencies are separated by band gaps. The low energy "valency" band, filled with electrons, is separated by a band gap from the "conduction" band containing mobile electrons, to which electrons can be promoted by absorbing energy such as heat or light. If the excited electron is not conducted away, it may relax back to the ground state (to recombine with the positively charged "hole" left behind) by radiating a photon, or nonradiatively as described above. Another source of photons in a solid state system is from "excitons," i.e., an excited electron-hole pair which can propagate over long distances within the system before giving up the energy by emitting a photon.

As we know, charge separation underlies the primary bioenergetic transduction processes associated with the mitochondria, and the formation of excitons and their propagation is believed to be widely involved in energy transduction and in biocommunication. In addition, Rattemeyer and Popp (1981) suggest that the DNA molecule is an excited duplex, or *exciplex*, in which photons are stored between the two DNA strands, and hence can be a source of emitted biophotons. So it is not surprising that living systems could emit light from processes taking place simultaneously all over the cell. And in the same way that the solid-state nature of the cell delocalizes electrons, the photons must also be delocalized. This is in accord with our observation that when single embryos were successfully imaged in the dark with a sensitive ccd camera and video-recorder after brief light exposure, photons were re-emitted from all over the embryo, although some areas may be emitting more strongly than others at different times (Ho et al. 1992b, Ho 1993). In effect, the organism stores its energy as a coherent photon (and phonon) field bound to its living matter (Ho 1993), and when stimulated, the stored energy is emitted from the coherent field as a whole.

We have no *direct* evidence that the light emitted is coherent. Nevertheless, the following considerations suggest that it might be so. The phenomenon of superdelayed luminescence bears a formal resemblance to superradiance (Gross and Haroche 1982) — the collective, phase-correlated emission of coherent light in a population of identical emitters. If this analogy is legitimate, it implies, first of all, that the embryos are identical; this is not too unreasonable, as they are actually synchronous to within one minute of developmental time. Second, and more importantly, each embryo must be seen as a coherent emitter with a definite phase, even though the light emitted from individual embryos, as indeed from all organisms, is in a broad band spread out evenly throughout the range of sensitivity of the photon detector (200–900 nm) and possibly beyond. In other words, there is a collective phase of many coupled modes, *which is only possible if the light from individual embryos is coherent.*

That one can regard each embryo as having a collective phase of many coupled modes may seem very strange at first. Actually, it is no stranger than the phenomenon which is familiar to everyone: that our heart — an extremely complicated organ — should beat so regularly as a whole, day-in day-out, as long as we are alive and well. The heart, too, has a collective frequency and phase because all the cells, and all the astronomical numbers of individual molecular motors within each cell, are working coherently.

Superdelayed luminescence can be modeled (Zhou, Ho, Popp and Gu 1994; Zhou et al. 1996) essentially as a critical phenomenon involving nonequilibrium phase transition in a population of nonlinear oscillators, that evolves from an initially random, nonsynchronous regime to a phase-locked, self-synchronized state (Strogatz and Mirollo 1988). Our procedure is comparable to that employed by Collins and Stewart (1992) who model quadrupedal locomotion in terms of four distinct but symmetrically coupled nonlinear oscillators — one for each limb connected to a central pattern generator, a putative network of neurons capable of producing oscillatory behavior. A large body of fascinating experimental and theoretical work on "neurobehavioral dynamical systems" (Kelso 1991) similarly is based on the assumption that the rhythmic activities of whole organs or functional systems — lungs, heart, limb, respiratory system, locomotory system and so on — are, each of them, a coherent, nonlinear oscillator which can be phase-locked one to another.

9 Global Birefringent Patterns Suggestive of Coherent Phase Ordering of Liquid Crystals

Finally, we show that coherence may be directly observable. We have alluded to the solid-state like nature of the cell or cytoplasm; a more accurate description is liquid-crystalline. Biological molecules, especially macromolecules, are not only giant dipoles, but also liquid-crystalline at physiological temperatures. And if there is anything like a coherent electrodynamical field in living organisms, would that not be reflected in the phase coherence of these liquid crystals? That prompted the recent discovery of a novel noninvasive imaging technique (Ho and Lawrence 1993, Ho and Saunders 1994) which enables us to see the internal tissues of living, moving organisms in specific interference colors, rather in the way that one might observe the colors of different rock crystals. The important distinction is that unlike the crystalline order in rocks, living crystalline order is *dynamically* generated and maintained. The intensity and hue of the colours are therefore dependent on both the structure or *birefringence* of the molecules *and* the degree of phase coherence existing among the molecules making up the tissues. Furthermore, whereas ordinary, nonliving liquid crystals are subject to equilibrium phase transitions dependent on temperature or concentration, living crystals are phase ordered by

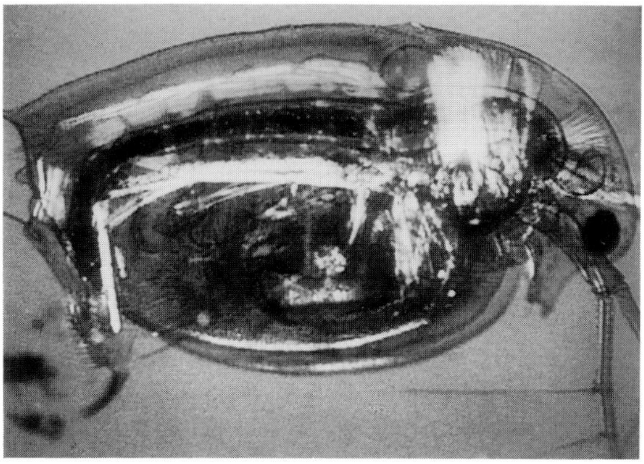

Figure 2: *Daphnia* under Interference Color Vital Imaging.

energy-transforming activities which they themselves are carrying out. Indeed, the colors change with the state of activity of the organism, becoming more intense as the level of activity increases, suggestive of a greater degree of coherent order as random thermal motion is suppressed. Conversely, when the organism dies, or when it is chilled, dehydrated, or otherwise unwell, the colors will fade accordingly as random thermal motion begins to dominate. Thus, the technique is unique in that it not only gives anatomical details noninvasively but, in addition, tells us whether the molecules of the tissues are in a coherent, energized state.

A remarkable feature which came to our attention is that for all animals examined, unicellular and multicellular species without exception, the anterior-posterior axis is the major polarizing axis, so that when it is aligned, all the tissues are maximally colorful, precisely as one would expect if the whole organism is one oriented, coherent field (Figure 2). With recent improvements in the technique, we can now see dynamic, birefringent, liquid crystalline patterns in early embryos long before cells or differentiated structures are formed (Ho et al. 1996) (Figure 3). These patterns have never been seen before, and are revealing to us the otherwise cryptic activities of the morphogenetic field, perhaps in the manner of a liquid crystal display reflecting underlying changes in electrical polarization patterns. We are currently mapping the evolution of these patterns in early development to see if they reveal something of the otherwise cryptic electrodynamical processes of the morphogenetic field.

Our tentative hypothesis is that electrodynamical processes give rise to polarization waves in the early embryo, probably following a successive bifur-

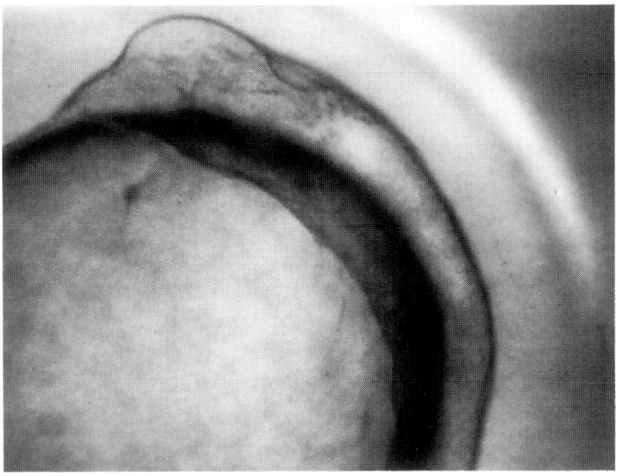

Figure 3: Birefringent, periodic patterns (light bands) on the undifferentiated head region of the day-old zebra-fish embryo, visualized with Interference Color Vital Imaging. (Obtained in collaboration with Philippe Herbomel, Pasteur Institute, Paris during his recent visit to the Bioelectrodynamics Laboratory.)

cation sequence (Ho 1990). These waves may arise out of the intrinsic dynamics of the coupled electric and mechanical interactions of membrane lipid and proteins much as Fröhlich has suggested. Or they may involve polymerization/ depolymerization reactions of cytoskeletal proteins which have been studied extensively in vitro (Tablony and Job 1992). These would then result in phase-ordered patterns of constituents of the cell membrane and/or peripheral cytoskeletal elements, and in turn, lead to a cascade of additional biochemical effects such as opening voltage-gated ion (calcium) channels and stimulating the transcription of special genes, ultimately resulting in overt pattern formation. According to our scheme, the electrodynamical processes precede specific gene expression, and may be responsible for establishing the patterns of gene expression themselves. This is suggested by previous observations that transformation of body pattern by physical agents, such as ether vapor (Ho et al. 1987), tend to be successfully induced before the corresponding pattern of gene expression appears in the embryo. Similarly, the characteristic transformation of the segmental body pattern into a helical configuration by weak static magnetic fields — which have yet to be found as genetic mutants (see above) — can be induced very early in development, before the "morphogenetic genes" are expressed. The problem of how pattern emerges in development must begin with the dynamics of the physico-chemical mechanisms generic to all living systems, that provide the effective forces and flows required for *patterning.*

That problem is simply not addressed in the standard explanation in terms of "morphogenetic genes" that turn one another on and off.

10 Conclusion

We have presented some theoretical arguments and empirical observations in support of the idea that coherence is the basis of living organization, and that coherent electrodynamical activities may be involved in laying down the basic body plan of all organisms. It is clear that we have only scratched the surface of what the coherence of organisms entails, or how it might organize the body plan. But we hope there is enough to stimulate the interest of biologists, physicists and mathematicians alike whose concerted effort is required for our further understanding of the coherence of organisms. For that is the most captivating problem of the coming century.

11 Epilogue

Lynn Trainor is a pioneer in applying physical and mathematical ideas to biology, and in attracting many brilliant young physicists to work with him. We were not fortunate enough to be Lynn's students, but nevertheless, owe a great deal to Lynn, for inspiring us, for encouraging M.W. Ho's excursion into physics, and Y.M. Zhou's excursion to Milton Keynes, and most of all for being always a kind and gentle friend.

References

Atkinson, D.E. (1977) *Cellular Energy Metabolism and Its Regulation.* New York: Academic Press.

Becker, R.O. (1962) The direct current control system. A link between environment and organism. *New York State Journal of Medicine* **62**, 1169–1176.

Burr, H.S. and Northrup, F.S.C. (1935) The electro-dynamic theory of life. *Quarterly Review of Biology* **10**, 322–333.

Clegg, J. S. (1984) Properties and metabolism of the aqueous cytoplasm and its boundaries. *American Journal of Physiology* **246**, R133–151.

Collins, J.J. and Stewart, I.N. (1992) Symmetry-breaking bifurcation: a possible mechanism for 2:1 frequency-locking in animal locomotion. *Journal of Mathematical Biology* **30**, 827–838.

Driesch, H. (1891) Entwicklungsmechanische Studien I-II. *Zeitschrift für wissenschaftliche Zoologie* **53**, 160–182.

Duffield, N.G. (1988) Global stability of condensation in the continuum limit for Fröhlich's pumped phonon system. *Journal of Physics A: Mathematical and General* **21**, 625–641.

Fröhlich, H. (1968) Long range coherence and energy storage in biological systems. *International Journal of Quantum Chemistry* **2**, 641–649.

Fröhlich, H. (1980) The biological effects of microwaves and related questions. *Advances in Electronics and Electron Physics* **53**, 85–152.

Goodwin, B.C. (1995) Biological rhythms and biocommunication. In S. Rose, ed., *Biocommunication, S327 Living Processes, Book 3*. An Open University Course. Milton Keynes: Open University Press.

Gross, M. and Haroche, S. (1982) Superradiance: An essay on the theory of collective spontaneous emission. *Physics Reports* **93**, 301–396.

Ho, M.W. (1989) Coherent excitations and the physical foundations of life. In B.C. Goodwin and P.T. Saunders, eds., *Theoretical Biology: Epigenetic and Evolutionary Order from Complex Systems*. Edinburgh: Edinburgh University Press, pp. 162–176.

Ho, M.W. (1990) An exercise in rational taxonomy. *Journal of Theoretical Biology* **147**, 43–57.

Ho, M.W. (1993) *The Rainbow and the Worm — The Physics of Organisms*. Singapore: World Scientific.

Ho, M.W. (1994) What is (Schrödinger's) negentropy? *Rivista di Biologia* **87**, 149–172.

Ho, M.W. (1995a) How to be a vibrant coherent whole? In M.W. Ho, ed., *Bioenergetics, S327 Living Processes, Book 2*. An Open University Course. Milton Keynes: Open University Press.

Ho, M.W. (1995b) Bioenergetics and the coherence of organisms. *Neural Network World* **5**, 733–750.

Ho, M.W. (1995c) Choreographer and dancer. In M.W. Ho, ed., *Bioenergetics, S327 Living Processes, Book 2*. An Open University Course. Milton Keynes: Open University Press.

Ho, M.W. (1995d) Enzymes — flexi-molecular energy machines. In M.W. Ho, ed., *Bioenergetics, S327 Living Processes, Book 2.* An Open University Course. Milton Keynes: Open University Press.

Ho, M.W. (1996a) Bioenergetics and biocommunication. In R. Paton, ed., *Proceedings IPCAT 95*, in press.

Ho, M.W. (1996b) The biology of free will. *Journal of Consciousness Studies* **3**, 231–244.

Ho, M.W., French, A., Haffegee, J. and Saunders, P.T. (1994) Can weak magnetic fields (or potentials) affect pattern formation? In M.W. Ho, F.A. Popp and U. Warnke, eds., *Bioelectrodynamics and Biocommunication.* Singapore: World Scientific.

Ho, M.W., Haffegee, J., Newton, R., Zhou, Y.-M., Bolton, J.P. and Ross, S. (1996) Organisms as polyphasic liquid crystals. *Bioelectrochemistry and Bioenergetics*, in press.

Ho, M.W. and Lawrence, M. (1993) Interference colour vital imaging: A novel noninvasive microscopic technique. *Microscopy and Analysis* September, 26 21.

Ho, M.W., Matheson, A., Saunders, P.T., Goodwin, B.C. and Smallcombe, A. (1987) Ether-induced disturbances in segmentation in *Drosophila melanogaster. Roux's Archives of Developmental Biology* **196**, 511–521.

Ho, M.W. and Popp, F.A. (1993) Biological organization, coherence and light emission from living organisms. In W.D. Stein, and F. Varela, eds., *Thinking About Biology.* New York: Addison-Wesley.

Ho, M.W., Popp, F.A. and Warnke, U. eds. (1994) *Bioelectrodynamics and Biocommunication.* Singapore: World Scientific.

Ho, M.W. and Saunders, P.T. (1994) Liquid crystalline mesophases in living organisms. In M.W. Ho, F.A. Popp and U. Warnke, eds., *Bioelectrodynamics and Biocommunication.* Singapore: World Scientific.

Ho, M.W., Stone, T.A., Jerman, I., Bolton, J., Bolton, H., Goodwin, B.C., Saunders, P.T. and Robertson, F. (1992a) Brief exposure to weak static magnetic fields during early embryogenesis cause cuticular pattern abnormalities in *Drosophila* larvae. *Physics in Medicine and Biology* **37**, 1171–1179.

Ho, M.W., Xu, X., Ross, S. and Saunders, P.T. (1992b) Light emission and rescattering in synchronously developing populations of early *Drosophila* embryos: evidence for coherence of the embryonic field and long range cooperativity. In F.A. Popp, K.H. Li and Q. Gu, eds., *Recent Advances in Biophoton Research and its Applications.* Singapore: World Scientific, pp. 287–306.

Jaffe, L.F. (1982) Developmental current, voltages and gradients. In S. Subtelny and P.B. Green, eds. *Developmental Order: Its Origin and Regulation.* New York: Alan R. Liss, pp. 183–215.

Kacser, H. (1987) On parts and wholes in metabolism. In G.R. Welch and J.S. Clegg, eds., *The Organization of Cell Metabolism.* New York: Plenum.

Kell, D.B. and Westerhoff, H.V. (1985) Catalytic facilitation and membrane bioenergetics. In G.R. Welch, ed., *Organized Multienzyme Systems.* Orlando, FL: Academic Press, pp. 63–139.

Kelso, J.A.S. (1991) Behavioral and neural pattern generation: The concept of neurobehavioral dynamical systems. In H.P. Koepchen and T. Huopaniemi, eds., *Cardiorespiratory and Motor Coordination.* Berlin: Springer-Verlag, pp. 224–234.

Liburdy, R. (1994) Electromagnetic fields and biomembranes. In M.W. Ho, F.A. Popp and R. Warnke, eds., *Bioelectrodynamics and Biocommunication.* Singapore: World Scientific.

Lumry, R. (1994) The new paradigm for protein research. In R.B. Gregory, ed., *Protein-Water Interactions.* New York: Marcel Dekker.

Nucitelli, R. (1988) Ionic currents in morphogenesis. *Experientia* **44**, 657–66.

Popp, F.A. (1986) On the coherence of ultraweak photoemission from living tissues. In C.W. Kilmister, ed., *Disequilibrium and Self-Organization.* Drodrecht: Reidel, pp. 207–230.

Popp, F.A. and Li, K.H. (1992) Hyperbolic relaxation as a sufficient condition of a fully coherent ergodic field. In F.A. Popp, K.H. Li and Q. Gu, eds., *Recent Advances in Biophoton Research and Its Applications.* Singapore: World Scientific, pp. 46–58.

Popp, F.A., Li, K.H. and Gu, Q. eds. (1992) *Recent Advances in Biophoton Research and Its Applications.* Singapore: World Scientific.

Prigogine, I. (1967) *Introduction to Thermodynamics of Irreversible Processes.* New York: John Wiley & Sons.

Rattemeyer, M. and Popp, F.A. (1981) Evidence of photon emission from DNA in living systems. *Naturwissenschaften* **68**, S572–573.

Schrödinger, E. (1944) *What is Life?.* Cambridge: Cambridge University Press.

Strogatz, S.H. and Mirollo, R.E. (1988) Collective synchronisation in lattices of non-linear oscillators with randomness. *Journal of Physics A: Mathematical and General* **21**, L699–L705.

Tablony, J. and Job, D. (1992) Microtubular dissipative structures in biological auto-organization and pattern formation. *Nanobiology* **1**, 131–147.

Wu, T.M. (1994) Fröhlich's theory of coherent excitations — a retrospective. In M.W. Ho, F.A. Popp and U. Warnke, eds., *Bioelectrodynamics and Biocommunication.* Singapore: World Scientific.

Zhou, Y.-M., Ho, M.W., Bolton, J.P., Milani, M., Costato, M. and Musumeci, F. (1996) Biophoton emission in synchronously developing populations of *Drosophila* embryos. *Proceedings BiOs96*, in press.

Zhou, Y.-M., Ho, M.W., Popp, F.A. and Gu, Q. (1994) An oscillator cellular automaton model of superdelayed luminescence in synchronously developing populations of early *Drosophila* embryos. In L. Beloussov and F.A. Popp, eds., *Non-Equilbrium and Coherent systems in Biophysics, Biology and Biotechnology.* Moscow, in press.

Is Spatial Pattern Formation Homologous in Unicellular and Multicellular Organisms?

Joseph Frankel

1 Introduction

One of the most important problems in modern biology is that of understanding how spatial complexity arises during development. All embryos, including human embryos, possess the genetic information needed to specify a complex organism. However, genes work within a preexisting cell, the activated egg; DNA cannot create an organism outside of this structural context.

What is the nature of this structural context and what is its relationship to the activity of genes? There are two general perspectives toward this question. One perspective is that there are specific aspects of pre-existing structural patterns and/or general rules of structural order which severely constrain what genes can accomplish. The view that emphasizes general rules of structural order is a manifestation in biology of a more general philosophical attitude known as "structuralism," whose best known modern exponents are the psychologist, Jean Piaget, with his laws of human cognitive development, and a linguist, Noam Chomsky, with his idea of a universal generative grammar. The biologists who embrace this view believe that there exist "generative laws of form," embodied during development in "embryonic fields" that have certain general ("generic") properties that are independent of the consequences of gene action (Goodwin 1985).

The contrary view, that everything interesting about development is ultimately encoded by genes, is by far the most popular among working developmental biologists. Although there are disagreements concerning the specific mechanisms by which DNA generates form, the theoretical framework that offers the best contrast to the "generative" view of developmental fields is that of "positional information," proposed by Lewis Wolpert (1969). Wolpert's idea is that developing animal embryos specify their spatial organization by "reading" a spatial grid. To visualize the essentials of this idea, think of road maps in a road atlas. All of the maps have similar coordinates on their margins; one can use these coordinates to locate towns in each state or country: In the Rand-McNally road atlas of the states of the United States, my home town, Iowa City, happens to be located at I-16 on the Iowa map, whereas on the Missouri map I-16 is the location of Farmington, Missouri. The Iowa and Missouri maps both are laid out using the same "universal grid," which is "interpreted" in one way in Iowa, in another way in Missouri. In an embryo, that grid is the positional information (which, like the grid in a road atlas, gives information *only* about position), and the different ways of reading out the grid are the different interpretations of that information. Wolpert further, and very boldly, held that all embryos of animals (at least) have the same universal positional information, which is read out differently by nuclei of embryos of different species depending on their inherited genotype and on the imprints of previous developmental events upon the capacity for expression of that genotype.

Wolpert's idea at first glance looks like a "universal grammar" of development, but is in fact quite the opposite. First, the way in which positional information is interpreted in any given system is entirely contingent on the evolutionary and developmental history of the system and is specified by the genotype; it obeys no general law whatsoever. Returning to our road atlas analogy, there is no a priori way of telling what type of town will be found at I-16 in any given state map. The same principle — or rather absence of principle — holds for fly and mouse embryos. Second and more dramatically, *the positional information (the grid) itself is the result of evolved gene activity.* It was found out — *after* Wolpert proposed his theoretical framework — that a set of genes originally known only in *Drosophila*, the "homeotic genes," are regionally expressed and utilized in all animals: They specify proteins with a special DNA-binding region (the "homeobox") that controls the expression of other genes found in nuclei situated in localized regions of the embryo (review: Kenyon 1994); utilization of these genes to specify axial position has even been held to be a defining criterion of what it means to be an animal (Slack, Holland and Graham 1993). Which genes are expressed by each localized nucleus of an embryo depends upon the combination of homeobox gene-products made in the region where that nucleus is found. In this view, all patterning during development of animals (and probably also of plants) boils down to action of genes, all is subject to natural selection, and therefore all is ultimately contingent. The logical conclusion of Wolpert's outlook is that a "universal grammar" of development, if it exists, is embodied only in the logic of gene-switching networks (Kauffman 1993).

My favorite organisms, the ciliated protozoa, are "exceptions" that can be used to "prove" (i.e. test) this notion of ultimate contingency of developmental mechanisms. Ciliates are large and complex unicellular organisms. They are covered by an organized lattice of ciliary structures used for swimming, with structures specialized for feeding, defecation, and osmotic regulation located at unique sites. Ciliates express — and transmit to their offspring — a degree of complexity similar to that of many multicellular organisms. Yet ciliates are single cells, with one central somatic nucleus. Therefore, if positional information in multicellular organisms is generated and interpreted as described in the previous paragraph, ciliate complexity must be achieved by entirely different means: Ciliate positional information (if any) cannot be embodied in DNA-binding proteins and cannot be read out by differential expression of genes located in different regions, as there are no (relevant) genes anywhere except in the central nucleus. Positional specification would instead be based on the differential localization and assembly of proteins on the cell surface: an entirely different mechanism. If this view is correct, ciliates and animals must generate their complex organization in different ways, and any resemblances must be results of evolutionary convergence.

Yet the resemblances are substantial. Ciliates have been subject to many experimental manipulations over the course of the last century, and it can fairly be said that these manipulations reveal that ciliates obey many of the phenomenological rules that hold for development of embryos. These include the capacity of a (nucleated) part to regulate to form a whole, the positioning of certain parts in a proportional manner irrespective of the size of the whole, and the induction of one specialized part by another (reviewed by Frankel 1989, Chapter 6). So we have an apparent paradox: As indicated in the previous paragraph, the probable mechanisms of development are different in ciliates and animal embryos, yet some major phenomena of development are remarkably similar. Is this simply a dramatic example of evolutionary convergence, or is there perhaps some "universal grammar" of development that underlies both the "genetic switch" mechanisms of specification and readout of positional information in multicellular organisms and the putative protein localization/assembly mechanisms in ciliates?

I will return to this question after describing some relevant phenomena of patterning in my favorite ciliate, *Tetrahymena thermophila*.

2 Nested Order in *Tetrahymena*

Tetrahymena is simple for a ciliate but complex for a cell. In addition to displaying the characteristic nuclear dualism of ciliates, with both a diploid germinal nucleus (micronucleus) and a complex polygenomic somatic nucleus (macronucleus), this ciliate possesses a nested hierarchy of cell-surface structures, with unique structural landmarks superimposed on a repeating cytoskeletal lattice. The positioning of these unique structural landmarks is emphasized here, but I will introduce this in the context of the structural lattice within which the unique landmarks are generated.

These dual levels of geometric order are illustrated, for *Tetrahymena*, in Figure 1. The upper-left diagram (Figure 1A) is a sketch of the ventral surface

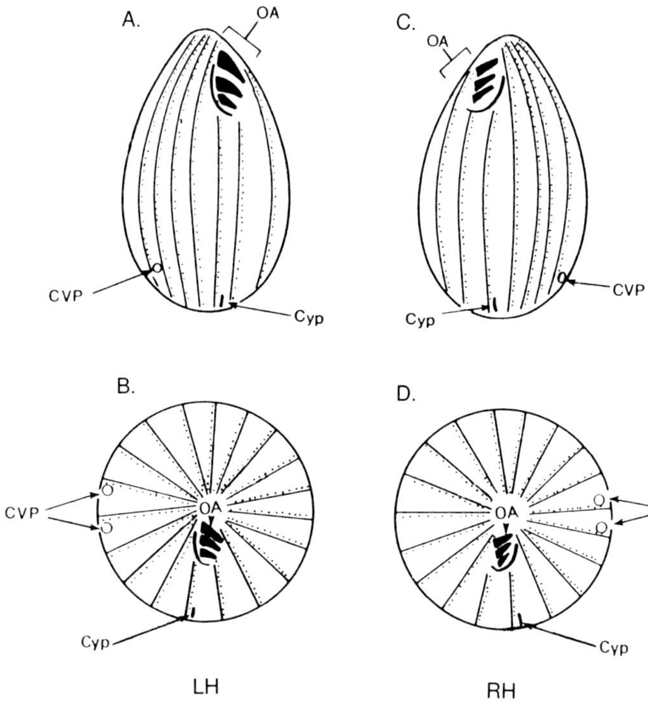

Figure 1: Diagrams of the cell surface organization of (A.,B.) right-handed (RH) and (C.,D.) left-handed (LH) *Tetrahymena* cells, drawn as ventral views (A.,C.) and as polar projections (B,D.). The oral apparatus (OA), cytoproct (Cyp), and contractile vacuole pores (CVP) are labelled. For further explanation, see the text (Note: Only one of the two CVPs is visible in A. and C.). Slightly modified from J. Frankel (1990) *Journal of Cell Science* **97**, 205–211, The Company of Biologists Ltd., with permission.

of a normal, "right-handed" (RH) *Tetrahymena* cell. The small dots are ciliary basal bodies, organized in longitudinal ciliary rows. The lines to the viewer's left (cell's right) of the dots represent longitudinal bands of microtubules which are parts of a polarized and asymmetrical cytoskeleton that makes up the structural lattice that is embedded within the cortical layer of the cell.

Within this lattice, three unique cortical landmarks develop within every ciliate generation: The most prominent is the *oral apparatus* (OA), located near the anterior end of the cell. This is made up of closely-spaced basal bodies assembled into four compound structures (membranelles and undulating membrane, labelled in Figure 2). In addition, a cytoproct (Cyp) and *contractile*

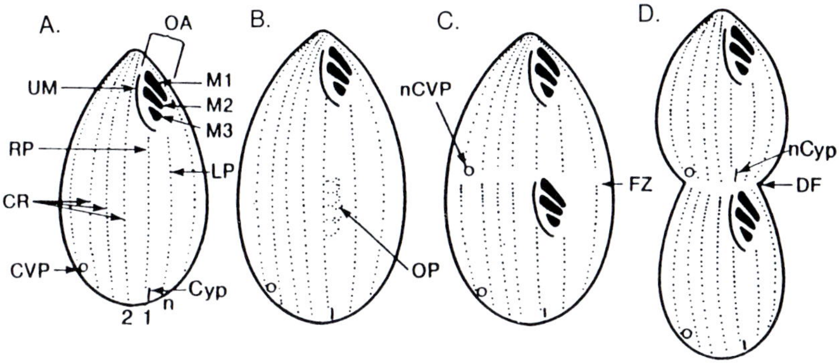

Figure 2: Structural features of the ventral surface of *Tetrahymena thermophila* at different developmental stages. A. A cell that has not begun cortical development. Seven longitudinal ciliary rows (CR) are shown, including the right-postoral (RP) and left-postoral (LP) rows. The ciliary rows are numbered from cell's left to right, starting with the right-postoral ciliary row (in this and all other figures, the cell's left corresponds to the viewer's right). The labelled structures include the three membranelles (M1, M2, M3) plus the undulating membrane (UM) of the oral apparatus (OA), the cytoproct (Cyp), and a contractile vacuole pore (CVP) (normally two CVPs are present, but only one of the two is shown). B. Early oral development. An oral primordium (OP) appears at midbody adjacent to the RP ciliary row. C. Late oral development. The OP has differentiated to form membranelles and an undulating membrane. A fission zone (FZ) has formed just anterior to the new OA, and new CVPs (nCVP) appear. D. Cell division. The division furrow (DF) constricts along the fission zone. A new cytoproct (nCyp) becomes apparent. From J. Frankel, L.M. Jenkins, E.M. Nelsen and C.A. Stoltzman (1993) *Developmental Biology* **160**, 333–354, Academic Press, with permission.

vacuole pores (CVP) are located near the posterior end. The geometrical relationships of these cortical landmarks are seen more clearly in a polar map (Figure 1B), with the anterior end of the cell at the center and the posterior end on the circumference. The OA and Cyp together define the *reference meridian* of the cell, with the CVPs (normally two, next to adjacent ciliary rows) located to the cell's right of this reference meridian.

In every cell cycle, new structures are formed along the same longitudes as old structures of the same kind but at different latitudes, i.e. near the prospective fission zone; only after these new structures are formed does the cell divide transversely (Figure 2).

What determines the distance of the CVPs from the reference meridian? The founder *of Tetrahymena* genetics, David Nanney, posed this question in a remarkably simple way. The cells shown in Figure 1 have 19 ciliary rows. What if the *number* of rows varies? Do the CVPs form at a fixed number of row-intervals to the right of the reference meridian, or a fixed *proportion* of the total number of rows? Answer: a nearly fixed proportion (Figure 3A, B). If the reference meridian is, like Greenwich, at longitude zero, then the CVPs are centered at about 80° west longitude (Nanney 1966). This implies that the cell cannot simply use a molecular or structural ruler to measure the distance from the reference meridian to the CVPs; it must also "know" some more global parameter.

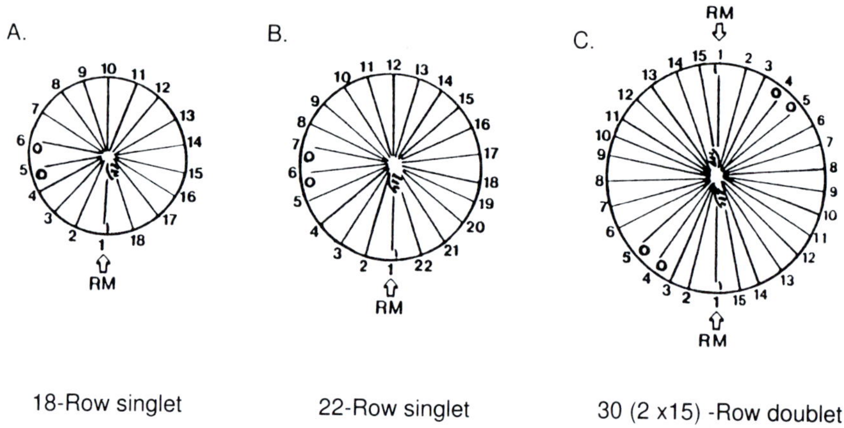

Figure 3: Schematic diagrams representing the geometry of CVP formation around the cell circumference of *Tetrahymena*. The diagrammatic conventions are the same as in Figure 1B., except that the dots representing basal bodies are omitted. Panels A. and B. depict singlet cells with 18 and 22 ciliary rows respectively, whereas C. depicts a siamese-twin doublet with reference meridians directly opposite one another and a total of 30 ciliary rows, 15 in each cellular field. All of the ciliary rows are numbered following the convention explained in the Figure 2 legend. Row 1, next to which the new oral apparatus and cytoproct develops (Figure 2), is called the reference meridian (RM). The doublet in C., with two reference meridians, has two tandem sets of numbers.

In the simplified presentation given above, the relevant global parameter appears to be the cell circumference; however, from the start Nanney realized that the true global parameter is the distance between reference meridians; in back-to-back "siamese-twin" doublet cells, the cells center their new CVPs at 22 per cent of the distance between reference meridians and not 80 degrees of a physical circumference (Figure 3C). The cells assess a surface field rather than using a compass and protractor to measure an angle (Nanney 1966).

If CVP positioning at "80°" (22%) to the cell's right of the reference meridian is possible, then why not 22% to the left? *Tetrahymena* cells have indeed been obtained in which the CVPs are located to the cell's left of the reference

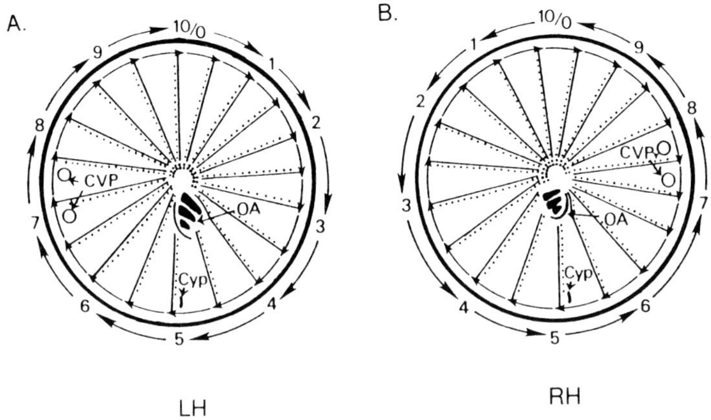

Figure 4: Polar representations of (A.) right-handed (RH) and (B.) left-handed (LH) *Tetrahymena* cells, with A. similar to Figure 1B. and B. similar to Figure 1D., but with indications of symmetry and postulated positional value superimposed. The arrows indicate symmetry: The inner small arrows show the *local* symmetry of the ciliary rows, the outer large arrows the *global* symmetry of the arrangement of the nonrepeating structures. The numbers indicate the postulated positional values. These values are *unrelated to the ciliary rows*; the 18-row and 22-row singlets in Figures 3A and 3B would have both have sets of circumferential positional values identical to those shown here; the doublet in Figure 3C would have two such sets of circumferential positional values in tandem. From *Pattern Formation: Ciliate Studies and Models* by Joseph Frankel (text) and Jane A. Otto (illustrations). Copyright ©1989 by Oxford University Press, Inc. Reprinted by permission.

meridian (which itself is shifted to the left rather than the right side of the OA). We call these "left-handed" (LH) cells (Figure 1C, D). This switch of handedness does not affect the relative distance between the CVPs and the reference meridian (Nelsen and Frankel 1989).

In these LH *Tetrahymena* cells, only the large-scale order of the cortical landmarks (symbolized by the large arrows and numbers outside the circumference of the polar maps in Figures 4A and 4B) is reversed. The local symmetry of the ciliary rows (symbolized by the small arrows inside the circumference) is the same in RH and LH cells (Nelsen and Frankel 1989). The discordance between the reversed global handedness and the normal local symmetry tends to disrupt the formation of normal new oral structures in LH cells (Nelsen, Frankel and Williams 1989) and probably is responsible for the slow division rate of such cells and hence their nonoccurrence (or extreme rarity) in nature.

It is likely that symmetry at the level of the cytoskeletal lattice associated with the basal bodies *cannot* be reversed, as it almost certainly is based on the complex asymmetrical geometry of the ciliary basal body (Fulton 1971). A deliberate effort to select mutants exhibiting such a reversal in an amenable unicellular organism (the algal flagellate *Chlamydomonas reinhardtii*) has failed (Holmes and Dutcher 1989). Global symmetry, however is more flexible, suggesting that it is based on dynamic relations among mobile molecules rather than on a fixed structural order; i.e. it is likely to be kinetically rather than structurally determined (Harrison 1982).

This conclusion was intuited by Nanney from the start. He wrote, "The concepts of fields and gradients so widely employed in a description of organismic integration in multicellular forms are equally as applicable (and equally as sterile perhaps) in a consideration of organization at the cellular level" (Nanney 1966, p. 317).

"Equally as sterile perhaps"? The parenthetical caveat was reasonable when written in 1966; by now, however, virtually everyone engaged in this subject would disagree with it — but perhaps for different reasons. For the molecular developmental biologist, fields and gradients in embryos are becoming reduced to cascading signal-transduction mechanisms and gradients of transcription factors (e.g. DeRobertis, Morita and Cho 1991; Lawrence and Sampredo 1993), whereas for some of the authors in this book, the reduction (if I may call it that) is to generic principles that also govern physical fields.

3 Reversals of Handedness and Developmental Fields in *Tetrahymena*

Left-handed (LH) tetrahymenas have not been found in nature. They are laboratory curiosities, obtained by special means and maintained by careful selection. The RH and LH cells that have been generated are vegetative variants that do not differ genically (Nelsen, Frankel and Jenkins 1989), yet can propagate their geometry. This propagation is possible because in ciliates growth is essentially longitudinal, as shown in Figure 2.

How is the transition from one global handedness to another achieved in a ciliate such as *Tetrahymena*? Detailed descriptions for *Tetrahymena* and other ciliates are provided elsewhere (Frankel 1989, Chapter 9; Suhama 1990; Frankel 1991, Shi et al. 1991; Yano and Suhama 1991), and in this volume (see the essay by Brandts and Totafurno). Briefly, the reversal of global symmetry in wild-type cells arises either from a spatial juxtaposition of parts that are not normally adjacent, or from the process of regulation of a doublet form such as the "siamese-twin" doublet alluded to earlier. Symmetry reversals have to my knowledge never been elicited by treatments with antimetabolites, inhibitors of macromolecular synthesis, or physical changes such as heat or cold shocks, except when the shock first causes a compound form to develop, for example as a consequence of arrest of cell cleavage (Nelsen, Frankel and Jenkins 1989). In other words, the proximate stimulus for a change of handedness is geometrical rather than chemical abuse.

A simple model might help to visualize what is going on during such geometrical transitions, and provide an intuitive background to the more precise and theoretical treatments of this subject by Brandts and Totafurno. To grasp the essence of the idea, first put your two hands together, palms facing you. Note that the little fingers of your two hands are next to one another; in terms of a numerical sequence (with thumb as 1, little finger as 5) what you see is 1-2-3-4-5-5-4-3-2-1. Every number is next to the same number or a consecutively adjacent number. This is positionally stable. Now imagine that you put your left hand, palm facing you, next to another left hand, palm also facing you (you will need to borrow a hand from someone else actually to do this). Now you have a numerical sequence 1-2-3-4-5-1-2-3-4-5. A little finger (5) is next to a thumb (1). Is this developmentally stable? *No.* Why not? Because two structures (the little finger and a thumb) that are ***not*** "normal(ly) neighbors" (cf. Mittenthal 1981) have become juxtaposed. What happens in a system capable of developmental regulation (which a fully formed human hand is not)? The following intercalation: 1-2-3-4-5-*4*-*3*-*2*-1-2-3-4-5 (numbers of intercalated structures italicized and underlined): A *right* hand will have been intercalated between two left hands in order to restore the continuity of successive integers. This type of left- right-left (or, equivalently right-left-right) triple-formation was first described (under the rubric of "secondary symmetry") in appendages of natural specimens of insects and crustacea a century ago by William Bateson in his 1894 book *Materials for the Study of Variation* (Bateson 1894, 1992) and has been since generated experimentally, and repeatedly modeled, in insects, crustacea and amphibians (see the essays by Brandts and Totafurno, and by Totafurno and Trainor, this volume). Even though the resulting monstrosities aren't very useful for their possessors, they inevitably develop once the appropriate surgical juxtapositions have been created; one simply cannot restore continuity between a "thumb" and "little finger" without a reversal of handedness in between. It is as if developing structures "need" to be near their normal positional neighbors, and if they are not, they form highly abnormal, nonadaptive, structures in order to restore "normal neighborhoods" everywhere. The nonadaptive nature of the structures is an important indication that a fundamental developmental mechanism is involved.

What we and others have found was that the same rule applies to intracellular development in ciliates. Bateson's empirical law and the subsequent models can be adapted (with modifications to accommodate the unique surface topology of growing and dividing ciliates) to explain numerous otherwise bizarre reversals of symmetry that we and others had observed in ciliates. This, then, brings us back to the notion of a "universal grammar" of development. Not only are there similarities between development in ciliates and embryos with regard to proportionality, induction, etc., but there is also the same tendency to form potentially maladaptive symmetry reversals when positional continuity is disrupted. The fact that the rule of continuity operates within as well as between cells seems to point toward some deep general property of developing systems that transcends the major and obvious differences between unicellular and multicellular systems.

How should we interpret this general property? We can imagine, by analogy to the road atlas introduced at the beginning of this essay, that there is an intracellular positional grid embedded within the surface layer of this ciliate. The circumferential coordinates of this grid are illustrated by the rings of numbers shown around the outside of the polar maps of Figure 4. These numbers can be taken to represent intracellular positional values, collectively forming a continuous sequence analogous to a clockface, which can be arrayed in either a clockwise (Figure 4A) or counterclockwise (Figure 4B) order (note that these numbers are totally distinct from the numbering of the ciliary rows). Particular positional values determine the placement of specific structures (e.g. OA and Cyp at 5, CVPs at 7). These positional values are propagated longitudinally during growth of the ciliate clone. Rules concerning stable spacing of these values account for occasional reversals of handedness. This model, derived from the polar coordinate model that accounted for counterintuitive patterns of regeneration in multicellular organisms (French, Bryant and Bryant 1976), does the same for pattern regulation in ciliates (Frankel 1989, Chapter 10).

Note that this formulation is agnostic with respect to the fundamental issue raised at the beginning of this essay. I have presented the model with a Wolpertian slant by stating that the numbers represent "positional values." Then how are such values generated and how are they interpreted? There are no localized genomes that can provide a readout of graded concentrations of transcription factors or other molecular signals. Of course, the "readout" might occur at the level of assembly rather than synthesis of unique proteins. The formulation provided by Brandts in this volume, in her essay with John Totafurno (see also Brandts and Trainor 1990a,b) is both more precise and explicit mathematically and physically; it provides an interpretation of how the positional values may be generated, but still is noncommittal with regard to "interpretation," or perhaps better, "implementation" of the positional instructions.

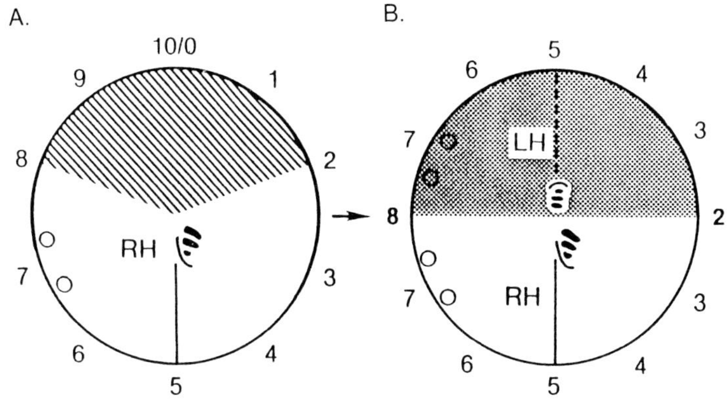

Figure 5: Simplified polar projections of *Tetrahymena* cells (A.) before and (B.) after expression of a *janus* mutation. Oral apparatuses, contractile vacuole pores, and only the right- postoral postoral ciliary row are shown. The numbers represent postulated circumferential positional values, as in Figure 4. The hatched region of the right-handed (RH) cell (A.) represents the subset of positional values that cannot be maintained by *janus* mutant alleles. This region then is replaced by a left-handed (LH) zone (shaded in B.). From J. Frankel (1990) *Journal of Cell Science* **97**, 205–211, The Company of Biologists Ltd., with permission.

4 Genic Mutations Affecting Global Patterning in *Tetrahymena*

The examples of reversal of global handedness considered thus far arise from responses to geometrical mishaps encountered by wild-type ciliates. There is also a genic route to reversal of cellular handedness; this involves expression of *janus* mutants in *Tetrahymena thermophila* (Frankel and Jenkins 1979; Frankel, Nelsen and Jenkins 1987). A homozygous *janus* cell (Figure 5B) possesses an extra abnormal OA plus a CVP set, organized in reverse of the normal arrangement, on what should be the dorsal surface of the cell. A detailed study of the phenotypic conversion of normal to *janus* cells after the macronuclear genetic transition from a $janA^+/janA^+$ to a $janA/janA$ genotype showed that the conversion took place while the number and configuration of ciliary rows remained unchanged. The CVP domain broadened and eventually split into two domains, and new oral primordia of the characteristic LH type later appeared within what previously was the mid-dorsal surface of the cell (Frankel and Nelsen 1986). The positional order changed profoundly while there was no apparent change in the structural lattice of the cell. In terms of the positional values presented in Figure 4, one could imagine that products of the jan^+ alleles are required for maintenance of a dorsal subset of the positional

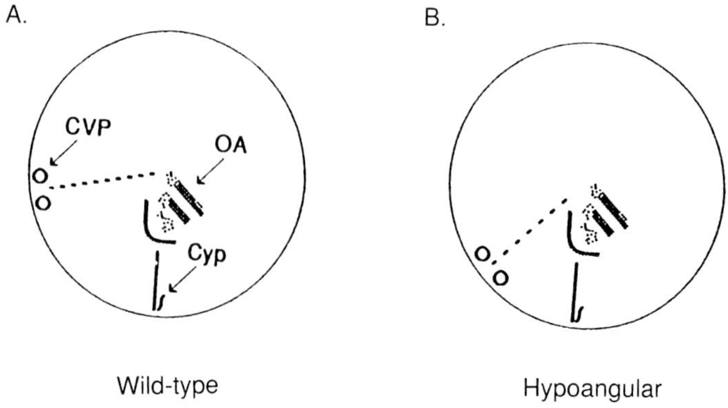

Figure 6: Simplified polar diagrams of (A.) wild-type and (B.) homozygous *hypoangular* cells, indicating typical relative positions of the oral apparatus (OA), cytoproct (Cyp), and contractile vacuole pore (CVP).

values (8-9-10/0-1-2)(Figure 5A). Following the withdrawal of a jan^+ gene product, the dorsal positional values would be lost, altering the spacing of values and thereby creating a potential major discontinuity at the former dorsal midline. To restore continuity plus near-normal spacing of positional values, the *janus* cell would then be forced to intercalate the permitted (ventral) subset of positional values, necessarily in reverse order (Figure 5B). This example of positional regulation has not yet been modeled in a manner comparable to that provided by Brandts and Trainor (1990a,b) for reversals of symmetry in wild-type cells, but presumably could be so modeled without much difficulty.

A mutant more challenging to our preconceptions is one called *hypoangular (hpo)*, so named because the "angle" between the reference meridian and the CVP midpoint is reduced, i.e. the CVPs are closer to the reference meridian than in wild-type cells (Figure 6). There are other changes as well, indicative of a general instability of cortical patterning (Frankel et al. 1993). The "clock-face" shown in Figure 4A, and the underlying energy functional in Brandts' model, would need to be altered in a profound manner in *hypoangular* cells.

Because of their instability of cortical phenotype, *hypoangular* cells can reverse their handedness more easily than wild-type cells. LH *hypoangular* cells have the expected LH organization, except that in *hypoangular* cells (unlike wild-type cells) the relative distance of the CVPs from the reference meridian is on the average smaller in LH than in RH cells (Frankel and Nelsen, unpublished observations). It is as if these LH cells somehow "recognized" the discrepancy between the symmetry of their large scale pattern and their structural lattice.

This observation fits in well with Brandts' idea of an asymmetrical energy functional (Brandts 1993 and this volume); it also suggests that the two nested positional systems, the local and the global, are in some kind of communication with one another.

5 Homology or Analogy?

Are the mechanisms of global patterning in ciliates homologous or analogous with corresponding mechanisms in multicellular organisms (animals in particular)? This question cannot be asked, let alone answered, before we have a clear understanding of the meaning of homology — which (in my view) leads us to the heart of the matter.

For a structuralist such as Goodwin, homology is a pre-evolutionary concept: "Homology is an equivalence relation on a set of forms which share a common structural plan and are thus transformable one into another," a relation which therefore is "independent of any historical or genealogical relationships which the actual structures may have" (Goodwin 1984, p. 101; see also his essay, this volume). For the majority of biologists, however, homology is what it was for Darwin, a record of inheritance from a common ancestor. From this latter perspective, relations of biological form are contingent outcomes of evolutionary tinkering; physical and chemical laws of course must be obeyed, but they do not illuminate similarities and differences among biological forms.

If we interpret homology in evolutionary terms, then both our question and the route to its answer are apparent. Comparative sequence analysis of ribosomal RNA molecules tells us that ciliates arose from an evolutionary lineage separate from those that gave rise to animals, plants, and fungi (Sogin, Ellwood and Gunderson 1986). The major evolutionary lineage that includes ciliates also includes certain flagellates (dinoflagellates) (Schlegel 1991). Dinoflagellates have a considerably simpler structural organization than ciliates, making it possible that the global patterning relationships discussed above arose within the ciliate sublineage. No clear indication of such global patterning relationships has yet been found in flagellates such as *Chlamydomonas* (Holmes and Dutcher 1989), nor in the most intensively investigated unicellular eukaryotes, budding yeast (Madden, Costigan and Snyder 1992). If, as is likely though not proven, the common evolutionary ancestors of ciliates and animals lacked any equivalent of the global patterning mechanisms of ciliates, then the similarities of global patterning relations in ciliates and multicellular organisms are likely to be convergent and hence at best analogous, not homologous.

The nonhistorical structuralist view of homology championed by Goodwin poses the question of homology in an entirely different manner: as one of transformability. If we were to interpret homology according to this view, our question would be whether pattern-forming mechanisms in ciliates and in some set of multicellular organisms (e.g. animals) belong to the same transformational series.

The existence in organisms of such transformational series is indisputable; the best example to my knowledge is phyllotaxis in plants: There exist three patterns of leaf arrangement generated at the shoot apex (distichous or two-ranked, whorled, and spiral; for further discussion see Goodwin's essay); transitions in the geometry of leaf initiation, either in naturally occurring phase-changes or in response to surgical, hormonal, or genetic intervention, can occur only between these distinct and well-defined patterns (Green 1987; Steeves and Sussex 1989, Chapter 7; Couder and Douady 1995). In the ciliate case, there is a restricted series of geometrical forms, and transformations of global handedness proceed through these forms (Figure 19 in Nelsen and Frankel 1989; Figure 3 in Brandts 1993). Furthermore, the more complex of these intermediate forms obey the geometrical rules of "secondary symmetry" first enunciated for arthropods by Bateson (1894, 1992: p. 479). From this perspective, the possibility that ciliates and animals may have evolved the same set of transformational rules independently of each other would be viewed as a contingent historical fact far less important than the commonalty of the rules governing the transformational series.

A corollary of this viewpoint is that the evolved developmental system is more like a discovery than an invention, as might be illustrated by the evolution of spiral phyllotaxis in a plant group that had not previously manifested it. This is an intellectually attractive idea, as it would give developmental biology a coherence lacking in a world view dominated by historical contingency. Nonetheless, the attractiveness of this viewpoint does not make it true. What is needed to validate this view is a successful comparison at a mechanistic level of systems that share formal properties in common yet look very different, e.g. ciliates and insect imaginal discs.

A transformational series of similar logical form in two entirely disparate groups of organisms will only be accepted by most biologists as homologous (in any sense) if the underlying molecular mechanisms show some important features in common. Hence in my own view, there is no substitute for finding the molecular mechanisms underlying the relevant patterns in ciliates, comparing these to the ones at work in animals and plants, and then (with luck) binding these together with a common formal basis.

Acknowledgements

I thank Dr. Anne Frankel for her comments on the manuscript. The research on which this essay is based was partially supported by NIH research grant HD-08485.

References

Bateson, W. (1894; 1992). *Materials for the Study of Variation, Treated with Especial Regard to Discontinuity in the Origin of Species.* Baltimore: Johns Hopkins University Press.

Brandts, W.A.M. (1993) A field model of left-right asymmetries in the pattern regulation of a cell. *IMA Journal of Mathematics Applied in Medicine & Biology* **10**, 31–50.

Brandts, W.A.M. and Trainor, L.E.H. (1990a) A non-linear field model of pattern formation: intercalation and morphallactic regulation. *Journal of Theoretical Biology* **146**, 37–56.

Brandts, W.A.M. and Trainor, L.E.H. (1990b) A non-linear field model of pattern formation: application to intracellular pattern reversal in *Tetrahymena. Journal of Theoretical Biology* **146**, 57–86.

Couder, Y. and Douady, S. (1995) Phyllotaxis: genetic determination and self organization. In D. Beysens, G. Forgacs and F. Gaill, eds., *Interplay of Genetic and Physical Processes in the Development of Biological Form.* Singapore: World Scientific, pp. 283–289.

DeRobertis, E.M., Morita, E.A. and Cho, K.W.Y. (1991) Gradient fields and homeobox genes. *Development* **112**, 669–678.

Frankel J. (1989) *Pattern Formation: Ciliate Studies and Models.* New York: Oxford University Press.

Frankel, J. (1990) Positional order and cellular handedness. *Journal of Cell Science* **97**, 205–211.

Frankel, J. (1991) Intracellular handedness in ciliates. In *Cell Patterning.* Ciba Foundation Symposium 62. London: Wiley, pp. 73–93.

Frankel, J. and Jenkins, L.M. (1979) A mutant of *Tetrahymena thermophila* with a partial mirror-image duplication of cell surface pattern. II. Nature of genic control. *Journal of Embryology and Experimental Morphology* **49**, 203–227.

Frankel, J. and Nelsen, E.M. (1986) How the mirror-image pattern specified by a *janus* mutation of *Tetrahymena* comes to expression. *Developmental Genetics* **6**, 213–238.

Frankel, J., Nelsen, E.M. and Jenkins, L.M. (1987) Intracellular pattern reversal in *Tetrahymena thermophila: janus* mutants and their geometrical phenocopies. In W.F. Loomis, ed., *Genetic Regulation of Development.* Society for Developmental Biology Symposium **45**. New York: Alan R. Liss, pp. 219–244.

Frankel, J., Jenkins, L.M., Nelsen, E.M. and Stoltzman, C.A. (1993) *hypoangular*: a gene potentially involved in specifying positional information in a ciliate, *Tetrahymena thermophila. Developmental Biology* **160**, 333–354.

French, V., Bryant, P.J. and Bryant, S.V. (1976) Pattern regulation in epimorphic fields. *Science* **193**, 969–981.

Fulton C. (1971) Centrioles. In J. Reinhardt and H. Ursprung, eds., *Origin and Continuity of Cell Organelles.* Results and Problems in Cell Differentiation **2**. Berlin: Springer-Verlag, pp. 170–221.

Goodwin, B.C. (1984) Changing from an evolutionary to a generative paradigm in biology. In J. W. Pollard, ed., *Evolutionary Theory: Paths into the Future.* New York: John Wiley and Sons, pp. 99–120.

Goodwin, B.C. (1985) What are the causes of morphogenesis? *BioEssays* **3**, 31–36.

Green, P.B. (1987) Inheritance of pattern: analysis from phenotype to gene. *American Zoologist* **27**, 657–673.

Harrison, L.G. (1982) An overview of kinetic theory in developmental modelling. In S. Subtelny and P. B. Green, eds., *Developmental Order: Its Origin and Regulation.* Society for Developmental Biology Symposium **40**. New York: Alan R. Liss, pp. 3–35.

Holmes, J.A. and Dutcher, S.K. (1989) Cellular asymmetry in *Chlamydomonas reinhardtii. Journal of Cell Science* **94**, 273–285.

Kauffman, S.A. (1993) *The Origins of Order: Self-Organization and Selection in Evolution.* New York: Oxford University Press.

Kenyon, C. (1994) Homeotic genes and evolution. *Cell* **78**, 175–181.

Lawrence, P.A. and Sampredo, J. (1993) *Drosophila* segmentation: after the first three hours. *Development* **119**, 971–976.

Madden, K., Costigan, C. and Snyder, M. (1992) Cell polarity and morphogenesis in *Saccharomyces cerevisiae. Trends in Cell Biology* **2**, 22–29.

Mittenthal, J.E. (1981) The rule of normal neighbors: a hypothesis for morphogenetic pattern formation. *Developmental Biology* **88**, 15–26.

Nanney, D.L. (1966) Cortical integration in *Tetrahymena*: An exercise in cytogeometry. *Journal of Experimental Zoology* **161**, 307–318.

Nelsen, E.M. and Frankel, J. (1989) Maintenance and regulation of cellular handedness in *Tetrahymena. Development* **105**, 457–471.

Nelsen, E.M., Frankel, J. and Jenkins, L.M. (1989) Non-genic inheritance of cellular handedness. *Development* **105**, 447–456.

Nelsen, E.M., Frankel, J. and Williams, N.E. (1989) Oral assembly in left-handed *Tetrahymena thermophila. Journal of Protozoology* **36**, 582–596.

Schlegel, M. (1991) Protist evolution and phylogeny as discerned from small subunit ribosomal RNA sequence comparisons. *European Journal of Protistology* **27**, 207–219.

Shi, X.-b., Lu, L., Qiu, Z., He, W., and Frankel, J. (1991) Microsurgically generated discontinuities provoke heritable changes in cellular handedness in a ciliate, *Stylonychia mytilus. Development* **111**, 337–356.

Slack, J.M.W., Holland, P.W.H. and Graham, C. F. (1993) The zootype and the phylotypic stage. *Nature* **361**, 490–492.

Sogin, M.L., Ellwood, H.J., and Gunderson, J.H. (1986) Evolutionary diversity of small subunit rRNA genes. *Proceedings of the U.S. National Academy of Sciences* **83**, 1383–1387.

Steeves, T.A. and Sussex, I.M. (1989) *Patterns in Plant Development.* Cambridge: Cambridge University Press.

Suhama, M. (1990) The regulation of homopolar doublets to singlets in *Glaucoma. Journal of Protozoology* **37**, 344–352.

Wolpert, L. (1969) Positional information and the spatial pattern of cellular differentiation. *Journal of Theoretical Biology* **25**, 1–47.

Yano, J. and Suhama, M. (1991) Pattern formation in mirror-image doublets of the hypotrich ciliate *Stylonychia pustulata*. *Journal of Protozoology* **38**, 111–121.

IV

Cellular and Organismic Biology

Statistical Mechanics of the Main Phase Transition in Lipid Bilayers

Francis P. Jones and Patrick Tevlin

1 Introduction

The cell membrane is a container for the cell's contents, giving it structure and mediating its interaction with the extracellular world. Cell membranes are today commonly understood in terms of the "fluid mosaic model" (Singer and Nicolson 1972; a recent introduction is Yeagle 1993), the main elements of which are lipids and proteins. The proteins are polypeptides of large molecular weight ($M_w \gtrsim 20,000$), while the lipids are small ($M_w \approx 800$) amphipathic molecules; one end of the lipid is polar and/or charged, while the other end is hydrophobic. The lipids form a bilayer in which the hydrophobic part is sequestered in the center of the bilayer, with the polar heads enjoying an aqueous environment (Gorter and Grendel 1925). Embedded within this two-dimensional structure are the proteins. The second major aspect of this model is that its constituent elements are, under the appropriate circumstances,"fluid," and free to diffuse laterally, in the plane of the bilayer. Although many other important aspects of cell membrane behavior are neglected in this description (see, e.g., McNeil 1991, Yeagle 1993), these two points give a general understanding of overall membrane organization.

The lipids in all membranes have great diversity and are comprised of a number of classes (Figure 1). Included here are the phospholipids, diacylglycerols, plasmalogens, sphingomyelins, and gangliosides. The major exception to the structural similarity of these classes is the sterol group. Within each lipid class, there may be many lipid *types*, each of which may contain many *species*. In this chapter we will consider only double-acyl chain phospholipids, and usually only the types denoted "PC" (phosphatidylcholine) and "PE" (phosphatidylethanolamine). In much of what follows, we will make comparison to the representative species DPPC (dipalmitoylphosphatidylcholine, with two 16-carbon acyl chains) and DPPE (dipalmitoylphosphatidylethanolamine, similar to DPPC but with the charged amino group N^+H_3 replacing the bulkier choline group $N^+(CH_3)_3$ in the lipid head region). However, our calculations and discussion will be quite general, and applicable to several lipid types.

Phospholipids in water show huge variations in behavior, form a large number of different structures, and can serve as templates for microstructure fabrication (e.g. Schnur 1993). Among these are micelles, hexagonally packed

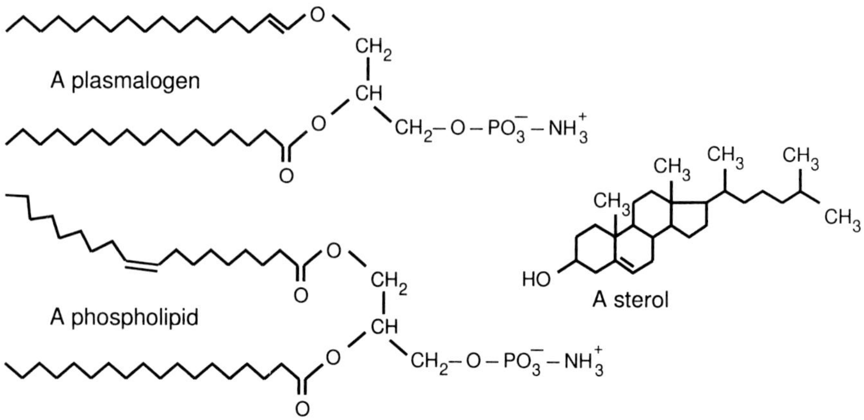

Figure 1: Structures of some common lipid types. The carbon backbones are indicated by line segments and most hydrogens are omitted for clarity.

cylinders, bilayers and monolayers. The phase diagrams for these systems are even now only partially known, and usually only as functions of water content and temperature (Figure 2, see also Janiak, Small and Shipley 1979; Sackmann 1982; Seddon, Cevc, Kayne and Marsh 1984). In excess water ($\gtrsim 30$ mol H_2O/mol lipid), the phases can be fairly simple for the molecules we consider, being basically lamellar dispersions (i.e., a planar lipid bilayer, with many parallel planes, repeated at a regular spacing). However, even for these simple phases, the addition of another species such as another phospholipid or cholesterol produces great complications in the phase diagram (see e.g. Lentz, Barrow and Hoechli 1980). Here we will be concerned with only the lamellar phases (both bilayer and monolayer) of a single lipid species in water as a first approximation to a cell membrane.

Heating a phospholipid dispersion in excess water can produce a number of phase transitions (termed "thermotropic") near room temperature. Those transitions move the system between various lamellar phases, depending on what species of lipid is involved.

A number of features characterize these phases and are embodied in their naming. The lamellar phases may have a flat surface (L), or have periodic ripples (P), of spatial period ≈ 10 nm. In addition, X-ray diffraction and NMR experiments (Tardieu, Luzzati and Reman 1973; Janiak, Small and Shipley 1976, 1979; Harlos 1978; Ruocco and Shipley 1982) show that the hydrocarbon chains may be ordered, with their ends packing on a (possibly distorted) two-dimensional hexagonal lattice. A system with chains in this state is in-

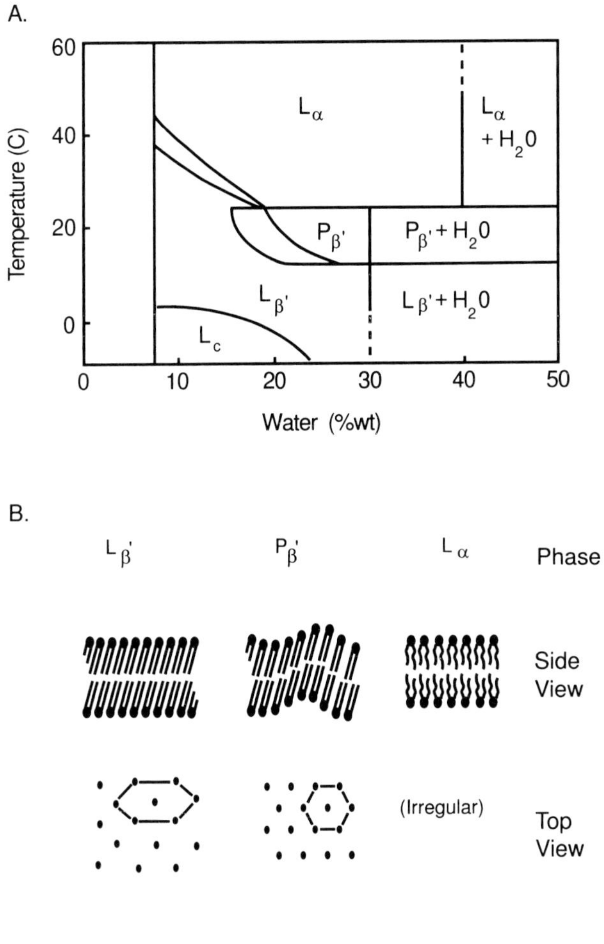

Figure 2: A. A phospholipid phase diagram as a function of water content and temperature. Only the more important phases are shown. B. Representations of the chain geometry (from above and from the side) in various lamellar phases. The head groups are represented by circles, the chains by lines. From M.J. Janiak, D.M. Small and G.G. Shipley (1979), *Journal of Biological Chemistry* **254**, 6068–6078, American Society for Biochemistry, with permission.

dicated by the subscript β. A crystalline phase (L_c) is also possible at lower temperatures, where the lipids pack much as they do in their anhydrous state. If the temperature is high enough, however, the chains become disordered or "melted." This phase is denoted by the subscript α. Furthermore, the lipid molecules are not necessarily perpendicular to the plane of the bilayer; this is indicated by a prime on the chain state. For example, DPPC in excess water goes from $L_{\beta'}$ to $P_{\beta'}$ at 35°C, and then to L_α at 41.5°C.

Near room temperature, the lamellar phases of a phospholipid dispersion in excess water may undergo a number of transitions known as the "sub-", "pre-", and "main-" transition, at temperatures T_s, T_p and T_m, respectively. Where these transitions exist, one finds $T_s < T_p < T_m$. Of these, the main transition is by far the most dramatic, with the lipid becoming fluid above the transition. The main transition is also the most common feature of the lamellar phases, in that the sub- and pre-transitions do not occur with all lipids. In this essay, we will only deal with the main transition. Nonetheless, all three of these transitions are ultimately of great interest in understanding the working of the lamellar phases since it appears that their driving mechanisms, and the resulting structures, are quite different.

The main phase transition is the most obvious feature in the thermotropic behavior of the lamellar phases. It is highly endothermic, with its transition enthalpy (for DPPC) of about 8 kcal/mol. This transition is known as the "gel-fluid transition," and is of the order-disorder type. The transition is also described as being first-order, since it exhibits a peak in the heat capacity, accompanied by a jump in enthalpy and sudden changes in molecular conformation.

Below T_m, the molecules exhibit a variety of symmetries. Scattering experiments with lipid monolayers show that the acyl chain ends usually display orthorhombic or hexagonal (two-dimensional) symmetries but *immediately* below T_m the ends exhibit a hexagonal or squashed hexagonal pattern. Also, the acyl chains forming the apolar lipid tails are extended, with few rotated C-C bonds. Above T_m the chain ends are "fluid" and disordered, but retain on average approximately 6 nearest neighbors. Internally, the chains are highly disordered and said to be in the "melted" state (Seelig and Seelig 1974; Seelig and Browning 1978; Israelachvili, Marčelja and Horn 1980), in response to a large number of rotated C-C bonds.

This appearance of disorder above T_m provides the key to understanding the main transition: It is a "chain melting" transition, and is a function of the acyl chain length n_c. In fact, the thermotropic behavior of the lamellar phases is quite regular. Within a type of lipid (such as PC), a homologous series shows trends in both T_m and the transition enthalpy ΔH as a function of the acyl chain length n_c. These are similar across many lipid types. This regular behavior is also exhibited by the heat capacities (Blume 1983). Thus, while the thermotropic behavior is a function of both the head group type and the acyl chain length, only the latter's effect on T_m and ΔH is currently predictable. We would expect that models of the lamellar phases which treat the acyl chains accurately ought to be able to predict the T_m, ΔH data for a homologous series within a lipid type. This is the approach we will follow.

We establish a Hamiltonian for the lamella so that equilibrium statistical mechanics can make predictions on the system's macroscopic behavior. The situation is complicated, since the inter-lipid interactions and the weighting of the various lipid conformations cannot be expressed in any useful closed form. This is largely because lipid geometry is complicated, and because there is currently no simple way to express molecular "shape" — and certainly no simple way of expressing the more complex *inter*-molecular steric effects — other than to list the positions of the constituent atoms. Because of these difficulties, we will approximate some terms in the Hamiltonian with a mean field contribution.

The accurate rendering of internal steric effects implies a fairly accurate consideration of a lipid's geometry in order to calculate the properties which are required by the Hamiltonian. We do this by calculating a function $\mathcal{D}$ with the following characteristics. $\mathcal{D}$ is a function of all the conformational quantities entering the Hamiltonian (such as the number of rotated bonds in a conformer); this set of quantities is denoted by $\{x\}$. Then $\mathcal{D}(\{x\})$ will be defined as the total number of states — out of all the possible conformers — which have this set of properties $\{x\}$. *In a slight departure from its usual statistical-mechanics meaning*, we will refer to this function $\mathcal{D}$ as the "density of states." While $\mathcal{D}(\{x\})$ *is not* the total number of states at a fixed energy, it is simply related to the usual density of states, and is central to our treatment of the main transition.

2 Molecular Interactions in a Lamellar Lipid System

Because of the large number Ω_0 of internal molecular states made possible by rotations around the $-C-C-$ bonds of their acyl chains ($\Omega_0 \sim 3^{2n_c} \sim 10^{15}$ for DPPC), it is currently very difficult to tackle lipid phases while accounting for correlations between these states — this would require specifying some Ω_0^2 interactions! And yet it is just these correlations which are important: on a purely geometrical basis, the steric restrictions imposed by a given molecule's neighbors are severe. The Hamiltonian which is chosen for describing lipids in a lamellar system needs to deal simultaneously with these two issues.

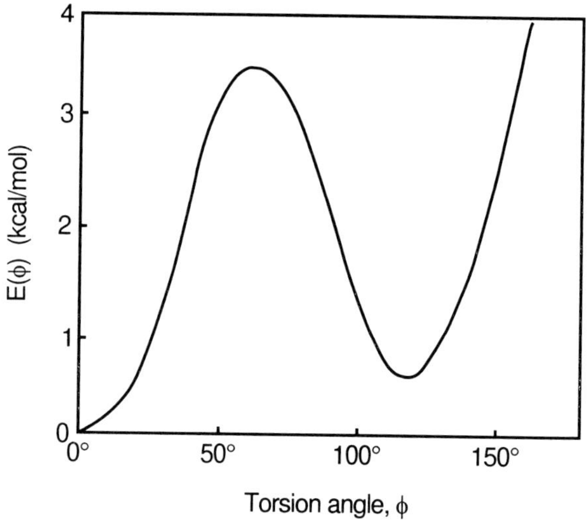

Figure 3: A sketch of the potential energy relative to the *trans* state, as a function of C–C bond rotation angle ϕ. The rotational isomeric approximation replaces the continuum of states (i.e., bond angles) with three discrete states: the *trans* state (t), and two rotated ($\phi = 120°$) *gauche* states ($g^{\pm}$) with energy $\approx$ 500 kcal/mol relative to the *trans* state. After Flory (1969).

One can think of the terms entering the Hamiltonian $\mathcal{H}$ as belonging to one of three categories, so that $\mathcal{H}$ is the sum $\mathcal{H}_{\rm id}+\mathcal{H}_{\rm int}+\mathcal{H}_{\rm ext}$. The first group, $\mathcal{H}_{\rm id}$, contains the "ideal gas terms," which include whole-body translation and rotation:

$$\mathcal{H}_{\rm id} = \mathcal{H}_{\rm id}^{trans} + \mathcal{H}_{\rm id}^{rot}. \tag{1}$$

These terms will not make a significant contribution to the behavior across the main transition.

The second group of terms, $\mathcal{H}_{\rm int}$, describes the internal state of each molecule. Included are the conformation-dependent terms, which we can further divide into those which exhibit nonbonded steric interactions ($\mathcal{H}_{\rm int}^{steric}$), and those which are bond rotation potentials ($\mathcal{H}_{\rm int}^{bond}$; Figure 3). The former are typically high-energy non-bonded interactions, while the latter are lower energy and of a local nature.

In addition, the internal kinetic term ($\mathcal{H}_{\rm int}^{kin}$) accounts for motions of the head group, chain splaying, and the propagation of bond rotations along the chains. These terms will be gathered into $\mathcal{H}_{\rm int}$:

$$\mathcal{H}_{\rm int} = \mathcal{H}_{\rm int}^{bond} + \mathcal{H}_{\rm int}^{steric} + \mathcal{H}_{\rm int}^{kin}. \tag{2}$$

Finally, the "external" terms are grouped into $\mathcal{H}_{\text{ext}}$. Included here are intermolecular steric repulsion terms ($\mathcal{H}_{\text{ext}}^{steric}$), which are complicated functions of the conformations of the two interacting molecules, their relative orientations, and their separation. There is currently no tractable expression for this important term, and so the "packing problem" must be treated with approximate methods. The difficulties in finding a reasonable $\mathcal{H}_{\text{ext}}^{steric}$ have given rise to the ubiquitous *pressure-area approximation* ($\mathcal{H}_{\text{ext}}^{steric} \sim \Pi \overline{A}$) first introduced by Marčelja (1974); an improvement to this model will be introduced here.

In addition to steric repulsion, lipid-lipid dispersion (i.e. van der Waals forces, $\mathcal{H}_{\text{ext}}^{vdW}$) and lipid-water ($\mathcal{H}_{\text{ext}}^{water}$) interactions must be accounted for. The van der Waals interaction depends on chain separation and conformation. Chain conformation can be described most simply through the nematic order of the chain (cf. Flory 1969). The nematic order S of a chain is defined as the average along the chain of $(3\cos^2\theta - 1)/2$, where θ is the orientation angle of a CH_2 unit relative to the plane of the lamella.

Finally, we note that for some systems a screened coulombic interaction between head groups ($\mathcal{H}_{\text{ext}}^{head}$) is also needed. The final form is

$$\mathcal{H}_{\text{ext}} = \mathcal{H}_{\text{ext}}^{steric} + \mathcal{H}_{\text{ext}}^{vdW} + \mathcal{H}_{\text{ext}}^{head} + \mathcal{H}_{\text{ext}}^{water}. \tag{3}$$

The "ideal gas" terms are essentially neglected, but in any "complete" theory — one which also describes the liquid-expanded to surface-gas transition of the monolayer — they would be essential. Also neglected are the "internal" kinetic terms, and the "external" lipid-water interaction. The kinetic terms are expected to make only a small contribution to the system's heat capacity. For the lipid-water terms any changes in the hydration shell resulting from expansion of the bilayer would be included *ad hoc*; the hydration energy will be assumed to be constant to avoid the introduction of such *ad hoc* parameters.

Our approach will be to approximate the exact Hamiltonian $\mathcal{H}$ through a mean field treatment. In mean field theory (MFT) the interaction of a particle with its neighbors is approximated by an interaction with a mean field, the mean field being determined self consistently. In the context of model membranes, MFT is used to simplify the lipid-lipid steric repulsion and van der Waals attraction terms.

Consider an effective lateral pressure Π introduced to mimic the repulsion between molecules. We noted above that this pressure can be coupled with an area (usually defined as some projection of the atoms in the molecule onto the plane of the bilayer):

$$\mathcal{H}_{\text{ext}}^{steric} \sim \Pi A. \tag{4}$$

Although area is the appropriate conjugate variable to lateral pressure, it is not obvious exactly what physical interpretation it should be given. Most area definitions arise from somewhat *ad hoc* physical scenarios which consider the time scale of molecular motions. The motivating idea is that lipid motion in a bilayer will affect how the lipids pack together. If whole-molecule rotation occurs on a timescale much faster than bond rotation, each (essentially rigid) molecule can be envisaged to sweep out a volume that largely excludes neighboring molecules. In this case, an appropriate area might be a circle whose radius is the perpendicular distance from the axis of rotation to, say, the carbon furthest from that axis; the lipid is effectively modeled as a rotating cylinder.

Such models will be rejected out of hand here, since they make a number of unreasonable predictions, such as: (1) an excessive number of conformers whose chain-chain separations are much larger than those observed (Day 1979) and (2) a very large volume change in going through the main phase transition.

Marčelja (1974) noted that the bilayer volume remains approximately constant (to within about 3%) across the main transition. Thus, if one can calculate the length l of a state in the direction perpendicular to the bilayer, it is a simple matter to define an effective area A for that state:

$$Al = A_0 l_0 \tag{5}$$

where A_0 and l_0 are the area and length of a reference state (usually the all-*trans* state), obtained from experiment. In this method, predictive power is sacrificed for a relatively easy calculation. We will favor an alternative approach.

We assume that the characteristic time for long-axis molecular rotation is much longer than that for rotation around carbon-carbon bonds within the lipid chains. The available data lean in this direction. In this "slow molecular rotation" scheme (Tevlin 1986), areas can be found by first projecting each carbon atom of the chain into the lamella plane, then assigning a circular area to each of these, and finally subtracting the overlapping portions of these circles. The resulting effective area is used below to evaluate intermolecular components of the Hamiltonian.

Pink, Green and Chapman (1980), as well as Tevlin, Jones and Trainor (1987) and Jones, Tevlin and Trainor (1989) have considered modifications of Marčelja's original model (Marčelja 1974). However, these authors did not model real lipids but treated individual acyl chains as the fundamental interacting units; the chains were assumed to be anchored in a plane, with the anchored points at the vertices of a two-dimensional triangular lattice. We will consider two-chain lipids, and will treat these as the fundamental interacting units.

In both the Pink and Tevlin models, the Hamiltonians are the same. The Hamiltonian used a repulsive ΠA term to model steric interactions between chains, and accounted for the internal chain conformation energy. The van der Waals interaction between chains was also accounted for, and contained a distance dependent term which varied $\sim 1/r^5$ in agreement with Salem (1960, 1962), in addition to the conformation-dependent factor used by Marčelja:

$$\mathcal{H}_{vdW} = -\frac{J_0}{2} \sum_{i\langle j \rangle} \frac{S_i S_j}{S_0^2} \left(\frac{\Delta_0}{\Delta_{ij}} \right)^5 \tag{6}$$

Here, i, j are molecules; J_0 is the interaction strength; S is the chain-averaged nematic order, with S_0 being the ground state (i.e. all-*trans* chains) nematic order; Δ_{ij} is the distance between nearest-neighbor chains i, j; Δ_0 is the distance between all-*trans* nearest-neighbor chains. The other terms in the Pink Hamiltonian account for chain internal energy, and the steric interactions between neighboring chains.

The Pink Hamiltonian exhibits all the main features necessary to model the main transition. In fact, its only unsatisfactory aspects are common to the MFT's treatment of the excluded volume effect through a ΠA term, and use of single chains to model two-chain lipids. We will modify the Pink Hamiltonian to model a two-chain lipid, and calculate a two-chain density of states. We will then use this density of states with the Hamiltonian to predict physical properties of the main phase transition of a phospholipid bilayer.

3 The Hamiltonian and Mean Field Theory

First, some notation. The thermodynamic average of a quantity A is

$$\langle A \rangle = \sum_s \exp(-\beta E_s) A_s,$$

where E_s is the energy of state s, found using the Hamiltonian $\mathcal{H}$. We will introduce a mean field Hamiltonian $\mathcal{H}_0$, for which thermodynamic averages will be written $\langle A \rangle_0$.

A single lipid molecule can take a state which we describe using the parameters n_g (the total number of *gauche* bonds in both acyl chains), S_1, S_2 (the nematic order of the first and second acyl chains), and A (the projected area of the molecule). To ease computations, we group these states into binned states α which are averages over small ranges of the continuous parameters S_1, S_2 and A (n_g is not binned). These states α may be degenerate, and we will denote this degeneracy by $D(\alpha)$.

For a collection of lipids such as a bilayer, we can regard the system state s as an ordered set of molecular states:

$$s = \{\alpha_1, \ldots, \alpha_i, \ldots, \alpha_N\} \equiv \{\alpha\},$$

where the subscripts $1, \ldots, N$ label the lipids in the system. The state of the ith lipid will be denoted α_i. The net degeneracy of the system when in state s is then just the product over the N molecules in the system,

$$\mathcal{D} = \prod_{i=1}^{N} D(\alpha_i).$$

The Hamiltonian (Pink et al. 1980) is

$$\mathcal{H} = \mathcal{H}_{\text{int}}^{bond} + \mathcal{H}_{\text{int}}^{steric} + \mathcal{H}_{\text{ext}}^{vdW} + \mathcal{H}_{\text{ext}}^{steric} + \mathcal{H}_{\text{ext}}^{head}, \tag{7}$$

which contains all the major terms necessary for the main transition. The pieces of $\mathcal{H}$ are

$$\mathcal{H}_{\text{int}}^{bond} + \mathcal{H}_{\text{int}}^{steric} = \sum_i E(\alpha_i) \tag{8a}$$

$$\mathcal{H}_{\text{ext}}^{vdW} = -J_0 \sum_{\langle ij \rangle} \Lambda_{ij} \tag{8b}$$

$$\mathcal{H}_{\text{ext}}^{steric} + \mathcal{H}_{\text{ext}}^{head} = \sum_i \Pi A(\alpha_i) \,. \tag{8c}$$

Here, i, j label molecules; α labels binned molecular states; and the summation notation $\langle ij \rangle$ indicates a sum over all *distinct* pairs of nearest neighbors (the term Λ_{ij} is defined below).

The "internal" terms are expressed using the rotational isomeric approximation and a steric conflict energy:

$$E(\alpha_i) = n_g(\alpha_i)E_g + E^{steric}(\alpha_i). \tag{9}$$

$n_g(\alpha_i)$ is the number of *gauche* bonds in conformer α_i and $E_g \approx 0.5$ kcal/mol is the energy due to a *gauche* rotation. Conformers exhibiting steric repulsion are modelled at $E^{steric}(\alpha_i) \sim \infty$, while those with no steric effects have $E^{steric}(\alpha_i) = 0$.

The van der Waals attraction between lipids i, j separated by a distance r_{ij} is similar to that found by Salem (1962), generalized for lipids containing two chains (c is an index over each chain in a pair of lipid molecules):

$$\begin{aligned}\Lambda_{ij} &= \sum_{c,c'=1}^{2} \frac{S_c(\alpha_i)S'_c(\alpha_j)}{4S_0^2}\left(\frac{2r_0}{r_i+r_j}\right)^5 \approx \sum_{c,c'=1}^{2} \frac{S_c(\alpha_i)S'_c(\alpha_j)}{4S_0^2}\left(\frac{r_0}{r_i}\cdot\frac{r_0}{r_j}\right)^{5/2} \\ &= \frac{(S_1(\alpha_i)+S_2(\alpha_i))/2}{S_0}\cdot\frac{(S_1(\alpha_j)+S_2(\alpha_j))/2}{S_0}\cdot\left(\frac{A_0^2}{A(\alpha_i)A(\alpha_j)}\right)^{5/4} \end{aligned} \quad (10)$$

Here, S_0, A_0 and r_0 are the nematic order, projected area and effective radius, respectively of the ground state (all-*trans* chains). The inter-lipid distance is taken to be roughly equal to the sum of the effective radii of the two molecules: $r_{ij} = r_i + r_j$.

Salem showed that the van der Waals interaction between two long, straight, parallel chains separated by distance r goes as r^{-5}. The product form for the distance dependence is simply an approximation which allows mean field theory to be applied more easily.

The "external" terms are of the Marčelja pressure-area form (Marčelja 1974) where Π is a lateral pressure to be fitted (typically, $\Pi \sim 30$ dyne/cm). For each state α, the projected area is $A(\alpha)$, and is calculated by the method of Tevlin (1986) described above, in which slow molecular rotation is assumed.

Putting the various pieces together, the Hamiltonian becomes

$$\mathcal{H} = -J_0\sum_{\langle ij\rangle}\frac{1}{4}\sum_{c,c'} I_c(\alpha_i)I_{c'}(\alpha_j) + \sum_i(\Pi A(\alpha_i) + E_g n_g(\alpha_i)), \quad (11)$$

where

$$I_c(\alpha_i) = (S_c(\alpha_i)/S_0)(r_0/r(\alpha_i))^{5/2} \quad (12)$$

is the van der Waals interaction factor. We also define $I(\alpha_i) = \frac{1}{2}\Sigma_c I_c(\alpha_i)$; the factor of 1/2 is included to allow a direct correspondence of the van der Waals coupling in this model with that of the single chain models such as those studied by Pink et al. (1980), and Tevlin et al. (1987).

The partition function can now be written in the usual way as a sum over the possible states of the system:

$$Z = \sum_s \mathcal{D}(s)\exp(-\beta E(s)) = \exp(-\beta G). \quad (13)$$

Here, the state of the system s is a set of molecular states α; the states α consist of sets of conformers with the same binned internal parameters n_g, A, S_1 and S_2. Because of this, the partition function is transformed into a sum over these internal parameters rather than the usual sum over system energies. Thus $\mathcal{D}(s)$ in the above equation is a generalization of the usual density of states.

If a reference system Hamiltonian $\mathcal{H}_0$ produces a free energy G_0, and if the total perturbation energy to $\mathcal{H}_0$ is $W_N = \Sigma\Sigma_{i<j}^N w(i,j)$, then we can write the true free energy G as a series (Zwanzig 1954)

$$\beta G = \beta G_0 + \beta\langle W_N\rangle_0 - \frac{1}{2}\beta^2(\langle W_N^2\rangle_0 - \langle W_N\rangle_0^2 + \ldots \qquad (14)$$

The deviation W_N is defined by

$$\mathcal{H} = \mathcal{H}_0 + W_N. \qquad (15)$$

An important feature of this expansion (sometimes referred to as the "λ-expansion") is that the first-order approximation gives an upper bound on the free energy of the system of interest, regardless of the choice of "reference" system $\mathcal{H}_0$. An upper bound G_1 for the true free energy G is

$$G \leq G_1 = G_0 + \langle W_N\rangle_0 \ . \qquad (16)$$

Since this relation holds for *any* reference system, we can get the best estimate of the free energy by parameterizing the reference system and minimizing the resulting G_1 with respect to the parameters. By making this parameterization sufficiently flexible, one can expect to bring the relation (16) closer to equality.

Once we choose a mean field reference system $\mathcal{H}_0$ for the lamella, the term W_N is fixed. Our choice is a parameterized mean field system:

$$\mathcal{H}_0 = X\frac{-J_0 z\Phi}{2}\sum_i I(\alpha_i) + \sum_i(E_g n_g(\alpha_i) + \Pi A(\alpha_i)) \qquad (16a)$$

$$W_N = \mathcal{H} - \mathcal{H}_0 = \frac{-J_0}{2}\sum_i I(\alpha_i)\left(\sum_{\langle j\rangle} I(\alpha_j) - zX\Phi\right) \qquad (16b)$$

where X is the parameter for minimizing G_1 and Φ is the usual molecular field (defined below). The new partition function Z_0 can be written

$$\begin{aligned} Z_0 &= \sum_{\alpha_1}\sum_{\alpha_2}\cdots\sum_{\alpha_N} D(\alpha_1)\ldots D(\alpha_N)\exp\left(-\beta\sum_i E_0(\alpha_i)\right) \\ &= \left[\sum_\alpha D(\alpha)\exp(-\beta E_0(\alpha)\right]^N \\ &= \mathcal{Z}_0^N \end{aligned} \qquad (17)$$

where $E_0(\alpha)$ is the energy of a single lipid in state α, and $D(\alpha)$ is the degeneracy of state α. $E_0(\alpha)$ depends on Φ:

$$E_0(\alpha) = X\frac{-J_0 z\Phi}{2} I(\alpha) + \Pi A(\alpha) + E_g n_g(\alpha), \tag{18a}$$

$$\Phi = \frac{1}{\mathcal{Z}_0}\sum_{\alpha} I(\alpha)\exp(-\beta E_0(\alpha)). \tag{18b}$$

To determine a *least* upper bound on the free energy we require

$$\frac{\partial G_1}{\partial X} = -\frac{NJ_0 z}{2}\Phi\frac{\partial\Phi}{\partial X}(X + 2(1-X)) = 0. \tag{19}$$

We set $X = 2$ in the formulae for $\mathcal{H}_0$ and W_N and obtain a free energy

$$G_1 = -kT\ln Z_0 + \frac{NJ_0 z}{2}\Phi^2, \tag{20}$$

from which we can find all the other thermodynamic functions of interest. For example the enthalpy change ΔH_1 across the transition is

$$\Delta H_1 = -T\Delta\left.\frac{\partial G_1}{\partial T}\right|_{\Pi} \tag{21}$$

with

$$-T\left.\frac{\partial G_1}{\partial T}\right|_{\Pi} = kT\ln Z_0 + NE_g\langle n_g\rangle_0 + N\Pi\langle A\rangle_0 - NJ_0 z\Phi^2. \tag{22}$$

4 The Density of States

The lipid chain lengths of interest are in the range $12 \leq n_c \leq 22$ with n_c even; unsaturated bonds will not be considered. Thus, the worst case density of states calculation would need to find the properties of some $3^{2\times 22} \sim 10^{20}$ states, which is far too many to examine individually. We will follow Tevlin (Tevlin 1986, Tevlin et al. 1987) and evaluate the properties of a randomly chosen fraction Ω/Ω_0 of these states, and from this estimate the density of states.

The central element in this approach is the diamond lattice approximation (the "DLA;" see Figure 4), under which computer modeling of acyl chains becomes a relatively fast process requiring mostly integer operations. The approximations are small distortions to the true acyl chain geometry, which cause the chain atoms to lie on the sites of a diamond lattice (hence the moniker).

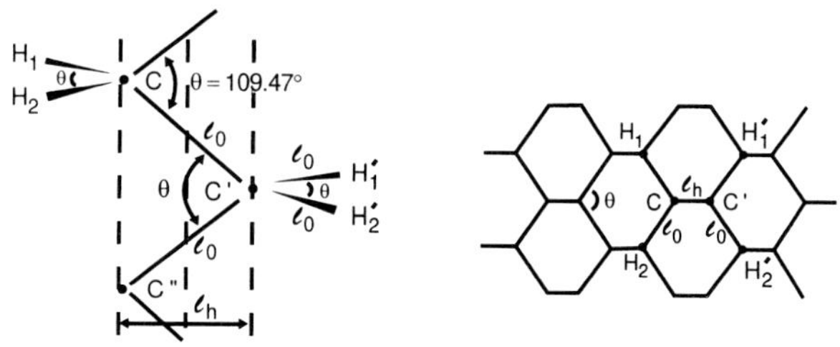

Figure 4: Chain geometry defining the DLA (Diamond Lattice Approximation).

The DLA is a modification of the rotational isomeric state scheme in which rotations around C–C bonds are restricted to three states $(g^{\pm}, t)$ separated by a rotation angle of 120°. Furthermore, all bond angles are set to the tetrahedral angle; this is a distortion of $\sim 3°$. The final alteration is to adjust the C–H bond length to equal the C–C bond length. This 50% increase in CH bond length simplifies the detection of states with steric conflict (see below), but does not affect the calculation of state properties (Tevlin 1986, Jones 1991).

With these approximations, all atoms of an acyl chain in any conformation coincide with the sites of an imaginary diamond lattice. Because of this, acyl chain states can be specified by their bond rotation sequence (e.g., $g^-ttg^+g^+tg^-tt\ldots$), and a simple table look-up (Tevlin 1986) translates this sequence into a set of sites on the diamond lattice. Steric conflict can then easily be found since it is signaled by more than one atom occupying the same site. This is especially important for multiple chain simulations such as ours, where no fast, simple criterion exists for determining the presence of chain-chain conflict.

In addition to this chain-chain conflict, simulations can also exhibit chain-head conflict, signaled whenever any of the atoms in a chain is placed above the plane of the glycerol group. Any state exhibiting steric conflict is rejected out of hand by our simulation procedure, effectively modeling such states at infinite energy. Consequently, terms in the Hamiltonian which are unique to such states can be dropped.

The DLA does not address the relative physical placement of the two acyl chains. For lipids in the crystalline state, atomic positions have been determined from X-ray scattering experiments and these have been used to model lipid structure near T_m (Elder, Hitchcock, Mason and Shipley 1977; Pearson and Pascher 1979). We use this data here to determine the relative placement of the acyl chains, with the result that only a slight adjustment is needed for the two chains to lie on the same imaginary diamond lattice. The geometrical distortion involved is of the same magnitude as the original DLA.

The DLA also does not address the modeling of the glycerol group. We assume that the glycerol group is completely rigid, in accord with experimental observations.

5 The Packing Problem

The calculation of the density of states accounts for all intramolecular (as opposed to just intrachain) steric conflicts. What about intermolecular steric hindrances as a function of the average spacing d between molecules? This requires that one calculate a function $f(\alpha_1, \alpha_2; d)$ of the molecular states α_i, and is clearly a much larger task than that of finding the density of states as described above. Its justification arises from its improvement over the somewhat *ad hoc* pressure-area repulsion term.

The general idea behind the function $f(\alpha_1, \alpha_2; d)$ comes from considering the following situation. Suppose one has two objects which may assume a variety of states α_i, and whose interaction depends on the states and the relative positions of the two objects. Further, each state α_i is composed of a (possibly large) number $D(\alpha_i)$ of degenerate molecular configurations; the states α_i are taken to be those of a single object in free space. Now we allow the objects to interact, with object 1 in a fixed orientation at the origin, and object 2 at a relative position $\vec{d}$ (and with relative orientation θ). We consider their steric conflicts only, leaving other interactions (such as the van der Waals term) to the Hamiltonian. A crude rendering of the two-dimensional picture applicable to a pair of lipids is shown in Figure 5.

If the objects did not interact, the net number of possible configurations would be $D(\alpha_1)D(\alpha_2)$. However, since they do interact sterically, they are able to assume only some fraction p of these configurations. In all, the two objects will have only

$$p(\alpha_1, \alpha_2; d, \phi, \theta)D(\alpha_1)D(\alpha_2)$$

states available in this configuration. Clearly, this "degeneracy reduction factor" p will have an effect on the statistical mechanics of N such interacting objects.

The fraction p defined above is too complicated to use and contains more information than needed in an initial calculation. We therefore reduce p to an averaged version

$$f(\alpha_1, \alpha_2; d) = C \int_\phi \int_\theta p(\alpha_1, \alpha_2; d, \phi, \theta) d\phi d\theta, \tag{23}$$

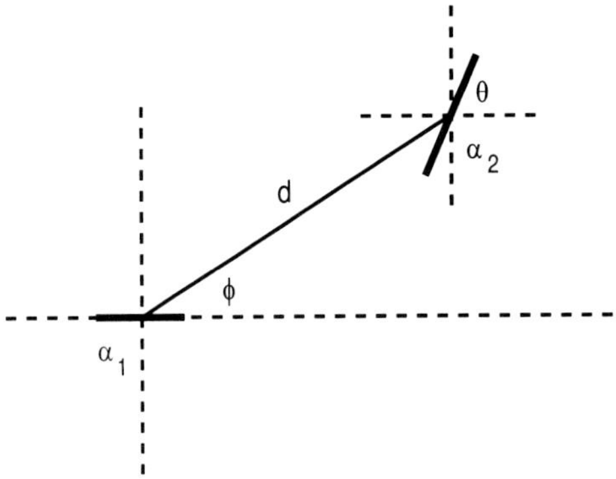

Figure 5: Relative positioning of objects, and definition of the geometrical quantities referred to in the text: the separation vector $\vec{d} = (d, \phi)$, and the relative orientation θ.

where the factor C is a normalization constant. From here on we will only consider the averaged function f and take the two interacting objects to be two-chain lipid molecules, with the degeneracy D being that of a two-chain lipid. The resulting degeneracy of a pair of molecules (separated by distance d) in state α_1, α_2 is then $f(\alpha_1, \alpha_2; d)D(\alpha_1)D(\alpha_2)$. Note that our calculation (in the previous section) of the density of states of a two-chain lipid accounts for intramolecular steric effects explicitly in D. Here, we add to that calculation to include intermolecular effects which can be accounted for *on average* through f.

We can remove the term $\mathcal{H}_{\rm ext}^{steric}$ from the Hamiltonian since the effect of intermolecular steric hindrance will now be accounted for by the degeneracy reduction function f. The energies of this new Hamiltonian $\mathcal{H}'$ will be denoted $E'(s)$, where s labels a state, as before. This still leaves the question of what to use for $\mathcal{H}_{\rm ext}^{head}$ (cf. equation (8c)). Certainly, for some lipid species with a net charge on the head group (such as phosphatidylserine), this term would be non-zero and dependent on the average lipid separation. However, for the sake of illustration we will ignore these effects, thus making the head group interaction energy independent of any bilayer expansion.

We rewrite the partition function (17) explicitly in terms of the states α:

$$Z_0' = \sum_{\{\alpha\}} \mathcal{D}(\{\alpha\}) \exp(-\beta\mathcal{H}_0') = \sum_{\{\alpha\}} \exp(-\beta\mathcal{H}_0' + \ln \mathcal{D}(\{\alpha\})) \qquad (24)$$

where

$$\mathcal{D}(s) = \mathcal{D}(\{\alpha\}) = \prod_i D(\alpha_i) = \prod_i \left(D(\alpha_i) \prod_{\langle j \rangle} D(\alpha_j) \right)^{1/(z+1)} . \tag{25}$$

$z = 6$ is the number of nearest neighbors of i in the $\langle j \rangle$ product. This form for Z_0' is — so far — similar to the Z_0 except that there is no pressure term. In both cases we can write the effective energy of state $s = \{\alpha\}$ as $\mathcal{H}_s^{eff} = \mathcal{E}_s - kT \ln D_s$, where $\mathcal{E}$ may stand for either $\mathcal{H}_0$ or $\mathcal{H}_0'$. The difference comes from modifying the system state degeneracy $\mathcal{D}(s)$ with the function f applied to each nearest-neighbor pair of molecules. Define

$$D_s' = \prod_{i\langle j \rangle} (D(\alpha_i) D(\alpha_j) f(\alpha_i, \alpha_j; d))^{1/(z+1)}, \tag{26}$$

and that the effective state energies in each case are now

$$\begin{aligned} \mathcal{H}_s^0 &= \mathcal{H}_s - kT \ln \mathcal{D}(s) \\ \mathcal{H}_s^{0'} &= \mathcal{H}_s' - kT \ln \mathcal{D}_s' \\ &= \mathcal{H}_s' - kT \ln \mathcal{D}(s) - \frac{kT}{z+1} \sum_{i\langle j \rangle} \ln f(\alpha_i, \alpha_j; d). \end{aligned} \tag{27}$$

The last term is positive, since $0 \le f \le 1$. This takes the place of the ΠA steric repulsion term (hidden in $E(s)$; see equation (8c)), and has the correct behavior since, as the lipids come closer together (at constant temperature), f decreases and the system energy increases.

The partition function Z_0' leads naturally to a Helmholtz (rather than Gibbs) free energy $F' = U' - TS'$:

$$F' = -kT \ln Z_0',$$

to first order. From this, the equivalent of a lateral pressure Π is found from

$$\Pi = - \left. \frac{\partial F'}{\partial A} \right|_T = - \left. \frac{\partial F'}{\partial d} \frac{\partial d}{\partial A} \right|_T . \tag{28}$$

Here, the average intermolecular distance is d, from which the average molecular area is approximated by a hexagon of area $A = \sqrt{3}/2d^2$. From equation (28), we see that this formulation of the effective state energy (equation (27)) is a natural description of the constant area (i.e., constant d) situation.

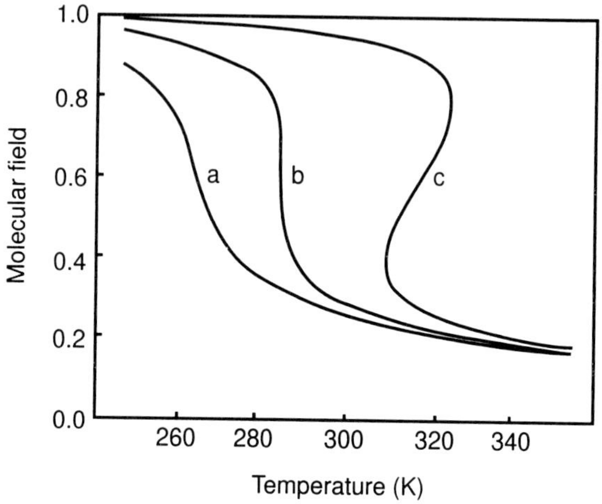

Figure 6: Plot of the molecular field Φ versus temperature T for three cases of interest, using the two-chain density of states for DPPC. Curve a shows a continuous transition from gel to fluid ($J_0 = 0.550 \times 10^{-13}$erg); curve b the critical case ($J_0 = 0.695 \times 10^{-13}$erg); curve c a first-order transition from gel to fluid ($J_0 = 0.940 \times 10^{-13}$ erg). In all cases, $\Pi = 44.0$ dyne/cm, and $E_g = 500$ cal/mol.

6 Results and Discussion

Once state occupancies are determined, all thermodynamic properties follow. How to determine the occupancies is not immediately obvious, but if we assume that they are known then we can make some observations concerning the behavior of the molecular field.

We note that $\Phi(T)$ can be multivalued, though the inverse function $T(\Phi)$ is always single valued. Three cases of interest are shown in Figure 6, though only the case of the first-order transition is of real interest here. In this case, Φ may be multivalued, depending on T. The multivalued regime deserves some comment, since only one solution can be correct: The solution with the lowest free energy is "physical." We note that both the extreme solutions Φ^g and Φ^f are "stable," in that they are always local minima of $G(\Phi)$, while the middle solution Φ^m is always a maximum. These points are illustrated in Figure 7.

Finding the state occupancies to get $\Phi(T)$ is done using an initial guess and an iterative set of equations for the molecular field. A self-consistent method for evaluating $\Phi(T; J_0, \Pi)$ is provided by equations (18b), (16a) and (17):

$$\Phi \rightarrow \mathcal{H}_0 \rightarrow Z_0 \rightarrow \Phi. \tag{29}$$

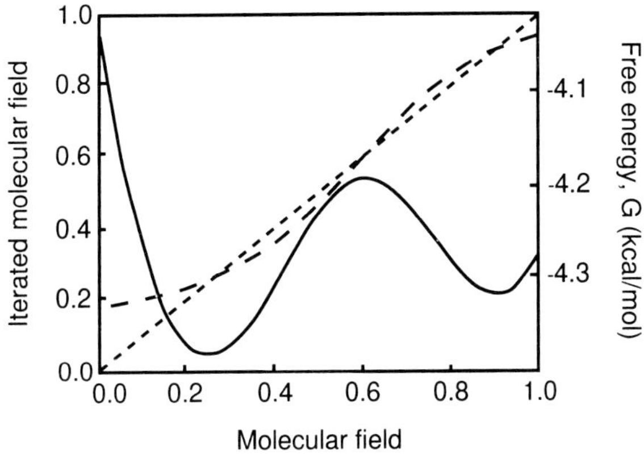

Figure 7: The metastable regime. The solid curve shows the free energy G versus the molecular field Φ in the mean field approximation. Note the curve's three extrema which correspond (left to right) to Φ^f, Φ^m, and Φ^g. Solution Φ^m is a maximum and so is unstable, whereas the other solutions are stable. In this case, the system is in the fluid state, since the Φ^f solution has a lower free energy. Also shown is the molecular field iteration of equation (29), demonstrated by the dashed (Φ vs. Φ) and dash-dotted (Φ_{iter} vs. Φ) The parameters used were for DPPC: $J_0 = 0.964 \times 10^{-13}$ ergs, $\Pi = 42.3$ dyne/cm, $E_g = 500$ cal/mol, $T = 316$K.

The mechanics of doing this, however, are best illustrated in the iterative map of Figure 7. This Figure shows Φ versus Φ (the straight dotted line) and Φ_{iter} versus Φ (dashed curve). The Φ_{iter} curve is derived from equation (29) by giving it a Φ as input, and then doing a single iteration to get Φ_{iter}. We can now use the result of a single iteration Φ_{iter} as the input to the next iteration; doing this repeatedly gives us a sequence of iteration pairs $\{(\Phi_0, \Phi_1), (\Phi_1, \Phi_2), \ldots\}$, with the second member of a pair being the iterated result of the first member. This sequence of pairs can be regarded as an ordered set of points in Figure 7; when these are joined by line segments, a series of horizontal and vertical lines results which are bounded by the dashed and the dotted curves. Furthermore, following this jagged path on Figure 7 (this path isn't drawn, however) leads to a fixed point on the curve. The solutions to equation (29) are then the values of Φ at the possible fixed point(s).

There are two cases to consider. The simple case is when conditions are such that there is only one solution (e.g. $J_0 = 0$, or $T \ll T_m$). In this case, we get an iterative map for which there is only one fixed point, and it doesn't matter where we start. However, in the second case, there are three possible solutions, two of which are stable. These two solutions can be found by starting the two iterations at $\Phi_0^g = 1$ (the gel state) and $\Phi_0^f = 0$ (the fluid state). For both of these cases, this two-iteration procedure will produce the stable Φ solution(s) (in the simple one-solution case the two iterations will converge to the same Φ).

Iteration of equation (29) is done until a stable point is reached, as determined by a change in Φ from one iteration to the next falling below a tolerance $\epsilon \simeq 10^{-4}$. This gives the molecular field Φ^g and Φ^f (corresponding to the gel and fluid initial conditions) which will be identical if the system is stable, and different if it is metastable. The fields are considered "identical" when $|\Phi^g - \Phi^f| \leq \delta$, where we typically set $\delta \simeq 10\epsilon$. If the molecular fields are not identical, the free energies corresponding to each phase are calculated, and the solution with the lower free energy is selected as the physical solution.

Thermodynamic properties were calculated for a suite of lipids. In each case, the density of states used was for a two-chain lipid with the second chain shorter than the first by just a single CH_2 group; we will therefore identify the lipids by the length n_{c1} of just the longer chain. The six lipids used were the even-n_{c1} lipids of biological importance: $12 \leq n_{c1} \leq 22$. However, because the density of states is built without regard for head group structure, these densities of states can be applied to the main transition of both phosphatidylcholines (PC's) and phosphatidylethanolomines (PE's).

On each graph of Figures 8 and 9 (corresponding to $n_{c1} = 12$ and 22), there are three points marked. The point "PE-Expt" (marked by $\bullet$) corresponds to the experimentally observed enthalpy and transition temperature for lipids with the PE head group. Similarly, the "PC-Expt" point (marked by $\otimes$) corresponds to the $\Delta H, T_m$ observed experimentally for PC's. The point "PC-DOS1" (marked by $\odot$) indicates the $\Delta H, T_m$ values which would be achieved using the 2-chain lipid model if the J_0, Π values found by Jones et al. (using a 1-chain density of states D_1) were used in the current calculation.

In order to compare these models, we note that the results presented here have all quantities on a *per molecule* basis. The parameter values corresponding to these points for $n_{c1} = 12$ to 22 are collected in Table 1 (PC-DOS1), Table 2 (PC-Expt), and Table 3 (PE-Expt). The dependence of the parameters on chain length is slightly nonlinear, if one is interested in an exact fit to the data. However, the nonlinearity is quite slight, as may be expected.

Anyhydrous lipids of different species but with the same chain length have nearly identical main transition properties (see Cevc and Marsh 1987, p. 241). This implies that the differences between PE's and PC's (PE's have a smaller headgroup (NH_3^+) than PC's ($N(CH_3)_3^+$)) are a result of headgroup-solvent interactions. Most of the transition enthalpy arises from "chain melting" and so ΔH should be relatively unaffected by changes in the headgroup, with the deviations from this behavior caused largely by differences in the amount of

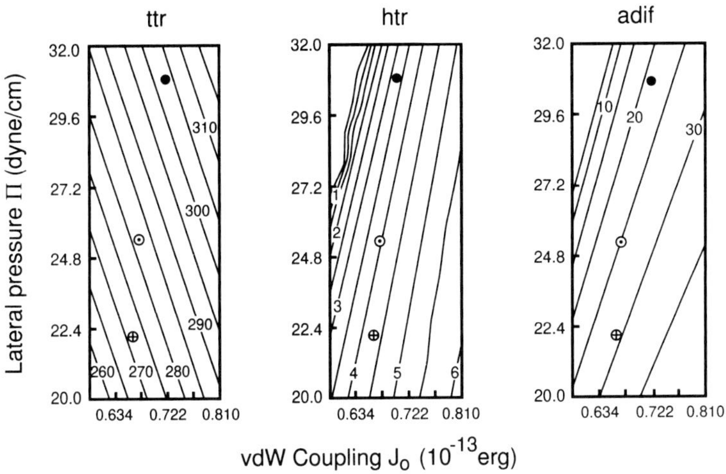

Figure 8: Plot of gel-fluid transition functions for $n_{c1} = 12$ carbons. Area change contours are in Å^2. Enthalpy contours are in kcal/mol, and transition temperature contours are in K. The symbols on each graph locate points having the following meaning: PC-DOS1, $\odot$; PC-Expt, $\oplus$; PE-Expt, $\bullet$.

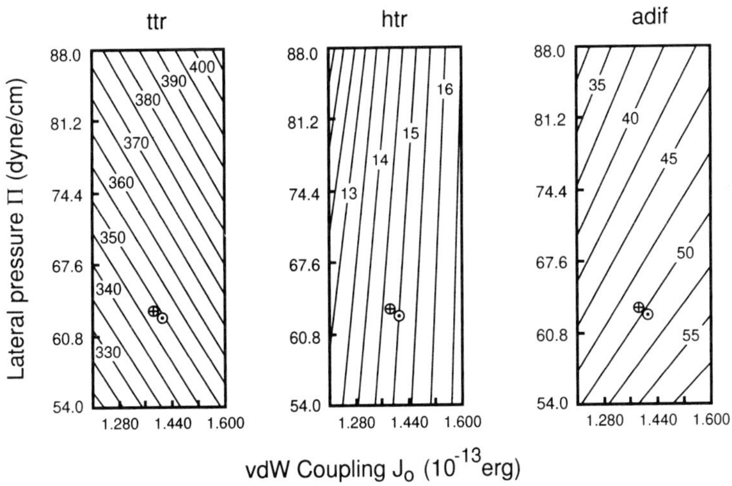

Figure 9: Plot of gel-fluid transition functions for $n_{c1} = 22$ carbons. Area change contours are in Å^2. Enthalpy contours are in kcal/mol, and transition temperature contours are in K. The symbols on each graph locate points having the following meaning: PC-DOS1, $\odot$; PC-Expt, $\oplus$.

Table 1: MFT Results for a Homologous Series of PC's. This table shows the parameter values necessary for the current model to reproduce the results achieved by Tevlin (1986) (J_0 in 10^{-13}erg, Π in dyne/cm).

n_c	J_0	Π	$T_m(K)$	ΔH (kcal/mol)
12	0.676	25.3	282.1	4.0
14	0.829	31.1	296.0	6.7
16	0.910	43.5	311.4	7.8
18	1.081	48.0	324.9	10.5
20	1.256	54.0	336.8	12.9
22	1.404	62.2	348.0	15.1

Table 2: MFT Results for a Homologous Series of PC's. This table shows the parameter values necessary for the two-chain lipid model presented here to reproduce the experimental results (J_0 in 10^{-13}erg, Π in dyne/cm).

n_c	J_0	Π	$T_m(K)$	ΔH (kcal/mol)
12	0.665	22.1	272.1	4.30
14	0.804	32.6	296.1	6.21
16	0.964	42.3	314.1	8.72
18	1.123	47.8	328.1	11.0
20	1.287	53.3	338.6	13.2
22	1.384	62.5	347.1	14.9

Table 3: MFT Results for a Homologous Series of PE's. This table shows the parameter values necessary for the two-chain lipid model presented here to reproduce the experimental results (J_0 in 10^{-13}erg, Π in dyne/cm).

n_c	J_0	Π	$T_m(K)$	ΔH (kcal/mol)
12	0.720	30.7	302.6	3.70
14	0.846	41.8	322.6	5.74
16	0.982	53.4	337.5	7.89
18	1.126	57.5	347.5	10.5
20	1.267	63.9	355.6	12.6

hydration between species. Furthermore, these changes have little effect on the fundamental nature of the van der Waals interactions between the acyl chains. For example, the hydrocarbon chains in both PE's and PC's see essentially no water, while the headgroups see differing amounts. Thus the van der Waals interactions will be essentially unmodified between PE's and PC's, and the differences observed experimentally are almost entirely reflected in the lateral pressure parameters.

We now look at the representative phospholipid dipalmitoylphosphatidylcholine (DDPC) in greater depth; results for the other lipids are similar. At the main transition the number of *gauche* bonds per chain undergoes a jump from roughly 0.3 to about 3.5. This is in fair agreement with the results of Seelig and Seelig (1974), and is somewhat less than the 3.65 *gauche* bonds calculated by Jones *et al.* (1989). The nematic order also shows the expected drop in magnitude at T_m, though the value just above the transition is somewhat higher than is measured (see, e.g. Seelig and Seelig (1974)); Jones et al. (1989) found a similar result. As in the earlier one-chain work, the two-chain predictions here likely suffer from neglecting whole-body rotation and collective chain tilting.

Finally, if we make a direct comparison between the single-chain and two-chain systems, the area change across the transition is roughly 95% for the two-chain density of states, and 105% for the single-chain density. While both of these are far greater than the 20% or so observed, it is apparent that the current system affords some improvement; this may be reflecting the more accurate accounting of intertwining of the acyl chains. An open question then is whether an $N \sim 6$ chain simulation would display even greater predictive power.

The magnitude J_0 of the van der Waals interaction is also open to some question. Estimates of J_0 have been made from various theoretical arguments (e.g., see Salem 1960, Marčelja 1974, Gruen 1979), and for DPPC these seem to fall in the range 1–4 $\times 10^{-13}$erg. The Jones et al. value for the single-chain system was $J_0 = 0.578 \times 10^{13}$erg; here, $J_0 = 0.964 \times 10^{-13}$erg, in closer agreement with the theoretical estimates. At the transition, the associated high chain areas lead to a large change in the system's net van der Waals energy, since this energy goes as $d^{-5/2}$. The somewhat better prediction of molecular area change across the transition in the two-chain system produces a more realistic J_0.

It is clear, however, that the areal expansion of the bilayer is still predicted to be much larger at the main transition than observed. It appears that the problem is still poor modeling of the cooperativity of the chains, and their propensity for entanglement. The model discussed here accounts for the geometry of only a given chain's molecular-partner chain; there are still — on average — five others to worry about. Because of this, we suggest that to make a good prediction of enthalpies, transition temperatures, *and* areal expansions, one must replace the pressure-area repulsion term. In its place, one needs a better method for treating the cooperative packing of the acyl chains. Though modeling such cooperative packing is not easy, the mean-field model presented here and the molecular dynamics studies of Jerala and his co-workers (Jerala, Almeida and Biltonen 1996) indicate that if cooperative packing were well treated, then only the simpler case of interacting single acyl chains would be needed to build a model capable of accurately predicting both mechanical and themodynamic properties of the lamellar phases.

Acknowledgement

The Natural Sciences and Engineering Research Council (NSERC) of Canada provided funding while much of the work discussed here was carried out. This work was inspired by Josef Locke and Count John McCormack. We would like to thank Flann O'Brien, P. McPhellimey, and Jem B.G. Casey.

References

Blume, M. (1983) Apparent molar heat capacities of phosopholipids in aqueous dispersion. Effects of chain length and head group structure. *Biochemistry* **22**, 5436–5442.

Cevc, G. and Marsh, D. (1987) *Phospholipid Bilayers: Physical Principles and Models.* Cell Biology: A Series of Monographs, Volume 5. New York: Wiley-Interscience.

Day, J. (1979) *Phase Transitions and Statistical Mechanics of Lipid Bilayer Systems.* Ph.D. Thesis, Boston University, Boston, Massachusetts.

Elder, M., Hitchcock, P., Mason, R. and Shipley, C.G. (1977) A refinement analysis of the crystallography of the phospholipid 1, 2-dilauroyl-DL-phosphatidyl ethanolamine, and some remarks on lipid-lipid and lipid-protein interactions. *Proceedings of the Royal Society of London, Series A* **354**, 157–170.

Flory, P.J. (1969) *Statistical Mechanics of Chain Molecules.* New York: Wiley-Interscience.

Gorter, E. and Grendel, F. (1925) On bimolecular layers of lipids on the chromocytes of the blood. *Journal of Experimental Medicine* **41**, 439–443.

Gruen, D.W.R. (1979) A statistical mechanical model of the lipid bilayer above its phase transition. *Biochimica et Biophysica Acta* **595**, 161–183.

Harlos, K. (1978) Pretransitions in the hydrocarbon chains of phosphatidylethanolamines. *Biochimica et Biophysica Acta* **511**, 348–355.

Israelachvili, J., Marčelja, S. and Horn, R.G. (1980) Physical principles of membrane organization. *Quarterly Reviews of Biophysics* **13**, 121–200.

Janiak, M.J., Small, D.M., and Shipley, G.G. (1976) Nature of the thermal pretransition of synthetic phospholipids: dimyristoyl and dipalmitoyllecithin. *Biochemistry* **15**, 4575–4580.

Janiak, M.J., Small, D.M. and Shipley, G.G. (1979) Temperature and compositional dependence of the structure of hydrated dimyristoyl lecithin. *Journal of Biological Chemistry* **254**, 6068–6078.

Jerala, R., Almeida, P.F.F. and Biltonen, R.L. (1996). Simulation of the gel-fluid transition in a membrane composed of lipids with two connected acyl chains: application of a dimer-move step. *Biophysical Journal* **71**, 609–615.

Jones, F. (1991) *Statistical Mechanics of the Main Phase Transition of a Lipid Bilayer.* PhD Thesis, Department of Physics, University of Toronto, Canada.

Jones, F., Tevlin, P. and Trainor, L.E.H. (1989) Phase transition of lipid bilayers. II. Mean field theory. *Journal of Chemical Physics* **91**, 1918–1925.

Lentz, B.R. Barrow, D.A. and Hoechli, M. (1980) Cholesterol-phosphatidylcholine interactions in multilamellar vesicles. *Biochemistry* **19**, 1943–1954.

Marčelja, S. (1974) Chain ordering in liquid crystals. II. Structure of bilayer membranes. *Biochimica et Biophysica Acta* **367**, 165–176.

McNeil, P.L. (1991) Cell wounding and healing. *American Scientist* **79**, 222–235.

Pearson, R.H. and Pascher, I. (1979) The molecular structure of lecithin dihydrate. *Nature* **281**, 499–501.

Pink, D.A. , Green, J.J. and Chapman, D. (1980) Raman scattering in bilayers of saturated phosophatidylcholines. Experiment and theory. *Biochemistry* **19**, 349–356.

Ruocco, M. and Shipley, G.G. (1982) Characterization of the sub-transition of hydrated dipalmitoylphosphatidylcholine bilayers. *Biochimica et Biophysica Acta* **691**, 309–320.

Sackmann, E. (1983) Physical foundations of the molecular organization and dynamics of membranes. In W. Hoppe, W. Lohmann, H. Markl and H. Ziegler, eds., *Biophysics*. New York: Springer-Verlag, pp. 425–457.

Salem, L. (1960) The calculation of dispersion forces. *Molecular Physics* **3**, 441–452.

Salem, L. (1962) Attractive forces between long saturated chains at short distance. *Journal of Chemical Physics* **37**, 2100–3223.

Schnur, J.M. (1993) Lipid tubules: a paradigm for molecularly engineered structures. *Science* **262**, 1669–1676.

Seddon, J.M., Cevc, G., Kayne, R.D. and Marsh, D. (1984) X-ray diffraction study of the polymorphism of hydrated diacyl- and dialkyl-phosphatidylethanolamines. *Biochemistry* **23**, 2634–2644.

Seelig, J. and Browning, J. (1978) General features of phospholipid conformation in membranes. *FEBS Letters* **92**, 41–44.

Seelig, J. and Seelig, A. (1974) The dynamic structure of fatty acyl chains in a phospholipid bilayer measured by deuterium magnetic resonance. *Biochemistry* **13**, 4839–4845.

Singer, S. and Nicolson, G. (1972) The fluid mosaic model of the structure of cell membranes. *Science* **175**, 720–731.

Tardieu, A., Luzzati, V. and Reman, F.C. (1973) Structure and polymorphism of the hydrocarbon chains of lipids: a study of lecithin-water phase. *Journal of Molecular Biology* **75**, 711–733.

Tevlin, P. (1986) *Statistical Mechanics of Model Biological Membranes*. PhD Thesis, Department of Physics, University of Toronto, Canada.

Tevlin, P., Jones, F.P. and Trainor, L.E.H. (1987) Phase transitions of lipid bilayers. I. Density of states calculation. *Journal of Chemical Physics* **87**, 5483–5496.

Yeagle, P.L. (1993) *The Membranes of Cells*. Second edition. New York: Academic Press.

Zwanzig, R.W. (1954) High-temperature equation of state by a perturbation method. I. Nonpolar gases. *Journal of Chemical Physics* **22**, 1420–1426.

Multi-neuron Interactions in Neural Network Models of Associative Memory

Allan E. Busch and L.E.H. Trainor

1 Introduction

The human brain's ability to store and retrieve innumerable events, motor movements, facts and other minutiae of every day life quickly, accurately and with relative ease is truly amazing! We can recognize a face in a crowd, a poem from a verse, and an entire song from just a few notes. Even though the brain is a highly complex system, many of these memory properties are already featured even in simple neural network models. An advantage of these simplified neural network models is that they resemble physical systems. Many of the tools of statistical mechanics that have been applied to such physical systems can thus be carried over in analyzing neural network models.

However, it is still important to capture as much as possible of the complexity of the brain's structure when creating these simplified models. To this end, we have studied the effects of incorporating multiple-order interactions by generalizing the Hopfield neural network model (Hopfield 1982). Inclusion of these interactions improves the correspondence with the way real neurons are connected in the brain and may give rise to interesting and more complex behaviour. Our approach includes both numerical computer simulations and an analysis using a mean field treatment from statistical mechanics.

2 Memory Models and Statistical Mechanics

One of the first applications of statistical mechanics to memory models was carried out by Little more than two decades ago (Little 1974). Over the subsequent twenty years the use of statistical mechanics in memory models became quite popular (for example, Peretto 1984; Amit, Gutfreund and Sompolinsky 1985; van Hemmen and Kuhn 1986; Busch and Trainor 1987; for a general overview see Arbib 1995).

In these models are built-in assumptions which make them resemble physical (non-biological) systems. First, with regard to the format, we consider the brain to have "content-addressable" memory as compared to the "location" addressable memory utilized by, for example, computers. Second, we assume only associative memory: Given partial information for a memory, the brain is capable of recovering the entire memory. Third, memory storage is by means of modification of connections between neurons in the brain, either by altering the number of processes between neurons or by altering the "strength" of interactions of individual processes one with another (from a modeling standpoint, these are equivalent). Although it is generally agreed that memory is stored via the modifications of the neuronal interactions, there is not yet any consensus on the actual "learning algorithm," i.e. the rules by which modifications take place in the brain. In fact, the main differences between various neural network models lies in their learning algorithms. It is also generally assumed in the modeling that the state of the brain is determined entirely by the states of all the neurons at any given time. Finally, we simplify by considering two state neurons: firing and not-firing. While this is a dramatic simplification, we take consolation from the observation that interesting global properties occur readily in physical systems containing large numbers of *simple* interacting elements, such as happens in spin glasses (see, e.g. Sherrington and Kirkpatrick 1975).

Within this framework, it makes sense to consider that memory states can be distinguished from non-memory states of the brain in that they are dynamically stable states which are easy to access and "recover." A memory state then is a point in phase space defined by the possible states of the neurons and acts like an attractor for the neuron dynamics. If started in a nearby state the system will tend to evolve toward the memory, and once there will remain there.

3 The Hopfield model

Mathematical work on neural networks was advanced dramatically by the work of McCulloch and Pitts in 1943, who presented a logical calculus of neural activity. The key assumptions made by McCulloch and Pitts (1943, p. 118) about model neurons remain influential today:

" 1. The activity of the neuron is an "all-or-none" process.

2. A certain fixed number of synapses must be excited at any time and this number is independent of previous activity and position on the neuron.

3. The only significant delay within the nervous system is synaptic delay.

4. The activity on any inhibitory synapse absolutely prevents excitation of the neuron at that time.

5. The structure of the net does not change with time."

The assumptions made by Hopfield in the initial presentation of his path-breaking model (Hopfield 1982; see also Hopfield 1994) are based on these McCulloch-Pitts assumptions. A Hopfield model is composed of a network of idealized binary state neurons, with each neuron connected to all the other neurons in the system. Although Hopfield's original formulation was in terms of neuron potential variables which could take on the values 0 and 1, we follow the current literature in presenting the model in terms of "spin" variables which take on the values of +1 or −1, corresponding to firing or non-firing, respectively.

The numerous synaptic connections from the axon branches of one neuron to the dendrites of another are modeled in a Hopfield net by a single effective weight factor, termed the synaptic interaction (or simply interaction strength). This synaptic interaction strength can be positive (excitatory) or negative (inhibitory). All delays in the transfer of a signal from one neuron to another are also modeled by a single parameter, W.

Let S_i denote the state of neuron i and J_{ij} denote the synaptic interaction strength from neuron j to neuron i. Using asynchronous[a] evolution, a randomly chosen neuron, i, is checked each time step and its state "aligned" to reflect its incoming potential, or "field", h_i, where

$$h_i = \sum_{j=1}^{N} J_{ij} S_j \ , \tag{1}$$

with N the number of neurons. If $h_i > U_i$, the neuron state is set to $S_i = +1$, and if $h_i < U_i$, the neuron state is set to $S_i = -1$, where U_i is the threshold potential of neuron i. For convenience U_i is usually taken to be zero. The evolution proceeds, one neuron at a time, until a steady state is reached in which each neuron remains in an unchanging state, i.e. $h_i S_i > U_i$, for all $i = 1, \ldots N$. Note that the value $1/W$ in this case is the mean attempt rate of the neurons, since this is the time delay for the change of state of one neuron to reach all the other neurons in the system.

To make a system like this into a usable memory model, it is desirable to have any chosen set of *memory states* become the equilibrium states of the system. This can be accomplished by setting the synaptic interaction strengths between the neurons via the algorithm

$$J_{ij} = T_{ij} \equiv \sum_{s=1}^{p} S_i^s S_j^s \ ; \quad T_{ii} = 0 \tag{2}$$

[a]As compared to "synchronous evolution" in which the state of *all* neurons are checked simultaneously each time step.

where S_i^s is the state of neuron i in the sth memory state and p is the number of memory states or patterns. Hopfield chose the interaction matrix to be symmetric between any two neurons. Equation (2) represents an averaging of the single memory prescriptions over the set of p memories, and is a generalization of Hebb's learning rule (Hebb 1949), which is based on the observed increase in the strength of synaptic connections between real neurons when they are firing simultaneously (within a certain time period). This algorithm for specifying the connection between neurons we shall term the *storage algorithm* because through it the system "learns" (stores) specific memory states.

The evolution algorithm and method of storage in Hopfield's model results in the minimization of a global "energy" function. To see this consider the quantity

$$E = -\sum_{i=1}^{N}\sum_{j=1}^{N} J_{ij} S_i S_j \ . \tag{3}$$

The change in E due to a change in state of neuron k is

$$\begin{aligned} \Delta E &= -\Delta S_k \sum_j (J_{kj} S_j) \\ &= -h_k \Delta S_k \end{aligned} \tag{4}$$

where h_k is the effective field acting on the k'th neuron. Thus E is a monotonically decreasing function, and reaches a value which is a local minimum when stability is reached in the network. In this sense, E is analogous to the energy of a dissipative system.

To illustrate the Hopfield model's associative memory properties, we define the state of the system to be an N dimensional vector whose components are the state values $(-1, 1)$ of the individual neurons, S_i. The learning algorithm serves to make the desired set of memory states, $\{S_i^s\}$, correspond to the set of local energy minima. We can thus construct a hypothetical energy contour plot in state space. If we then start the system near one of the assigned memories, corresponding to partial information input about the memory, the evolution of the system will be such that the memory is recalled in full (see Figure 1). The model is thus able to "recall" previously learned memories if it is given a "cue" or partial information. Furthermore, if the partial information is not enough to distinguish between two or more memories, the system will tend to make a decision and recall only one of the memories which it resembles — the probability of a given memory being chosen corresponding to its overlap with the initial (cue) state.

There is a restriction on the number of memories that can be stored such that all memories will be local energy minima, and thus useable memories. This

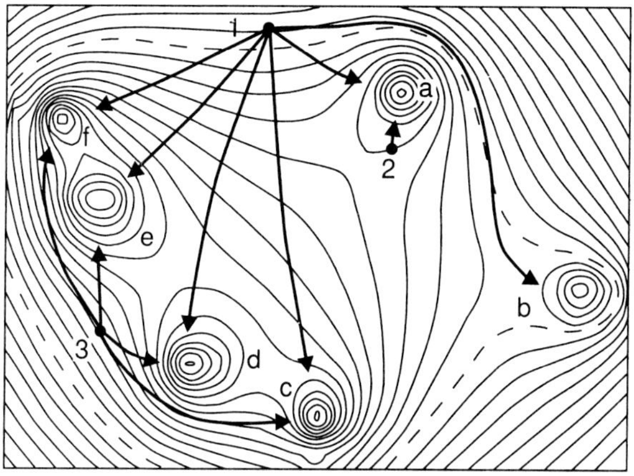

Figure 1: Hypothetical energy contour diagram illustrating the memory recall property of Hopfield's model. Letters a, b, c, d, e and f represent energy minima (memory states) while the numbers 1, 2 and 3 represent initial states that partly overlap one or more of the memories. State 2 then corresponds to an initial state which could evolve only to memory state "a." Initial states 1 and 3 represent "ambiguous" starting states, and can thus evolve to one of a number of states. The only restriction on the energy E is that it must decrease.

is referred to as the neural network *capacity*. Hopfield (1982) observed that the capacity of the network is about $0.15N$ memories. Above this number, the error when recalling memories becomes severe. Consider the fraction of *totally stable* memories (memories which are recalled exactly) as a function of both the total number of imposed memories, p, and the number of neurons, N. Figure 2 shows this for $N = 50, 100, 200$ and 1000 neurons, where the averaging has been done over all memories in a given memory set as well as over a number of randomly chosen memory sets. Plotting the average number of error-free memories versus the number of (chosen) memories, it can be seen (Figure 3) that there is an optimal number of memories for which we get a maximum number of stable ones. These "optimization" curves are beneficial as a criterion in comparing the effects on storage capacity of various modifications to a neural network model, such as the inclusion of higher order neural interactions (see section 5).

We also consider the retrievability of a Hopfield type model by starting the system at an initial system state which differs from a particular memory state by randomly changing a given number of neuron states in that memory. Figure 4 shows the fraction of these initial system states which returned to the memory state as a function of their Hamming distance from the memory

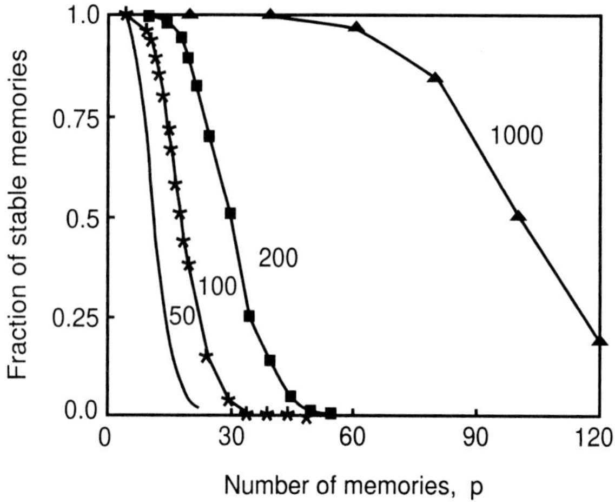

Figure 2: Fraction of totally stable memories versus the number of memories p for systems with 50, 100, 200 and 1000 neurons. Straight-line interpolation is done between points.

state for systems with 100 neurons and separately for 10, 15 and 20 stored memories.

4 The Hopfield Model, "Spin" Systems and Statistical Mechanics

There is a similarity between Hopfield's model and an Ising spin system. By replacing "neuron" with "spin" and "synaptic interaction" with "exchange interaction," the sum over the incoming potentials becomes a sum over local fields from surrounding spins. If the set of memory states $\{S_i^s\}$ is chosen randomly, then we have an infinite-range Ising spin model whose exchange interaction between spins is random and given by the prescription of equation (2). Hopfield alluded to this analogy in his claim that for a random distribution of memories the interactions are effectively random and thus the system becomes a spin glass.

With this similarity in mind, it makes sense to look at how the tools of statistical mechanics can be used to further study the neural network models and explore their pseudo-thermodynamic properties. As we noted above, this important question has been considered by numerous researchers, such as Peretto (1984), Little (1974), and Amit, Gutfreund and Sompolinsky (1985, 1987). The standard approach is to introduce "noise" into neural network models by turning the system into a system in contact with a heat bath at a

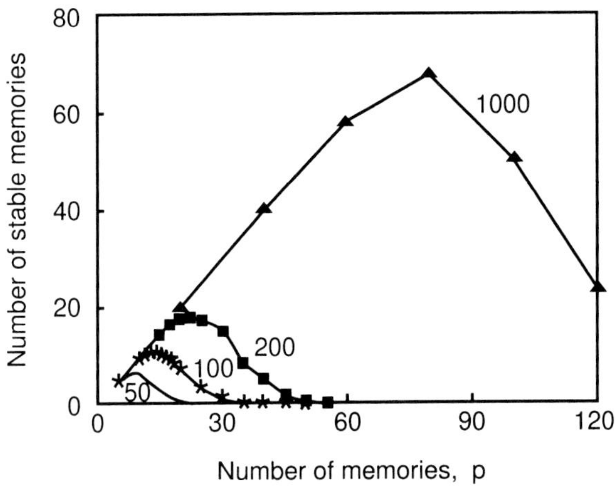

Figure 3: Optimization curve plotting the number of stable memories versus the number of memories for 50, 100, 200 and 1000 neurons. Straight-line interpolation is done as before.

"temperature" T and then using canonical ensemble methods in the analysis.[b] For example, since the Hopfield model has a global energy function, we can define an effective Hamiltonian for the system as

$$\mathcal{H} = -\sum_{i,j} J_{ij} S_i S_j \ . \tag{5}$$

Note that this is done for statistical purposes only, since Hopfield's system is not, in its own right, a Hamiltonian system dynamically. This will be discussed further in the section on the mean field analysis of the multiple-interaction model.

5 The Multiple Interaction Model

Extensive research has shown Hopfield's model to be a simple, elegant representation which captures essential elements of associative memory, such as memorization and recall. Its very simplicity, however, has led to important criticisms. Numerous researchers have developed modifications in an attempt to bring the model closer to physiological realism (Clark, Rafelski and Winston

[b]The terms "heat bath" and "temperature T" correspond metaphorically to an excitation background (bath) at level T which can exchange excitations with individual neurons. T is then a measure of the excitation level of the network.

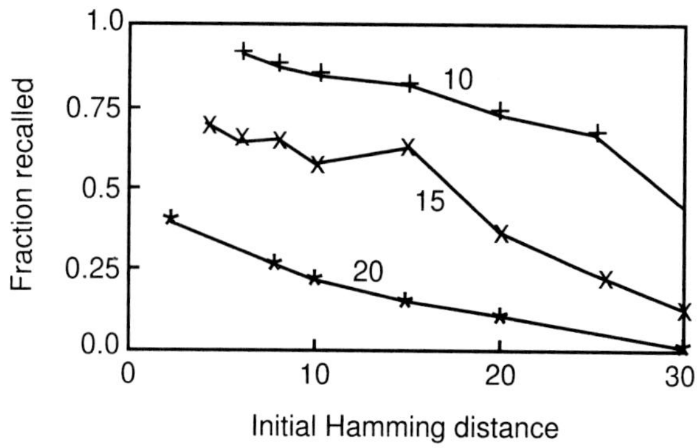

Figure 4: Fraction of states which returned to the memory states as a function of initial distance from the memory for $N = 100$, and $p = 10, 15$, and 20. Averaging was performed over all memories in a set as well as over different memory sets.

1985). We have been especially interested in studying the effects of *multiple neuron interactions* (MI's) on memory formation, capacity and recall.

Real neural tissue is very dense and has a high degree of connectivity — indeed the brain is mostly comprised of connections! Furthermore, the connections between neurons can be of very long range. These are some of the reasons that led us previously to consider so-called "higher-order interaction models" (Busch and Trainor 1987, Gardner 1987, Busch 1988). In a higher-order interaction, the state of any given neuron depends not only on the states of the other neurons taken each independently, but on the states of several neurons taken together (see also e.g. the idea of *heterarchy* and *heterarchical organization* of interactions in Lumsden's essay, this volume). We call such interactions Multiple Interactions and designate them as $MI(n)$ where n is the order of the interaction.

An example of how a triple-order ($n = 3$) interaction might be realized in real neural tissue is shown in Figure 5, which illustrates what is termed an axo-axonal interaction. In this case, the axon of one neuron, A, synapes directly on the axon of a second neuron, B. The state of both A and B together must be considered in determining how the axon from B affects a third neuron, C. Considering the density of neural connections, one can readily appreciate how higher-order interactions can play a significant role in determining the information processing behavior of such a complex network as the brain.

Multiple interactions are irreducible (see the essay by Trainor, this volume) in that they cannot be expressed or formulated in terms of lower-order interactions. For example, an irreducible 3-neuron interaction ($MI(3)$) cannot

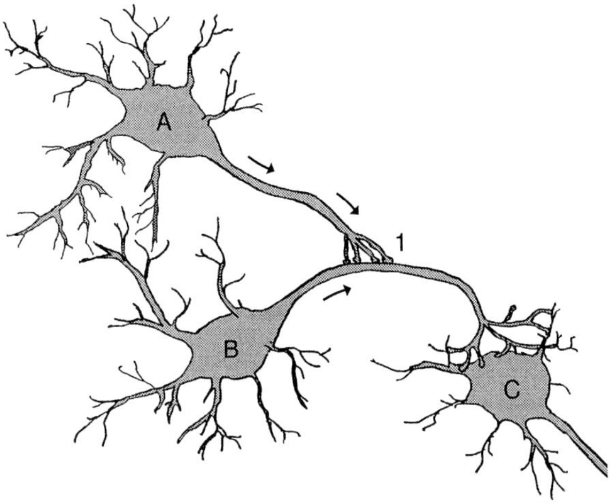

Figure 5: Illustration of how triple-order interaction can be realized in a neural network via axo-axonal interactions between nerve cells (neurons A, B, C and synapse at 1).

be expressed as a function of 2-neuron interactions ($MI(2)$).

In condensed matter and nuclear matter theory, 3- and higher-body influences on the forces between particles are usually ignored without serious effects on calculations. In both nuclear and condensed matter systems, the mediums are effectively of low density and the effective forces of short range, so that interactions of order greater than 3 make negligible contributions. Neural nets, however, are exceptional in this respect: Their nonlinear dynamics and complex interactions are at least suggestive of irreducible many-body interactions, with neurons playing the role of "particles" and the inter-neuron synapses supplying the interactions or "force." Because of the high inter-neuron connectivity the neural net is effectively a high density system with long-range interactions between neurons; thus it is appropriate that multiple-neuron interactions be assessed.

Accordingly, let us consider a population of binary (two-state) neurons, with the state of the ith neuron designated by $S_i = \pm 1$. We investigate the effects of higher order interactions. Each order n can be considered separately. For interactions of nth order, the state of $n-1$ neurons as a group affects any given neuron. We define the nth order contribution to the field at neuron i to be

$$h_i = \sum_{j_2=1}^{N} \sum_{j_3=1}^{N} \cdots \sum_{j_n=1}^{N} J_{i,j_2\cdots j_n} S_{j_2} S_{j_3} \dots S_{j_n} \ , \tag{6}$$

where $J_{i,j_2\cdots j_n}$ is the strength of the interaction between the group of neurons $j_2, j_3 \dots j_n$ and the neuron i.

The interactions are set by a Hebb-like learning algorithm. Let S_i^μ denote the state of neuron i in the μth memory, and p the number of memories the system is to learn. Then we chose for the elements of the interaction matrix

$$J_{i,j_2\cdots j_n} = \sum_{\mu=1}^{p} S_i^\mu S_{j_2}^\mu \dots S_{j_n}^\mu , \tag{7}$$

with $J_{i,j_2\cdots j_n} = 0$ if $j_k = i$ or $j_k = j_l$ for every k, l in the range 2 to n. Matrix elements with repeated indices are taken to be zero in order to completely eliminate lower-order contributions and self-stimulation. We consider all possible unique n-tuples of neurons in an $MI(n)$ network with all lower-order interactions eliminated, thus maximizing the effects of using orders of interaction greater than 2. In this way the effect of an n-order interaction alone can be better assessed than if a mixed order model were considered. Ultimately, in a real neural net, various combinations of doublet ($n = 2$), triplet ($n = 3$), quadruplet ($n = 4$), etc., would probably exist together but such a mixed order model will not be considered here. Note that in a system hardwired to perform multiple interactions automatically, as the brain presumably is, no extra processing time would necessarily be required over strictly pairwise interactions, regardless of the order of the interaction.

An MI Hopfield model is taken to be the same in all respects as the standard Hopfield model except with regard to the interactions between neurons, which are now extended to multiple-order interactions. The evolution is chosen to be asynchronous, with a random neuron i being chosen at each time step. The field at the neuron i is determined from equation (6) above, and the state adjusted to align with the field, such that $h_i S_i > 0$. This state change is communicated to every neuron in the next time step. Evolution occurs until all neurons are aligned with their local fields due to all other neurons in the net.

The "energy" for this system is taken to be

$$E = -C \sum_{i=1}^{N} \sum_{j_2=1}^{N} \cdots \sum_{j_n=1}^{N} J_{i,j_2\cdots j_n} S_i S_{j_2} \dots S_{j_n} \ , \tag{8}$$

where C is a positive constant expressing the overall synaptic strength. A change in the state of neuron i thus results in a change of energy

$$\begin{aligned} \Delta E &= -C \sum_{j_2=1}^{N} \cdots \sum_{j_n=1}^{N} J_{i,j_2 \cdots j_n} S_{j_2} \cdots S_{j_n} \Delta S_i \\ &= -C h_i \Delta S_i \ . \end{aligned} \tag{9}$$

If S_i changes from -1 to $+1$, then $\Delta S_i = 2, h_i > 0$ and so $\Delta E < 0$. If S_i changes from $+1$ to -1, then $\Delta S_i = -2, h_i < 0$ and again $\Delta E < 0$. Thus E is a monotonically decreasing function and behaves much like the "energy" for the ordinary Hopfield model. The energy is a global function describing the overall "goodness" of the state of the system. However, the energy landscape is much more complex. This will be discussed further when we explore the thermodynamic properties of the MI model.

6 Numerical Analysis of MI(3)

For the triplet, $MI(3)$, model, the field at neuron i is given by the double sum

$$h_i = \sum_{j=1}^{N} \sum_{k=1}^{N} J_{ijk} S_j S_k \tag{10}$$

and the synaptic interaction matrix is specified by the elements

$$J_{ijk} = \sum_{\mu=1}^{p} S_i^{\mu} S_j^{\mu} S_k^{\mu} \tag{11}$$

with $J_{iik} = J_{iji} = J_{ijj} = J_{iii} = 0$ for all i, j, k. The total number of unique pairs of neurons connected to neuron i is $(N-1)(N-2)/2$. Compare this to the $(N-1)$ individual neurons in the standard Hopfield model! There is an increase in the amount of information per imposed memory that a given neuron is receiving under $MI(3)$ over $MI(2)$.

Computer simulations were performed aimed at analyzing two properties of the $MI(3)$ model. First, the stability of the memories was investigated with the use of optimization curves. Second, the retrievability of the imposed memories was examined by assessing the ability of the system to recall a memory when started a given distance from that memory.

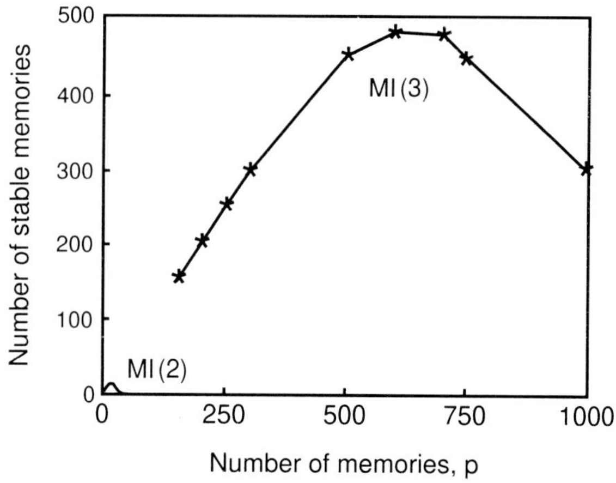

Figure 6: Optimization curves comparing the original Hopfield model ($MI(2)$) with a Hopfield model with interactions of order 3 ($MI(3)$) for $N = 100$.

6.1 Stability and MI(3) Storage Capacity

As in the Hopfield model, the stability of the imposed memories is determined by starting the system at one of the memory states and noting whether the system evolves away from it (in which case the memory state is unstable), or whether it remains in the memory state (in which case the memory is stable). This amounts to determining the number of memories which are local energy minima. The optimization method then involves plotting the number of stable memories as a function of the number of imposed memories, averaged over the number of memories in a memory set and over many randomly chosen memory sets. A memory set is chosen by randomly selecting a value of ± 1 for each of the N spin values S_i^m for each of the p memories.

The effect on the maximum number of stable memories m_s due to the triplet ($MI(3)$) interactions for a system with $N = 30$ is shown in Figure 6, with the optimization curve for the original Hopfield model ($MI(2)$) shown for comparison. It can be seen that m_s increases dramatically. For the $N = 100$ system with $MI(3)$, $m_s = 480$ as compared to $m_s = 11$ for $MI(2)$. This factor of 44 is somewhat better than the expected increase that results from the increase in the total number of unique connections: $(N - 2)/3 = 33$. Note that to obtain the same number of unique connections from strictly $MI(2)$ interactions would require a system with about 580 neurons, which, nevertheless, still has $m_s \sim 48$. Thus, the $MI(3)$ interactions increase by a factor of about

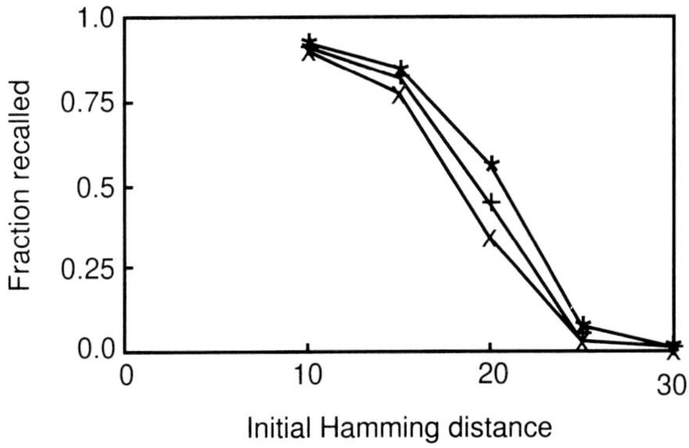

Figure 7: Retrievability of $MI(3)$ Model. The fraction of memories which are retrievable from a given initial Hamming distance is plotted for $N = 100$ and $p = 100, 150$ and 200 (right to left curves, respectively).

10 the storage capacity of a Hopfield-like network over the traditional $MI(2)$ interactions, with the same number of unique connections, and they do so with far fewer neurons!

We also analyzed the storage capacity for an $MI(4)$ model under the original Hopfield formalism (using potential variables $V_i = 0, 1$). For a system with $N = 30$, $m_s = 55$. One anticipates that for the spin glass ($S_i = \pm 1$) formalism, this will be increased by a factor of about 8, i.e. $m_s \approx 440$, compared to 60 for $MI(3)$.

6.2 Memory Retrievability

The widths of the memory energy minima were measured in a similar manner to that employed for the stability. After choosing the memories randomly and determining the interaction strengths, the system was started at a state close to one of the assigned memories. For a given initial Hamming distance (IHD), the system was allowed to evolve for a number of times steps, and the final HD from the memory calculated. Those memories that resulted in a zero final HD were said to be exactly retrievable from a given initial HD. Averaging was done over different initial states with the same HD, over all the memories in a set and over different memory sets.

The results for a $MI(3)$ model are shown in Figure 7, where the fraction of stable memories is plotted as a function of the IHD for various numbers of memories. An initial state was considered to have reached the chosen memory if it came within 1 HD of that memory. We see from the retrievability curves that starting the system with 20% of the neurons (20 neurons) in the wrong state gives a storage capacity of about 50%, which is half that of the stability result. Comparing Figure 7 for $MI(3)$ to Figure 4 for $MI(2)$ we see that energy basins, while narrower, are much more sharply defined. This means if a memory is recalled it is recalled more accurately under $MI(3)$ than $MI(2)$. The storage capacity depends on several factors: (1) the amount of error in recall of a memory that can be tolerated (i.e., the final HD from the memory); (2) the number or percentage of memories desired to be retrievable (the capacity required); and (3) the degree of retrievability required (the maximum IHD).

7 Mean Field Theory of the MI Model

The computer simulations show that an MI neural architecture greatly enhances the number of stable memories achievable as well as their retrievability in the extended Hopfield model. It is also clear from the definition of the energy that the energy landscape of an MI Hopfield model is different. What effect does this alteration have on other aspects of the MI model? In particular, are there other novel, global properties that emerge due to higher-order interactions? One needs to explore the properties of the energy landscape to answer such questions. In general this is difficult without using the averaging methods of statistical mechanics. All the thermodynamic properties would then follow from the free energy. To this end, we have applied the methods of Peretto (1984), and of Amit et al. (1985), which introduce noise by placing the network effectively in a heat bath at temperature T. The resulting MI model is restructured by the imposition of the heat bath, which allows for "thermal fluctuations" in "energy." (The neural network analog of temperature will be discussed further below.)

Consider a canonical ensemble with an effective Hamiltonian given by the MI exchange terms:

$$\begin{aligned}\mathcal{H} &= -\sum_{(i,j_1\cdots j_{n-1})} J_{i,j_1\cdots j_{n-1}} S_i S_{j_1} \ldots S_{j_{n-1}} \\ &= -\frac{N}{n!}\left[\sum_{\mu=1}^{p}\left(\frac{1}{N}\sum_{i=1}^{N} S_i^{\mu} S_i\right)^{n}\right] \qquad (12)\end{aligned}$$

where the sum is over all $N!/n!(N-n)!$ unique n-tuples, denoted by $(i, j_1 \ldots j_{n-1})$. The step to equation (12) involves the fact that we are constraining the terms with repeated indices to be zero (see equation (11)), and thus the restricted sum is proportional to the unrestricted sum. This is the

Hamiltonian for a particular memory set. The effective free energy f per "neuron" is

$$f[\mathcal{H}] \equiv \lim_{n\to\infty} -\frac{1}{\beta N}\ln \mathrm{Tr}\exp -(\beta\mathcal{H})$$
$$= \lim_{N\to\infty} -\frac{1}{\beta N}\ln \mathrm{Tr}\exp \frac{\beta N}{n!}\sum_{\mu=1}^{p}\left(\frac{1}{N}\sum_{i=1}^{N} S_i^\mu S_i\right)^n \quad (13)$$

where $\beta = 1/kT$, and the trace is the sum over all possible neuron states. The usual mean field theory method of completing the square using the Hubbard-Stratonovitch, or Gaussian identity and using the saddle-point approximation does *not* work with these $MI(n)$ systems since the potential is of the nth order rather than a simple quadratic. However, progress can still be made toward deriving the free energy by using a mean field approximation. This can be done as long as the interactions are separable, which is true if the number of memories is finite.

The mean field formulation developed here is slightly different in form than the standard mean field theory (see e.g. Stanley 1971) for Ising spin models, which involves writing the spin at each site as a mean spin plus a fluctuation. The formalism to be presented is in fact equivalent to the standard formalism. The reasons for using it will become evident.

We begin by writing the Hamiltonian in the form

$$\mathcal{H} = \frac{N}{n!}\sum_{\mu=1}^{p}(V_\mu)^n \quad (14)$$

where $V_\mu = \frac{1}{N}\sum_{i=1}^{N} S_i^\mu S_i$. We have switched to a "memory set" space rather than a spin space, where the potential operator V_μ is the contribution to the effective energy from the μth memory.

We now write the potential term V_μ as

$$V_\mu = m_\mu + (V_\mu - m_\mu) = m_\mu + \delta V_\mu \quad (15)$$

where δV_μ is a fluctuation term, $m_\mu = \langle V_\mu\rangle_{MF}$, and $\langle\rangle_{MF}$ is the thermodynamic average with respect to the mean field Hamiltonian. Substituting this into the Hamiltonian and ignoring terms quadratic and higher in the fluctuation gives the mean field Hamiltonian

$$\mathcal{H}_{MF} = \frac{N}{n!}\sum_{\mu=1}^{p}\left\{(n-1)(m_\mu)^n - n(m_\mu)^{n-1}V_\mu\right\} \quad (16)$$

The parameters m_μ can be written as

$$m_\mu = \langle V_\mu \rangle_{MF} = \frac{1}{N} \langle \sum_{i=1}^{N} S_i S_i^\mu \rangle_{MF} = \frac{1}{N} \sum_{i=1}^{N} \langle S_i \rangle_{MF} S_i^\mu \, , \tag{17}$$

and thus denote the overlap of the mean spin with the memory state μ. The reason for our formalism in "memory set" space now becomes clear: For a memory model we are fundamentally interested in whether the memory states are free energy minima or not, and how well a particular thermodynamic equilibrium state corresponds to such memory states. The order parameters m_μ describe exactly that. We write $\vec{m}$ for the array $(m_1, m_2, \ldots m_p)$.

Substituting the mean field Hamiltonian of equation (16) into the free energy expression of equation (13) gives

$$f(\vec{m}) = \frac{(n-1)}{n!} \sum_{\mu=1}^{p} (m_\mu)^n - \lim_{N \to \infty} \frac{1}{\beta N} \ln \operatorname{Tr} \prod_{i=1}^{N} \exp \frac{\beta n}{n!} \sum_{\mu=1}^{p} (m_\mu)^{n-1} S_i^\mu S_i \tag{18}$$

where the argument of the natural logarithm in equation (18) has the form

$$\operatorname{Tr} \prod_{i=1}^{N} \exp a_i S_i = \operatorname{Tr} e^{a_1 S_1} e^{a_2 S_2} \ldots e^{a_N S_N} . \tag{19}$$

The free energy is thus

$$f(\vec{m}) = \frac{(n-1)}{n!} \sum_{\mu=1}^{p} (m_\mu)^n - \lim_{N \to \infty} \frac{1}{\beta N} \sum_{i=1}^{N} \ln \left(2 \cosh \left(\frac{\beta n}{n!} \sum_{\mu=1}^{p} (m_\mu)^{n-1} S_i^\mu \right) \right) . \tag{20}$$

This free energy looks like a sum of free energies, one for each different memory in the particular set chosen, but weighted differently due to the S_i^μ factor. The first term is the mean field internal energy of a frozen lattice, in the "memory set" space, and the second is a kind of mean field of the neighboring memories in that space.

The extremum condition on the free energy requires that $\partial f / \partial m_\eta = 0$. Since

$$\frac{\partial f}{\partial m_\eta} = \frac{1}{(n-2)!} (m_\eta)^{n-2} \left[m_\eta - \lim_{N \to \infty} \frac{1}{N} \sum_{\mu=1}^{p} S_i^\eta \tanh \left(\frac{\beta n}{n!} \sum_{\mu=1}^{p} (m_\mu)^{n-1} S_i^\mu \right) \right] , \tag{21}$$

either $m_\eta = 0$ or

$$m_\eta = \lim_{n\to\infty} \frac{1}{N} \sum_{i=1}^{N} S_i^\eta \tanh\left(\frac{\beta}{(n-1)!} \sum_{\mu=1}^{p} (m_\mu)^{n-1} S_i^\mu \right) . \tag{22}$$

Equations (21) and (22) are the mean field equations for the MI model. It can be shown that ignoring the quadratic- and higher-order terms in the potential is quite justifiable and that the mean field equations are in fact exact (in the thermodynamic limit; see Busch 1988).

The solutions to equations (21) and (22) can be divided into three useful categories: retrieval states, which have an overlap with only one of the imposed memories; symmetric solutions, which have an equal overlap with several memories; and asymmetric solutions, possessing an unequal overlap with several memories. Here, we consider the retrieval state and symmetric solutions as they are the most relevant for memory modeling.

7.1 Retrieval State Solutions

Apart from the trivial solution, there are other solutions which are important at low temperatures. We first consider solutions which have only one non-zero m_μ component, the remaining $p-1$ being zero, i.e., solutions of the form $\vec{m} = (0, 0, \ldots, 0, m_\mu, 0, \ldots, 0)$ where $m_\mu \neq 0$. Let $m_\eta = m$ be the only non-zero m_μ. Then

$$m = \lim_{N\to\infty} \frac{1}{N} \sum_{i=1}^{N} S_i^\eta \tanh\left(\frac{\beta n}{n!} m^{n-1} S_i^\eta \right) . \tag{23}$$

Since $S_i^\eta = \pm 1$ and $\tanh(-x) = -\tanh x$, each term in the sum has the same value, so that

$$m = \tanh\left(\frac{\beta n}{n!} m^{n-1} \right) . \tag{24}$$

The free energy simplifies considerably for these solutions as well. Since $\cosh(S_i^\eta x) = \cosh(x)$,

$$f(m) = \frac{n-1}{n!} m^n - \frac{1}{\beta} \ln\left\{ 2\cosh\left(\frac{\beta n}{n!} m^{n-1} \right) \right\} . \tag{25}$$

These solutions are of great interest for memory modeling as they correspond to the overlap with a particular memory. Following Amit et al. (1987), we call them retrieval states. For $n = 2$ and finite p they are the Mattis states of the generalized Hopfield model studied by Amit et al. (1985, 1987). These

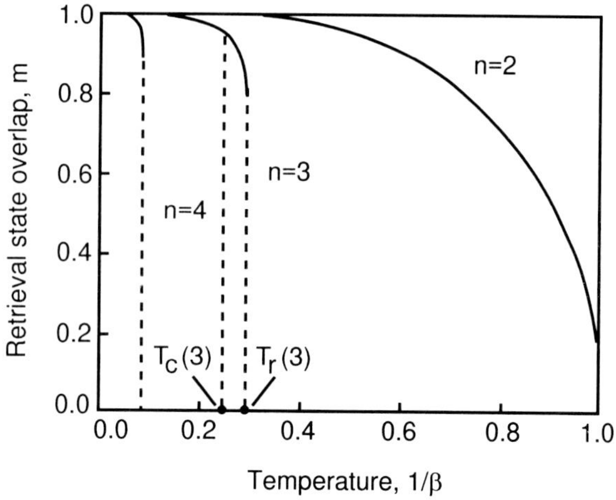

Figure 8: Retrieval state overlap, m, as a function of temperature for orders $n = 2, 3$ and 4 (right to left curves, respectively). The dashed lines for $n = 3$ and $n = 4$ indicate a discontinuity in the value of m.

solutions are equivalent, via a Mattis transformation, to the ferromagnetic solutions. There are p such solutions, one for each of the imposed memories. The behavior and properties of these states still needs to be determined from the minimum requirement on the free energy. The values of m satisfying equation (24) can be determined by plotting $\tanh\left(\beta n m^{n-1}/n!\right)$ as a function of m for a given temperature and noting where it crosses the $m = m$ line. We can then plot retrieval state overlap and network free energy as functions of temperature (Figures 8, 9). The trivial solution is the only solution above $T_r(n)$, and at $T_r(n)$ a non-zero solution appears discontinuously. Since the trivial solution is also stable past this "transition" temperature, where the retrieval state solution appears, we must chose the one which gives the minimum free energy. The non-zero m solution for order n gives the lowest free energy when below a temperature $T_c(n) < T_r(n)$. This can be seen in Figure 9, where the free energy is plotted as a function of temperature for $n = 2, 3$ and 4. At temperatures below $T_c(n)$, the retrieval state solutions are the correct thermodynamic solutions as they give the lowest free energy.

There are two important differences in the $n > 2$ solutions over the case $n = 2$. First, it appears that we no longer have a second-order transition, as there is in the doublet case ($n = 2$). This can be seen from the fact that the free energy is not smooth through the transition temperature $T_c(n)$. The second difference is that any inclusion of odd-order multiplets results in a

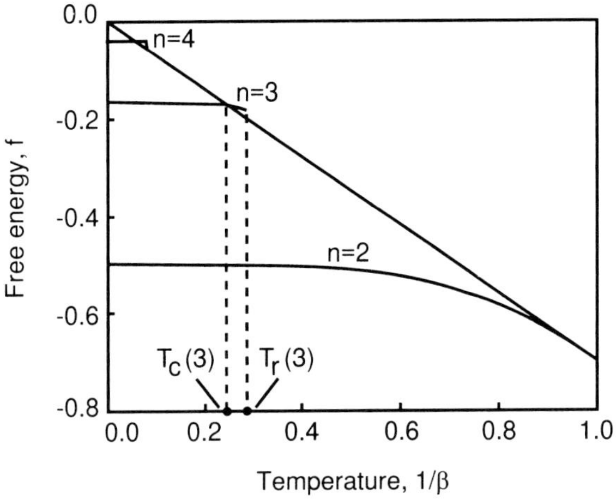

Figure 9: Retrieval state free energy versus temperature, orders $n = 2, 3$ and 4.

broken symmetry. Whereas in the doublet case, and for all purely even-order multiplets, for every $+m$ solution there is a $-m$ solution, for odd orders this is no longer the case. For purely odd orders, only the $+m$ solution (and the zero solution) exists. For a mixed multiple-order model containing odd orders the positive and negative solutions are no longer equal in magnitude.

7.2 Symmetric Solutions

Solutions of the form $\vec{m} = m_s(1, \ldots, 1, 0, \ldots, 0)$, where there are n_s 1's ($n_s \leq p$) correspond to states that are equal mixtures of n_s memories. For the MI model, the order parameters for these solutions are

$$m_\nu = m_s = \lim_{N \to \infty} \frac{1}{N} \sum_{i=1}^{N} S_i^\nu \tanh\left(\frac{\beta n}{n!} m_s^{n-1} X_i\right), \quad \forall \nu = 1, \ldots n_s \tag{26}$$

where $X_i = \Sigma_\mu^{n_s} S_i^\mu$. The free energy is

$$f(m_s) = \lim_{N \to \infty} \left\{ \frac{n-1}{n!} n_s m_s^n - \frac{1}{\beta N} \sum_{i=1}^{N} \ln\left(2 \cosh\left(\frac{\beta n}{n!} m_s^{n-1} X_i\right)\right)\right\}. \tag{27}$$

Since the equation for the parameter m_s must be true for any value of ν in the range 1 to n_s, in particular for $\nu = \mu$ and $\nu = \eta$, then $m_\mu = m_\eta$. Since the system is self-averaging in the thermodynamic limit (Busch 1988):

$$n_s m_s = \left\langle X_i^\mu \tanh\left(\frac{\beta n}{n!} m_s^{n-1} X_i\right)\right\rangle_D . \tag{28}$$

Consider the special case of $n_s = 2$. For any pair of memories, labeled μ and η, we have $m_\mu = m_\eta \equiv m_2(\mu, \eta)$ and

$$X_i = S_i^\mu + S_i^\eta \begin{cases} = 2S_i^\mu & \text{if } S_i^\eta = S_i^\mu \\ = 0 & \text{if } S_i^\eta = -S_i^\mu \end{cases} . \tag{29}$$

Reindexing the neuron states such that $S_j^\mu = S_j^\eta$ for $j = 1, N_{\mu\eta}$, where $N_{\mu\eta}$ is the number of spins that are the same in the μ and η memory states, gives

$$m_2(\mu, \eta) = \lim_{N\to\infty} f_{\mu\eta} \tanh\left(\frac{2\beta n}{n!} m_2^{n-1}\right) \tag{30}$$

where $f_{\mu\eta} = (N_{\mu\eta}/N)$ is the limit of the fraction of spins which are the same in memories μ and η. $f_{\mu\eta}$ is also called the *fractional overlap* between the μ and η memory states. Averaging over memory sets yields $f_{\mu\eta} = 0.5$ for randomly chosen memory sets. If, however, we were to consider a finite-N model, then in principle $f_{\mu\eta}$ could take on other values. In Figure 10 the behavior of m_2 is plotted as a function of temperature for $f_{\mu\eta} = 1.0, 0.75$ and 0.5. For a given set of memories, there are $p(p-1)$ of these pair solutions, one for each pair of memory states in the memory set. Each pair will have a different transition temperature for m_2 since the overlaps of each pair will be different.

The finite-N free energy for the m_2 solution is

$$\begin{aligned} f_N\left(m_2(\mu, \eta)\right) &= \frac{n-1}{n!} m_2^n - \frac{1}{\beta}\left[f_{\mu\eta} \ln 2 \cosh\left(\frac{2\beta n}{n!} m_2^{n-1}\right)\right. \\ &\left. \quad + (1 - f_{\mu\eta}(1 - f_{\mu\eta}) \ln 2)\right] . \end{aligned} \tag{31}$$

The finite-N free energy is plotted in Figure 11 for $f_{\mu\eta} = 1.0, 0.75$ and 0.5. For all orders n the transition temperature for the averaged pair solutions (i.e., $f_{\mu\eta} = 0.5$), is less than the transition temperature for the retrieval states. This is in agreement with the earlier results for $n = 2$ (Amit et al. 1987) that there is a temperature range (in the thermodynamic limit) for which the retrieval states are the only non-zero mean field theory solutions.

One can see from Figure 10 that the symmetric states with $f_{\mu\nu} \geq 0.68$ have higher transition temperatures and lower free energy than the retrieval states. This suggests that strongly correlated memories are harder to distinguish but more easily recalled.

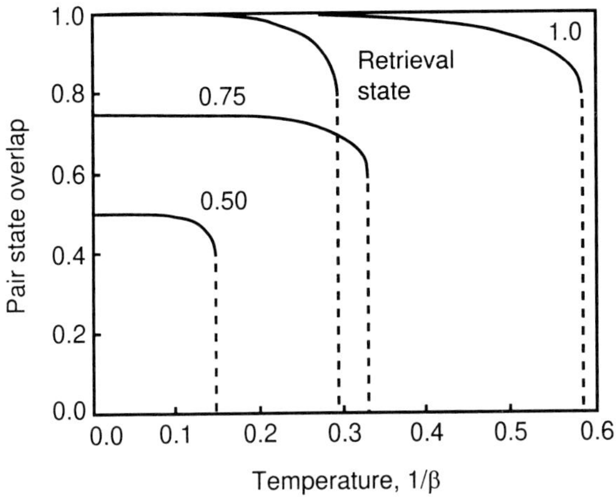

Figure 10: Symmetric solution overlap for $n_s = 2$ as a function of temperature for $f_{\mu\eta} = 1, 0.75$ and 0.5, with the retrieval state overlap shown for comparison.

8 Discussion

In light of the numerical and theoretical investigation of multiple-order interaction neural network models, we conclude with several observations:

First, the numerical simulations show that higher-order neuron interactions allow for a substantial increase in memory storage capacity through an increase in the number of the interactions, without requiring a non-efficient increase in the number of neurons. A past criticism of multiple-order interaction models with $n > 2$ has been that although the storage capacity increases with increasing interaction order, the "number of stored bits per interaction" actually decreases. In the model above, however, the number of stored bits per unique interaction increases slightly with increasing order of interactions (as seen from the numerical simulations). Moreover, even if the storage per interaction were to decrease under specific circumstances, multiple interactions may be of great importance for systems in which the "cost" per element is much greater that the "cost" per interaction. This is most likely the case for real neural nets, where the cost per neuron is greater than the cost per synapse. In brain tissue, multiple interactions are manifested in the impressive arborization of neuron processes and the intensive synaptic activity between processes. The number of neurons in the brain is fixed in early development. The massive growth that occurs subsequently is in the connections between neurons, and it seems likely that these include some amount of higher-order interactions.

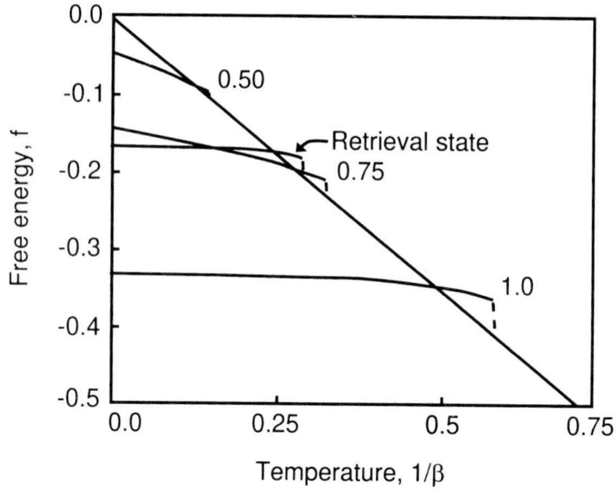

Figure 11: Symmetric solution free energy for $n_s = 2$ as a function of temperature for $f_{\mu\eta} = 1, 0.75$ and 0.5 with the retrieval state free energy shown for comparison.

The second observation is that some interesting features arise from the mean field theory analysis discussed above: the idea of transitions which result from changing a "temperature" type parameter and the non-second order of these transitions for models with interactions of order greater than two.

In order to apply the mean field theory familiar to statistical mechanics, our neural network model was assumed to be in contact with a "heat bath" (excitation bath) at temperature (level) T (generalized Hopfield model, Peretto 1984). This parameter presumably describes or represents a mechanism which enable neurons to spontaneously "change state" or exchange "excitations." One can then see how a system with interacting elements may operate in terms of memory recall as follows: To begin, consider the theory of ferromagnetism. Below a transition (Curie) temperature, there are two solutions to the mean field theory equations. Actual ferromagnetic systems are in only one of these states when below the critical temperature. Above the Curie temperature, there will be an excess of spins in one direction. In practice, a non-zero magnetic field is also not obtainable. If the temperature of a real system is lowered through the Curie point, the system will enter one of the two magnetization states. Thus, in order to perform correct thermodynamic averages (to match reality), a small biasing field must be introduced to force the model system into either one of the positive or negative magnetization states. A ferromagnetic system could behave, then, like a memory system with only two memories (one with all the spins up and one with all the spins down). The system can recall one of two states by lowering the temperature from above to below the Curie temperature. Which state is recalled depends on which "memory" state the initial state resembles the most or if there is a "biasing field" for that state. By controlling the temperature and applying biasing fields, or fixing spins, we can alternatively "recall" (by lowering the temperature) and "clear" (by raising the temperature) the "memory".

One could envision a similar mechanism for a neural network system, including ones with higher-order interactions, in contact with a heat bath, and possessing states which have been memorized by a learning algorithm that makes these memory states the stable states of the system. Several differences arise compared to the ferromagnetic "memory" viz. in the number of memories, in the nature of the external "field" and in the order of the transition for higher-order interactions (discussed further below). For neural network models the biasing "fields" correspond to afferent input "fixing" some of the neurons in states which would represent part of a memory (a cue, or certain known elements). Again by controlling the temperature T (now the excitation level), we could alternatively "recall" and "clear" a memory. A single parameter, the "temperature," would control whether the system has remembered (or is in the process of remembering), or is ready to remember, another state. The excitation level parameter is thus the memory driver! It seems likely that this level parameter T would correspond to the overall spontaneous firing rate of the neurons, which may be linked to the global rate of production of neurotransmitters in the brain.

This makes a certain amount of sense based on the analysis of the retrieval and the symmetric state solutions: "Strong" memories can be viewed as memorizing the same state multiple times and thus correspond to symmetric states with $f_{\mu\nu} = 1$. These states are more easily recalled at higher temperatures (greater neural activity). Less strong memories have a lower transition temperature and thus require a lower overall firing rate for recall. To carry this further, tranquilizers and other drugs could seriously affect such a system by altering the excitation level, which may disrupt memory recall by preventing the change needed to get to or get out of a memory, or it may enhance memory by allowing lower excitation level transitions to be reached.

With this in mind, let us now consider the order of the phase transition for models with interaction orders greater than $n = 2$. We have seen that the transition in these cases is discontinuous and not of second order as for the $n = 2$ case. The recall of a memory for an $n = 2$ model is much more gradual and occurs over a longer range of excitation levels than the higher order models. One could conjecture that higher-order models thus capture the "Aha!" experience of memory recall, in which a memory is quite suddenly recalled. One could speculate whether observations on neuron firing rates and times for memory recall could, in principle, be used to determine the extent of multiple neuron interactions in real brains.

References

Amit, D.J., Gutfreund, H. and Sompolinsky, H. (1985) Spin glass models of neural networks. *Physical Review A* **32**, 1007–1018.

Amit, D.J., Gutfreund, H. and Sompolinsky, H. (1987) Statistical mechanics of neural networks near saturation. *Annals of Physics* **173**, 30–67.

Arbib, M., ed. (1995) *The Handbook of Brain Theory and Neural Networks.* Cambridge, MA: MIT Press. See as well http://www-mitpress.mit.edu/mitp/recent-books/neuro/arbib.html

Busch, A.E. (1988) *Mean Field Theory of Disordered Neural Networks with Multiple Neuron Interactions.* PhD Thesis, Department of Physics, University of Toronto, Canada.

Busch, A.E. and Trainor, L.E.H. (1987) Neuron multiplets and storage capacity. *Physics in Canada* **43**, 72.

Clark, J.W., Rafelski, J. and Winston, J.V. (1985) Brain without mind: computer simulation of neural networks with modified neuronal interactions. *Physics Reports* **123**, 215–273.

Gardner, E. (1987) Multiconnected neural network models. *Journal of Physics A* **20**, 3453–3464.

Hebb, D.O. (1949) *The Organization of Behavior: A Neuropsychological Theory.* New York: John Wiley and Sons.

Hopfield, J.J. (1982) Neural networks and physical systems with emergent collective computational abilities. *Proceedings of the U.S. National Academy of Sciences* **79**, 2554–2558.

Hopfield, J.J. (1994, February) Neurons, dynamics and computation. *Physics Today* **47**, 40–46.

Little, W.A. (1974) The existence of persistent states in the brain. *Mathematical Biosciences* **19**, 101–120.

McCulloch, W.S. and Pitts, W.H. (1943) A logical calculus of the ideas immanent in nervous activity. *Bulletin of Mathematical Biophysics* **5**, 115–133.

Peretto, P. (1984) Collective properties of neural networks: a statistical physics approach. *Biological Cybernetics* **50**, 51–62.

Sherrington, D. and Kirkpatrick, S. (1975) Solvable model of a spin-glass. *Physical Review Letters* **35**, 1792–1796.

Stanley, H.E. (1971) *Phase Transitions and Critical Phenomena.* Oxford: Clarendon Press.

van Hemmen, J.L. and Kühn, R. (1986) Nonlinear neural networks. *Physical Review Letters* **57**, 913–916.

Network Hierarchies in Neural Organization, Development and Pathology

Jeffrey P. Sutton

By comparing the morphology and relative abundance of axonic and dendritic collaterals of the cerebral pyramids in the vertebrate scale, the conclusion was reached that intellectual power, and its most noble expressions, talent and genius, do not depend on the size or number of the cerebral neurons, but on the richness of their connective processes, or in other words on the complexity of the association pathways to short and long distances.

Santiago Ramón y Cajal[a]
(1852–1934)

1 Introduction

The human brain is enigmatic. Many consider it the most complex and mysterious structure within nature, and while a myriad of its secrets may forever defy our curiosity, the past one hundred and fifty years have been a remarkable time for the brain sciences. Growth in knowledge about nervous system structure and function has been escalating at an enormous pace. New insights into development, molecular neuroscience, neural circuitry and cognitive neuroscience are revealing interesting and profoundly intricate mechanisms of brain and behavior. At one time, such knowledge was completely inaccessible. Progress has been made using an abundance of innovative techniques which record brain activity across many orders of magnitude, from the gene to the entire brain and beyond. There have been marked advances concerning normal function as well as disorders affecting the nervous system. Common neurological and psychiatric diseases afflict one in three individuals, and patients of all ages are often helped by modern treatments (Adams and Victor 1989, Kaplan and Sadock 1989).

With the accumulation of scientific information about basic and clinical neuroscience, there is a pressing need to integrate knowledge acquired at different scales of neural organization. This is an important problem since links between different scales, such as the molecular, cellular and systems levels, help to bridge the apparent dichotomy between brain and mind. Historically, the problem is an old one. It dates to antiquity and occupies a central place in philosophical inquiries (Churchland 1986). However, the ability to probe

[a]From Cajal (1937; 1991) p. 459.

the inner workings of the brain has begun to shed new insight into how neural processes mediate mental functions, including memory, learning, attention and perception. Models are playing an increasingly important role since they are capable, in principle, of bridging scales or levels of neural organization. Modeling complements experimentation, which is typically restricted to one level of analysis (e.g. electrophysiological recording from a single neuron).

The scientist interested in modeling biological neural networks must effectively combine neuroscience and computational principles. Computers have revolutionized the field of theoretical neuroscience, and there is an intriguing relationship between neural networks and computer science (Sutton 1995). Not surprisingly, some computer components have analogies with brain circuitry. The very same computers are also used to help investigate neural information processing, which leads to discoveries in neuroscience and to the development of better computers.

Moreover, there is an underlying notion in neuroscience today that the brain utilizes computational strategies to encode and handle information, and that a study of these strategies, especially between levels of neural organization, leads to a better understanding of the apparent gap between brain and mind. Several research programs in computational neuroscience are testament to how successful bridges can be built between brain and mind (e.g. Anderson and Rosenfeld 1988; Anderson, Pellionisz and Rosenfeld 1990). While many discoveries have been guided by specific neuroscience questions, others have been motivated by less biological factors but have given rise to interesting engineering and mathematical applications. Indeed, both software and hardware implementations of neural networks are areas of active development.

In this chapter, a mathematical approach to studying an enormous neural network, the neocortex, is described. The neocortex is arguably the most sophisticated structure within the mammalian brain. Judged on basis of sheer size, it is our largest brain structure. Its properties endow us with qualities that are unique to our species. To develop an approach that is both tractable and useful, enormous anatomical and physiological simplifications must be made. It is also necessary to focus on specific aspects of function, such as associative memory and learning, at the expense of many other important characteristics. Such simplifications and focus are required in any model of the brain.

Since the model examines different scales or levels of neural organization within the neocortex, there is considerable discussion in this chapter about how neurons cluster together. It is proposed that nested clusters of neurons are fundamental to the structural and functional integrity of the neocortex. The organization of the clusters changes across many different time scales, and the connections amongst individual neurons appear disordered. Individually and together, these features give rise to unusual computational and physical attributes. There are experimental data supporting the approach and verifying predictions of the neocortical model, and the dynamic features are beginning to lead to new ways of performing extremely rapid parallel and associative computations.

The chapter is organized as follows. Section 2 summarizes some of the neurobiology relevant to the theoretical research. In section 3, a hierarchical cluster model of neocortical networks is outlined. The model has multi-leveled memory and learning capabilities that are mathematically and computationally described in section 4. The relationship of these models to development is presented in section 5. Simulations of neurodegenerative disorders are described in section 6. Finally, in section 7, the importance of integrative neural modeling is discussed from the perspectives of complex systems, neural computation and applications to technology and neuroscience.

2 Neurobiological Considerations

2.1 Overview

The human brain consists of approximately 10^{11} neurons connected to each other in a complex manner. In the human, most of the cells and connections are in the forebrain (Figure 1A). The largest forebrain structure, the neocortex, is a convoluted mantle of tissue that, if flattened, would have the surface area several times the size of this page. Approximately 99% of the connections between cells in the neocortex arise from other cells within the neocortex. Only 1% of the connections originate from or terminate onto non-cortical structures (Shepherd 1990). Thus, the neocortex is largely a self-contained structure.

There is considerable heterogeneity among the types of neurons in the neocortex. Some cells have relatively few connections, and these are confined to distances of less than 100 microns (μm). Other cells, such as the pyramidal cells, have upwards of 10^5 connections located throughout an arborization extending from several centimeters to more than a meter (Brodal 1981). It is estimated that if all the connecting fibers in the neocortex were extended end to end, they would reach from the earth to the moon and back again.

A tremendous amount of knowledge has been acquired about the anatomy and physiology of the human brain (e.g. Kandel, Schwartz and Jessell 1991). It has been known for almost a century that different regions of the neocortex have cytoarchitectural differences (Figure 1B). Functional correlates of many regions have also been determined. For example, Brodmann area 17, located in the occipital pole, or back part of the brain (Brodmann 1909), is associ-

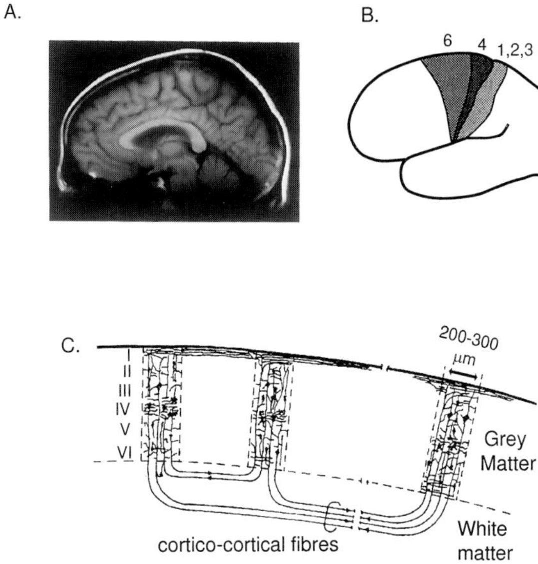

Figure 1: A. Magnetic resonance image of the human brain. A mid-sagittal section is shown which reveals the convoluted structure of the neocortex. The neocortex occupies a significant portion of the brain. The skull is outlined by a bright white line, and the front portion of the brain is on the left of the image. Several prominent structures are visible, including the corpus callosum (which connects the two cerebral hemispheres), the cerebellum and the brainstem. B. The neocortex is subdivided into regions based on microscopic differences in the ways cells are arranged. A commonly used map is the one developed by Brodmann (1909), parts of which are shown in the Figure in highly schematized form. Panels A. and B. show the brain in the same orientation. C. This diagram illustrates the layering of cells in the neocortex. Layer I is the most superficial layer, and it lies just underneath the surface of the neocortex shown at the top of the figure. Layer VI is the deepest layer of cells. Cell bodies form a considerable portion of the gray matter. The white matter consists of axons, which are cable-like projections emanating from the neurons. The axons are encased in an insulating substance known as myelin which has a white appearance. Axons linking different parts of the neocortex are called cortico-cortical fibers. In addition to anatomical details based on structure, the neocortex can be subdivided based on function. Three functional units (cortical columns) are depicted. After Sutton, Beis and Trainor (1988b).

ated with primary visual processing. While some morphological features of cell organization may vary from region to region, many structural details are preserved between regions. In particular, almost all areas of the neocortex have six layers of cells arranged parallel to the cortical surface (Figure 1C).

At the level of individual nerve cells, electrochemical signals are received by neurons and transduced in a variety of ways. For example, in the so-called "classical neuron doctrine" (which is now outdated), the output of a neuron is the result of changes induced by cumulative inputs to the neuron (Shepherd 1990). When these inputs cause the membrane threshold of the neuron to be exceeded, and the neuron is not refractory to signal generation, then an action potential is triggered. The action potential behaves as a traveling voltage pulse. It propagates along the neuron to the axon terminal, where it releases chemical substances into a 20-nanometer synaptic cleft for diffusion to receptors on other neurons. The chemical neurotransmitter may be either excitatory or inhibitory in its effect on the other neurons. The cycle from input to output is commonly referred to as the "all or none" response of individual neurons. This will be more fully characterized in section 3.

When two neurons connect, there is a transfer of information that may be unidirectional or bidirectional. If neuron A originates an action potential that propagates to neuron B, then neuron A is said to send a feedforward connection to neuron B. Neuron A becomes the afferent cell and neuron B becomes the efferent cell. If an action potential generated by neuron A travels down an axon with a collateral projection that synapses back onto neuron A, then feedback occurs. Feedforward and feedback loops are common and involve many cells within the neocortex.[b] There are also loops between the neocortex and other structures, such as the thalamus.

Neurons connect to each other via synapses. The notion that synapses between neurons may be a site of plasticity and learning dates back more than one hundred years to the works of Cajal (1909) and Freud (1897). Several researchers, including Hebb (1949), have proposed learning rules based on correlated activities between pre-synaptic and post-synaptic neurons. Positive correlations between neurons strengthen the likelihood of simultaneous neuron firing; negative correlations weaken the probability of synchronous fir-

[b]There are many different types of connections in the neocortex. For columns separated by more than 1 mm, layer II and III cells project to layer IV cells, and layer III, V and VI cells project to layer I, V and VI cells (e.g. Figure 1C). These two sets of projections mediate information loops between columns separated by distances up to many centimeters (Maunsell and Van Essen 1983). There is another source of feedforward and feedback connections among layer I cells, which typically have projections limited to 3 mm. In section 6, the importance of large layer III cells that link different cortical regions together, via cortico-cortical fibers, is discussed with respect to Alzheimer's disease.

ings. These theoretical predictions have only recently been experimentally confirmed. The mechanisms mediating positive and negative correlations are referred to respectively as long term potentiation and long term depression (Brown, Kairiss and Keenan 1990).

2.2 Functional Anatomy

Establishing linkages between neurobiological structure and function is a fundamental challenge in neuroscience. It is also a motivating factor in the development of any model of the brain. Functional anatomy is a discipline that attempts to elucidate function based on structural details. For example, neurons are complex and heterogeneous cells whose anatomical and physiological features suggest that they are not simple summing devices for incoming information. They modify inputs in a plethora of ways that depend upon cell morphology and the relative positions and strengths of the synaptic connections. The details link structure and function, and they associate a repertoire of computational features to individual cells (McKenna, Davis and Zornetzer 1992).

Similarly, clusters of neurons that constitute cortical columns can be treated as functional or computational units (Szentágothai 1977). Columns are oriented perpendicular to the surface of the neocortex and they contain on the order of 10^2 to 10^4 neurons. Each is approximately 200 to 300 μm in diameter and each column traverses all six cellular layers (Figure 1C). There are no obvious structural boundaries that demarcate a column from its neighbors. However, boundaries are clearly demarcated by physiological means, such as by the electrophysiological recording from electrodes positioned obliquely through visual cortex during sensory stimulation of the eyes. There is also evidence for columnar organization of the neocortex based on developmental grounds. During ontogeny, each column arises from a single clone cell that develops into a network of radially projected cells (Rakic 1981). The cells generate a column with the inner aspects being constructed before the outer aspects. Increases in size of the neocortex witnessed in phylogeny are likely the result of increased numbers of columns, without striking change in their vertical organization (Rockel, Hiorns and Powell 1980).

The neurons within columns are densely connected to each other. As the distance between the columns grows, the density of connections between cells in different columns decreases. As a result, the neocortex as a whole is sparsely connected. This is important to realize since many models of neural networks assume that there is relatively full connectivity among the neurons in the network. Such models generally apply to relatively small neural networks.

2.3 Nested Distributed Systems

While the brain is sparsely connected, in that every cell is not connected with every other cell, there are nevertheless organizing principles that cluster neocortical cells together. The concept of the cortical column is one such example. More generally, it has been suggested that groups of neurons organize themselves into networks across many scales (Mountcastle 1978). Neurons cluster to form small networks, which in turn cluster to form larger networks, and so on. Some of the early evidence for nesting among neural networks arose from motor cortex data (Phillips and Porter 1978). Small functional units, such as mini-columns, were seen to coalesce to form columns, which clustered to form overlapping motor regions (Figure 2C). The resulting architecture is schematically shown in Figure 2D. It consists of a network of networks that links neurons together into distributed systems of varying size.

The notion of nested distributed systems within the neocortex includes but transcends early work on how to integrate structure and function. Hughlings Jackson (1884), for instance, proposed the concept of distributed networks linked by reflex loops within the neocortex. The idea was tied to an evolutionary view of the brain. Specifically, the neocortex was seen as a highly sophisticated processor sitting atop a stack of other structures containing distributed networks of neurons. Each structure performed successively more complicated tasks, and information was fed up and down the hierarchy (Figures 1A, B). It is now know from both theoretical (Freeman 1975, Edelman 1978) and experimental grounds (Van Essen, Anderson and Felleman 1992) that the neocortical processing picture is much more complicated than this simple scheme suggests. In particular, parallel processing involves networks at many scales, and it is ubiquitous in the neocortex.

Moreover, one of the interesting realizations to emerge in recent years has been that functional groups of neurons *dynamically* organize themselves into spatial clusters (T'so, Gilbert and Wiesel 1986; Eckhorn et al. 1988). Patterns of correlations among neurons evolve within fractions of a second, and they occur in systematic relation to behavioral events (Vaadia et al. 1995). The patterns also evolve over much slower time courses and in relation to learning (Karni et al. 1995) and neocortical injury (Merzenich et al. 1983). What emerges is a picture of distributed networks linked in space and time throughout the neocortex (Goldman-Rakic 1988; Donoghue, Leibovic and Sanes 1992; Singer 1994).

In the model described in the next section, a theoretical approach is described to formally integrate ideas about parallel processing, distributed neural networks and hierarchy. The model was motivated by anatomical and phys-

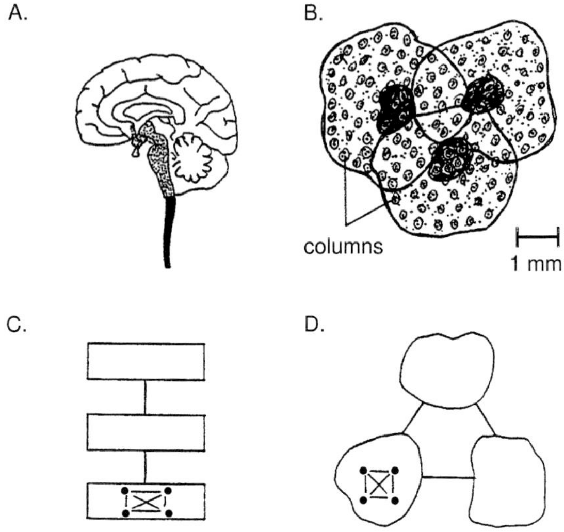

Figure 2: A. Schematic representation of the human brain showing the spinal cord (black), brainstem (stippled) and cerebrum (white), including the cortex and cerebellum. B. A layered network showing three sub-networks in sequence. C. A diagram looking down onto the surface of the cortex. Nested distributed networks of neurons are shown. Note the overlap of the three large networks. D. The topology in panel C. is illustrated for three sub-networks. (Compare this architecture with that shown in panel B.) After Sutton and Trainor (1990).

iological data about the neocortex that were available in the mid-1980's. It has since been put to the test by new research supporting the functional organization of nested networks, especially across a range of temporal scales. Fortuitously, the approach also comes at a time when neuroscientists are striving to bridge gaps and traverse scales of understanding in the 1990's, "Decade of the Brain."

3 Hierarchical Cluster Model

3.1 Architecture

The hierarchical cluster model (HCM) of Sutton, Beis and Trainor (1988a) is an artificial neural network model consisting of nested distributed networks. Its architecture is shown in Figure 3. The model is based in the tenets of

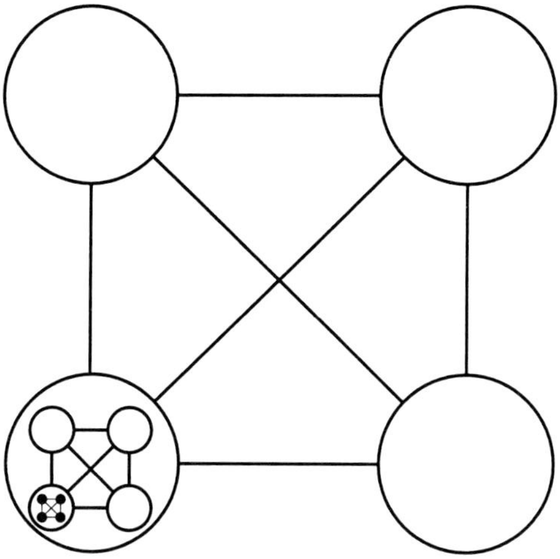

Figure 3: Fractal-like design of the hierarchical cluster model (HCM) proposed by Sutton and co-workers. A large network of successively smaller networks are organized in a nested manner. After Sutton, Beis and Trainor (1988a).

neocortical anatomy and physiology, and it can be viewed as a first order approximation to the posited neural architecture outlined in the previous section.

There are two ways of constructing the model, and they are equivalent. The first way is to consider individual model neurons linked together into local networks. The neurons form a trivial level of cluster organization, and the local networks comprise the first level of cluster organization. Coalescence of several first-level clusters leads to the formation of second-level clusters, and so on. Another way of viewing the nesting is to consider the entire network as a single cluster at the highest level. This cluster is partitioned into several smaller but still large clusters. Each of these large clusters is further partitioned, and so on, down to the level of individual neurons.

Regardless of which view is adopted, the properties of individual neurons are maintained at all levels of organization. Developmental considerations are also important in the construction of the model. Plasticity in the neural connections is ongoing, and the model has the capacity to continually redefine its structure. Certain entities, such as individual neurons and cortical columns, have close associations between structure and function. These are natural levels of cluster organization. Cytoarchitectural differences between different brain regions provide further boundary constraints. These regions may delineate higher levels of clustering. Other levels of organization are based on other spatial and temporal criteria, such as those discussed in the next subsection.

As in any hierarchical structure, the details of how the levels or boundaries are determined are critical. The boundaries in the HCM are determined by several factors. These include the variations among neuron types, the differences in the connection patterns, the sign and strength of synaptic weightings and the inputs to the system. Some neurons only have connections to other neurons in their immediate vicinity. Other neurons propagate great distances within the neocortex and have many collateral connections (Figure 4A). Some strategically positioned neurons may have high level functions, such as those associated with psychophysical phenomena in biological networks (Newsome, Britten and Movshon 1989). It is the heterogeneity in individual neurons and neural populations that is critical to how cluster boundaries emerge.

3.2 Dynamic Form

In the brain, learning and adaptation are ongoing processes that persist throughout a life span. Synapses alter with learning (Kandel et al. 1987), and on a much larger scale, the cortex reorganizes itself with learning (Karni et al. 1995). Within the HCM, clusters organize based on learning rules which operate at the synaptic level. Regardless of where an individual model neuron projects, all neurons in the model are assumed to obey local learning rules. That is, neurons perform local computations without special knowledge of the global properties of the neocortex.

Changes in the weights among model neurons change cluster boundaries. This will be discussed more fully in section 5. However, there is another mechanism for altering boundaries. Co-activation of adjacent clusters yields transiently larger clusters (Figure 4B). The correlated activity has the potential to encode information on a higher level than that associated with individual clusters. This becomes apparent when the transient boundaries propagate throughout the entire network and provide a means of rapid signal transfer across widespread distances.

3.3 Assertions of the HCM

The main assertions and implications of the HCM as originally proposed and then modified by Sutton and co-workers are as follows:

1. The neocortex uses nested clusters of neurons as an organizing principle.

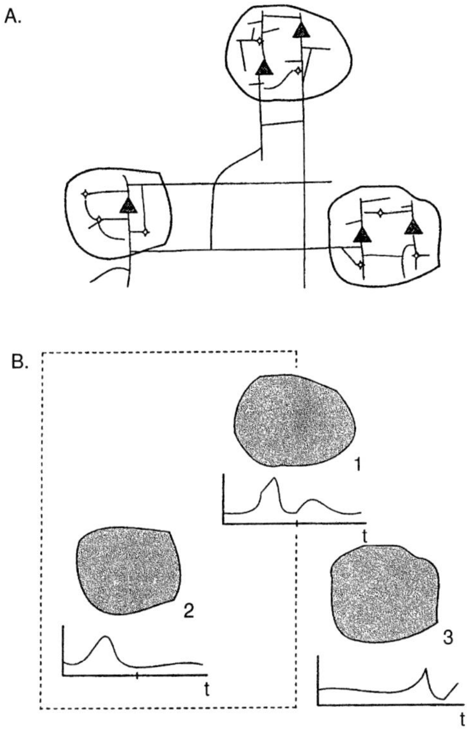

Figure 4: A. Three clusters illustrating intra-cluster and inter-cluster connections among heterogeneous model neurons. B. The activation profiles of three clusters are shown below each cluster as a function of time. During the first half of the time courses, clusters 1 and 2 are essentially synchronized, whereas cluster 3 is inactive. The simultaneous activity in adjacent clusters 1 and 2 forms a transient boundary (dotted line). The boundary delineates a functionally larger cluster, relative to the size of either cluster 1 or cluster 2. This larger cluster has the capacity to encode correlated information between clusters 1 and 2. After Sutton and Anderson (1994).

2. Cluster boundaries are initially determined by structural and functional criteria laid down in development.

3. The boundaries are plastic and are governed by the same rules dictating individual neuron and synaptic changes.

4. The boundaries are also determined by the simultaneous activation of linked clusters.

5. Patterns of neuron activity within a cluster are stored in stable ways. The stability endows the network with multileveled memory properties.

6. Higher-level memories are spatiotemporal correlations among lower-level memories.

7. Memory storage scales exponentially with the number of levels.

8. The model is massively parallel, with scaling between levels that is vector-valued. This is a result of preserving individual neuron properties across all levels. Traversing levels does not necessitate scalar averaging the activity of many neurons.

9. In principle, the boundaries between clusters do not have to be discrete. They vary in time and may be continuous in space.

10. The speed of computation is linearly related to the number of neurons (compare with item 7. above).

11. Within a level of cluster organization, stable memory patterns may represent information that has *meaning* at different levels. For example, clusters in the dorsolateral prefrontal cortex, which are important in working memory, may perform more sophisticated tasks relative to similarly sized clusters in primary sensory cortex.

12. The model is readily testable using current techniques which record spatiotemporal signals across scale (e.g. multiple unit recordings, functional magnetic resonance imaging).

13. Conceptually, the HCM provides a framework for integrating knowledge about some neuropsychiatric diseases (see section 6).

14. The architecture of the HCM can be applied to practical problems for *real time* processing, and its design is also suitable for hardware implementation (i.e. silicon).

3.4 Limitations of the HCM

The HCM approach to neocortical organization suffers from several limitations. Some of these are listed below.

1. Considerable anatomical detail at all levels of description is ignored.

2. The evidence for nested clustering is limited to relatively small regions of neocortex.

3. It is not clear what constitutes intermediate levels of nesting. Spatial and temporal properties are clearly important, but it is difficult to delineate the relationship between these properties and levels of neural clustering.

4. The approach is hard to validate and refute, despite being a simple postulate of brain organization.

4 Multilevel Memory Properties

4.1 Model Implementation

There are several ways to implement the hierarchical cluster model (Sutton et al. 1988a; Anderson, Gately, Penz and Collins 1990; Guan 1994). While different implementations vary in their computational features, they share properties of hierarchical and parallel distributed processes. These will be emphasized in this section.

According to Salthe (1985), at least three levels of structure are required for the analysis of any evolving hierarchical system. The self-similarity aspects of the HCM, as shown in Figure 3, allow an examination of a simple three-level version to be extended to an arbitrarily large network. Sutton (1988) has rigorously shown how to generalize to the n-level HCM.

There are many different mathematical formalisms and artificial neural network approaches that can be used in analyzing the HCM. For convenience, the methods employed by Hopfield (1982) are used here since the properties of the Hopfield model neural network are extremely well understood (Amit 1989, Peretto 1992). It should not be construed that the use of nested Hopfield networks implies that the neocortex is actually composed of Hopfield networks. It is the general architectural principles of nesting and parallelism that are important. Furthermore, it is not necessary that clusters in the HCM behave as Hopfield networks in order for the HCM to work. There are reasons to believe that alternative computational approaches may actually be more appropriate and yield similar results (Sutton and Anderson 1995).

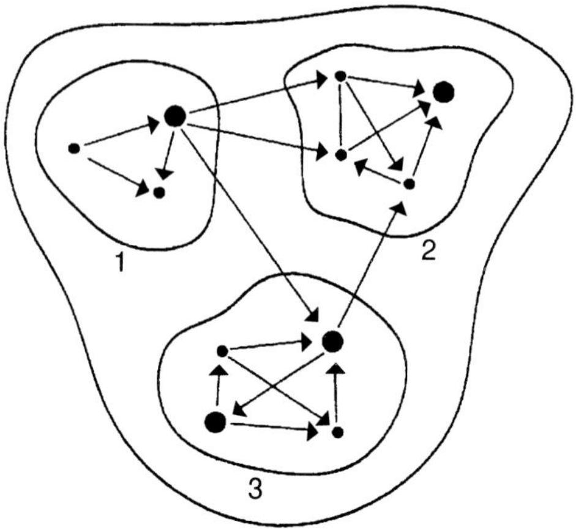

Figure 5: A simple nested network with three levels of structure. Individual neurons are shown as solid dots. They come in two "sizes." Neurons represented by small dots send and receive unidirectional connections to other neurons within each of the three small networks. Small dotted neurons may also receive connections from other networks but these neurons cannot project to other networks. Inter-cluster projections can only emanate from neurons represented by large dots. Each small network is also termed a first-level cluster. In this terminology, the neurons form trivial (or zeroth-level) clusters. The entire network, made up of the three first-level clusters, constitutes a second-level cluster. Only some neurons and connections are shown for simplicity. After Sutton and Trainor (1991).

4.2 Mathematical Formulation

4.2.1 Spatial features

A network consisting of three first-level clusters is illustrated in Figure 5. Within each such cluster, indexed by $k = 1, 2, 3$, there are N neurons or zeroth-level clusters. The system as a whole forms a second-level cluster. That is, there are three levels of structure: neurons, first-level clusters and the entire system.

The neurons have different firing rates and properties that convey interesting and important information. In the simplest scenario, each neuron is taken to be a binary state unit (after McCulloch and Pitts (1943)). The state values are given by $S_{ik} = \pm 1$, $i = 1, ..., N$, where the value S_{ik} characterizes the firing state $(+1)$ or the resting state (-1) of the ith neuron in the kth cluster (Figure 6).

A memory consists of an array of length N with elements ± 1 (Figure 7). Each first level cluster has the capacity to store p memory patterns $\left\{\xi_{ik}^{\mu}\right\}_{i=1}^{N}$,

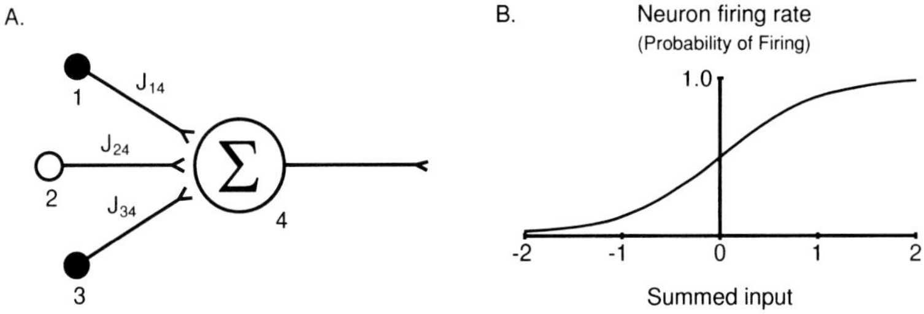

Figure 6: A. Diagram of four highly simplistic neurons. Neurons 1, 2 and 3 project to neuron 4, which effectively sums the input from these three neurons. The inputs are determined by the state of the pre-synaptic neurons (black = firing state, white = quiescent state) and the weights J_{ij}. The weights may be either positive or negative. The subscript on J_{ij} indicates that neuron i projects to neuron j. B. Response curve of a neuron as a function of its summed input. The firing threshold is assumed to be zero. Inherent noise in the system prevents the response curve from being a Heaviside function. Nevertheless, the "all-or-none" response is a common assumption in many neural network simulations. (See Sutton, Mamelak and Hobson (1992) for a discussion of firing probabilities as a function of neuromodulation.)

$\mu = 1, ..., p$. In practice, the ξ_{ik}^{μ} are chosen randomly. Surprisingly, this simple choice leads to some interesting properties.

By pairing the neurons together within a first-level cluster according to the rule

$$J_{ik,jk} = \frac{1}{N} \sum_{\mu=1}^{p} \xi_{ik}^{\mu} \xi_{jk}^{\mu}, \qquad i \neq j, \tag{1}$$

a learning scheme similar to Hebb's rule is created (Hebb 1949). This rule correlates pre-synaptic and post-synaptic firing activities. Positive correlations occur when pre-synaptic and post-synaptic neurons are either both firing or both are quiescent. This increases the value of $J_{ik,jk}$. Negative correlations, or mismatches in pre-synaptic and post-synaptic firings, decrease the strength of $J_{ik,jk}$. The geometry is such that every neuron effectively connects with every other neuron within a first-level cluster.

Hopfield (1982) demonstrated that when patterns of memory are stored in the symmetric connections according to equation (1), and when model neurons asynchronously update their firing states, then partial input of a memory pattern retrieves the entire memory. The Hopfield network is one of several artificial neural networks displaying this content-addressable memory property. Another popular approach has been proposed by Kohonen (1987).

In the context of the HCM version examined here, the first-level clusters are similar to Hopfield networks. In addition, subpopulations of model neurons

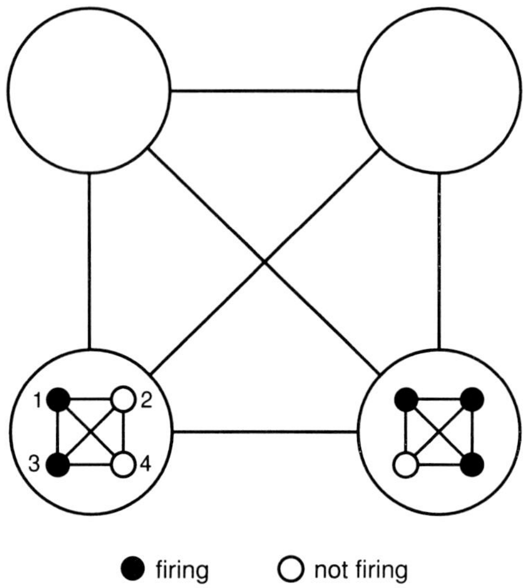

Figure 7: Expanded view of the bottom left portion of Figure 3. Four first-level clusters are shown. In the bottom two clusters, the firing and non-firing states of representative neurons are shown. Each first-level cluster has 2^N firing states. Some of these states are designated as memory patterns. Note that the firing configuration illustrated in the bottom left cluster corresponds to the neural connectivities shown in Figure 6A.

within each first-level cluster also connect with neurons in other first-level clusters. Not only is the symmetry among all the neurons broken by this feature, but it is also assumed that inter-cluster connections are asymmetric. This is more biologically plausible than assuming symmetry in the connections of neurons having long range fibers.

As illustrated in Figure 5, the model has two populations of neurons. One population forms connections with every neuron within its own first-level cluster, but does not form connections with neurons in the other two clusters. This population receives connections from neurons located anywhere in the system. The other population of neurons forms and receives connections both within and between the first-level clusters. While grossly oversimplifying the relationship to neurobiology, the arrangement is not unlike the populations of stellate cells and pyramidal cells in the neocortex. Stellate cells have local connectivities only, while pyramidal cells have both local connectivities and projective branchings that extend for considerable distances.

Mathematically, neuron connections between the clusters are given by

$$J_{ik,jl} = \frac{\gamma}{N} \sum_{\langle \mu\mu' \rangle} \xi_{ik}^{\mu} \xi_{jl}^{\mu'}, \qquad k \neq l. \tag{2}$$

The connection linking the ith neuron in cluster k to the jth neuron in cluster l is given by $J_{ik,jl}$. Since the network is not fully connected, many of the $J_{ik,jl}$ are equal to zero. The number of non-trivial connections between the first-level clusters is proportional to the ratio, x, of the number of neurons with projection-like (i.e., pyramidal neuron) features to N, the total number of neurons in a cluster. The parameter γ in equation (2) expresses the average strength of inter-cluster connections relative to intra-cluster connections. This measure allows for possible differences in the types and quantities of neuro-transmitters, the densities of projection neurites and synaptic terminals, and the receptor characteristics in the connections between clusters relative to those within clusters.

In the simple model under study, memory is encoded at two levels (Figure 7). At the first level, memory patterns are the $\left\{\xi_{ik}^{\mu}\right\}_{i=1}^{N}$, $\mu = 1, ..., p$, arrays stored in a distributed manner within each first-level cluster $k = 1, 2, 3$. Memory organization at the second level spans the entire network. The second-level memories consist of unique triplets of first-level memories, each one being from a different first-level cluster. That is, second-level memories are correlated triplets of first-level memories.[c]

If (ν, ν', ν'') denotes a specific second-level memory, where ν, ν' and ν'' represent individual first-level memories in clusters 1, 2 and 3 respectively, then $J_{i1,j2}$ contains a contribution of $\xi_{i1}^{\nu}\xi_{j2}^{\nu'}$. The sum over all such predetermined pairs $\langle \mu\mu' \rangle$ is weighted by γ/N. This sum gives the value of $J_{i1,j2}$ shown in equation (2).

By combinatorial analysis, we find that p^3 second-level memories are possible. It is a general property of the model that the number of memories grows exponentially with the number of levels. Many higher-level memories therefore arise from lower-level memories, and this is extremely advantageous for learning (section 5) and for engineering applications (section 7).

4.2.2 *Temporal features*

While the storage capacity of higher-level memories grows exponentially with the number of levels, the time required to compute at the systems level only

[c]Content-addressable memory features are assured for each first-level cluster when $p/N < 0.14$ (Amit, Gutfreund and Sompolinsky 1985). The conditions for stability of memories in the second- and higher-level clusters as a function of storage capacity have not been well characterized.

grows linearly with the number of clusters (Anderson and Sutton 1995). There are many dynamics that demonstrate this property, and for convenience a stepwise updating rule is used here.

It is assumed in the dynamical equations that the time scale for the model to evolve toward a memory pattern is much shorter than the time scale for learning. That is, equations (1) and (2) are taken to be time independent. Then, at each time step t, a neuron is chosen at random, and it assesses and modifies its state value according to

$$S_{ik}(t+1) = \text{sgn}\left[h_{ik}(t) - \theta_{ik}(t)\right]. \tag{3}$$

In this equation, $h_{ik}(t)$ is the instantaneous membrane potential. Evaluation of any state change involves a comparison between this potential and the membrane threshold, $\theta_{ik}(t)$. In the general case, the different values and time dependency of $\theta_{ik}(t)$ lead to interesting biological and computational properties (Horn and Usher 1989). For simplicity, the thresholds for all the neurons are set to zero. That is, it is assumed that $\theta_{ik}(t) = 0$.

The membrane potential is given by

$$h_{ik}(t) = \sum_{\substack{j=1 \\ j\neq i}}^{N} J_{ik,jk} S_{jk}(t) + \sum_{l\neq k} \sum_{j=1}^{N} J_{ik,jl} S_{jl}(t-\tau) + I_{ik}(t). \tag{4}$$

The first term reflects intra-cluster contributions, the second term gives inter-cluster effects, and the third term arises from external inputs, $I_{ik}(t)$. To simplify the discussion, it is assumed that $I_{ik}(t) = 0$. The τ in equation (4) is a temporal (or phase) lag measure. It is integer valued and corresponds to propagation delays between neurons in different first-level clusters. Delays reflect not only spatial distances between the clusters, but they are also used to incorporate different connection properties among intra-cluster and inter-cluster linkages.

The form of equation (4) is similar to the membrane-like equations found in some fully-connected associative networks with temporal sequencing capabilities (Sompolinsky and Kanter 1986; Gutfreund and Mezard 1988; Sutton, Mamelak and Hobson 1992). Indeed, for large τ values, the HCM operates as a temporal sequencing network (Sutton and Trainor 1991).

Although equation (3) describes serial updating of individual neurons, each step in the computation derives input from the entire system. When $\tau = 0$, there is instantaneous input from all the neurons connected to the selected neuron. For $\tau \neq 0$, first-level processes within a cluster are modified by second-level processes delayed in time. Evolution proceeds toward independent first-level memory patterns until second-level effects become apparent. These second-level effects potentially modify the network's global memory recall. Regardless of τ, the number of time steps required for convergence of the model scales with the total number of neurons.

4.2.3 Computational energy

Sutton et al. (1988a) have shown that when the HCM dynamics are governed by the evolution equation (3), there is a decrease in the value of a "computational energy function"

$$E(t) = -\frac{1}{2}\sum_{k=1}^{3}\sum_{i=1}^{N} h_{ik}(t)S_{ik}(t). \tag{5}$$

It is useful to introduce an energy-like term since there is a close association between the local minima and the memory patterns in an energy representation of attractor neural networks (Amit 1989). However, the computational energy in equation (5) is not a true energy in the thermodynamic sense. In order for this to occur, all of the connections must be symmetric. This is the case in the Hopfield model, and the associated computational energy constrains the system in a manner not unlike the role played by the potential energy function in a Hamiltonian system (Hopfield 1982).

While the asymmetry of the connections between the clusters in the HCM prevents its E function from operating as a Hamiltonian, there are certain conditions whereby equation (5) maintains some general energy-like characteristics. The main constraint is that the maximum value of γ not be much greater than unity. When γ is very large, the inter-cluster connections significantly distort the attractor properties of the first-level clusters and the network as a whole. This impedes the associative memory capabilities of the system, as verified in the simulations in the next subsection.

As discussed in section 2, there is biological evidence that the number of connections between different levels of neocortical organization is limited. This property is incorporated into the HCM, where the fraction of inter-cluster projection neurons, x, is inversely related to the level of cluster organization. There are no biological data, to the author's knowledge, to suggest that γ should vary inordinately (on average) from unity. Furthermore, the time delay, τ, should be short, which is consistent with very small propagation delays across the neocortex. Under these aforementioned conditions, the HCM has been shown to behave as a complex attractor neural network with energy-like properties associated with the E function of equation (5). Moreover, the reasons why this occurs have been investigated in detail in the mean field approximation (Sutton 1991). Some of the computer simulations demonstrating the hierarchical memory performance of the HCM are described in the next subsection. The interested reader is referred to the appropriate reference for a discussion of the mean field analysis.

4.3 Simulations

4.3.1 Overview

Computer simulations are helpful in gaining insight into how particular parameters and variables alter the performance of a neural network. Simulations are also useful in characterizing the limitations of a model. When the HCM was first developed, simulations were used to demonstrate how the hierarchical structure affected the parallel distributed properties of the network as a whole (Sutton et al. 1988a). In the results presented here, the focus will be on three parameters. These are (1) the fraction, x, of inter-cluster connections relative to intra-cluster connections, (2) the relative strength, γ, of these connections and (3) the temporal lag, τ, of inter-cluster events relative to intra-cluster events. The results are discussed in the $\tau = 0$ case and in the $\tau \neq 0$ case. To simplify the presentation, attention is placed on second-level memory recall only.

4.3.2 Method

In the version of the HCM that was analyzed, each of the three first-level clusters had 30 neurons and three stable memory patterns. These patterns were stored distributively among the intra-cluster connections, as shown in Figure 7. The second-level memory patterns were combinations of first-level patterns. For convenience, only three such combinations were taken as second-level memories. (There were $3 \times 3 \times 3 = 27$ combinations in total.) The determination of the second-level memory patterns was made according to equation (2).

The second-level memory performance of the network was calculated by initially placing (i.e., cueing) one of the first-level clusters in a memory pattern forming part of a second-level memory. Then, with the other two clusters in random initial patterns, the system was allowed to evolve according to equation (3) until an equilibrium state was reached. This process was repeated thirty times for each combination of x, γ and τ being investigated. The repetition was required to obtain averages over the initial conditions and over the choices of first- and second-level memories.

One way of expressing the performance of the network was to determine the per cent correct correlation among first-level memories making up a second-level memory pattern. Suppose that the network was initiated (1) with one cluster in a memory pattern forming part of a second-level memory and (2) with the other two clusters in random patterns. If the network evolved to the second-level memory pattern associated with the cued first-level pattern, then the network was said to have correctly recalled the second-level memory pattern. If the network evolved to some other pattern, even if a significant portion of that pattern was a second-level memory, then the network was said not to have recalled the second-level memory on that trial.

The results were expressed as a percentage of correct recall of second-level memory patterns over all the trials. Thus, if the network correctly recalled the partially cued second-level memory on every trial, then the per cent correct correlation at the second level was 100%. In practice, this value could not be achieved because of inherent noise and frustration in the connections amongst the neurons (Sutton 1988). The result was that the network evolved to spurious patterns that were either quasi-stable or the inverse of memory patterns. Inverse memories were represented by arrays where every $+1$ was replaced by -1 and vice versa.

While the network was not expected a priori to perform at 100% even under the most ideal conditions, it was similarly not anticipated to yield 0% output under random conditions. In theory, the chance occurrence of relaxation to a second-level memory pattern was 3 out of 27, or 11%. The actual probability was less than this value since the evolution of the non-cued clusters sometimes resulted in spurious local minima. The time for convergence to memory patterns was approximately $12N\tau$, or three times the estimated $4N\tau$ convergence time for a single Hopfield network (Hopfield 1982).

4.4 Results

4.4.1 $\tau = 0$ conditions

With instantaneous updating of neurons throughout the network, the dependence of second-level memory recall on x is as shown in Figure 8. The range of correct correlations as a function of x for selected values of $0.5 \leq \gamma \leq 3.0$ was 7.3% to 94% Most of the gain in second-level memory association occurred for $x < 0.4$ when $\gamma \geq 1.0$. This was not surprising since the number of terms contributing to the membrane potential equation (4) scaled as $N(= 30)$ for intra-cluster connections; whereas, the number of terms effectively increased as $3N \cdot x$ for inter-cluster connections. For weak inter-cluster connectivity values of $\gamma < 0.5$, the recall ability of second-level memory patterns deteriorated

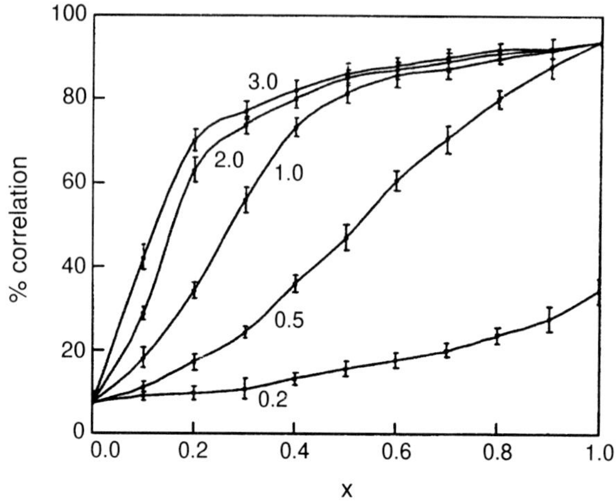

Figure 8: Plots of second-level memory recall expressed as a function of the inter-cluster connectivity fraction, x. As x increases, the performance of the network increases. There is a trade-off with time (not shown), since each additional connection adds $N-1$ calculations to determining the % correlation at each time step. The plots are shown for $\gamma = 0.2, 0.5, 1.0, 2.0$ and 3.0. After Sutton et al. (1988a).

dramatically.

As $x \to 0$, the system became dissociated into disjoint clusters. The inter-cluster correlations of first-level memory patterns were no longer stored. In this limit, the correlation as a percentage was 7.3%. The effect of spurious local minima partially explained the non-zero value and the less than perfect correlation of first-level memory patterns that was seen as $x \to 1$. The relatively large number of second-level connections in this limit occasionally caused the cued memory pattern to become unstable. When random initial states of the non-cued clusters were associated with a deep valley on the E surface that excluded the cued memory pattern, the model tended to come to rest in either some combination of first-level memory patterns not representing a second-level memory pattern or some other spurious pattern (Sutton et al. 1988a).

The dependence of second-level memory recall on γ is as shown in Figure 9 for $\tau = 0$. Second-level memory recall was optimized in a range of γ values that varied with x. When $\gamma < 0.2$, the second-level effects of the model for $x \leq 0.4$ were essentially negligible. On the other hand, for strong inter-cluster connectivities of $\gamma > 7$, the second-level correlations tended to dominate the first-level events. This caused the model to lose the fine detail of the computa-

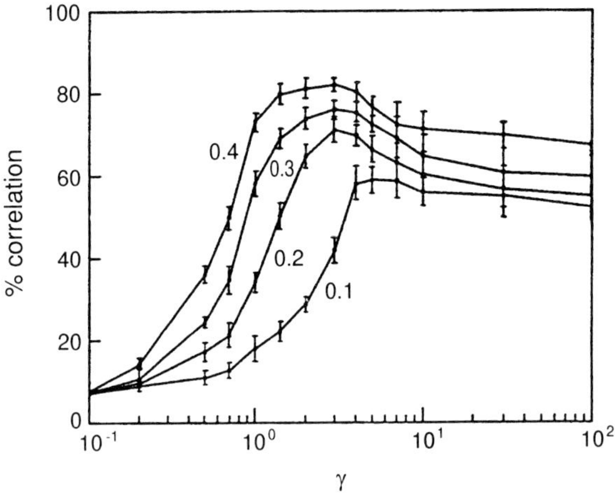

Figure 9: Graph of % correlation as a function of the relative inter-cluster connectivity strength, γ, for x=0.1, 0.2, 0.3 and 0.4. There is a maximum in each curve, such that very large γ values destroy the performance of the network. After Sutton et al. (1988a).

tional surface, and the efficiency of the memory storage and recall capabilities diminished. Relatively large variability in second-level recall, on the order of 15%, occurred in this range. In summary, peak second-level associations occurred for $2 < \gamma < 5$ and $0.1 \leq x \leq 0.4$.

4.4.2 $\tau \neq 0$ *conditions*

Figure 10 shows the average retrieval of second-level memories as a function of τ for $\gamma = 1.0$ and $x = 0.4$. When $\tau = 0$, the system evolved to a second-level memory pattern approximately 75% of the time.

Average correlations among first-level memory patterns forming second-level memories decreased monotonically with τ to a value of approximately 19%. This occurred for $\tau > 400$. Under such conditions, there was serial input of first and then second level contributions to the membrane potential equation (4). Following relaxation to random triplets of first-level memory patterns, the system rearranged itself via input from the inter-cluster connections. This process yielded a greater percentage of second-level memory patterns than that due to chance alone. The effect of the second-level architecture was made apparent by the improved value of 19% compared to the 7–11%, corresponding to the expected performance of the network when second-level inputs were absent and the first-level clusters acted independently. Thus, there was a

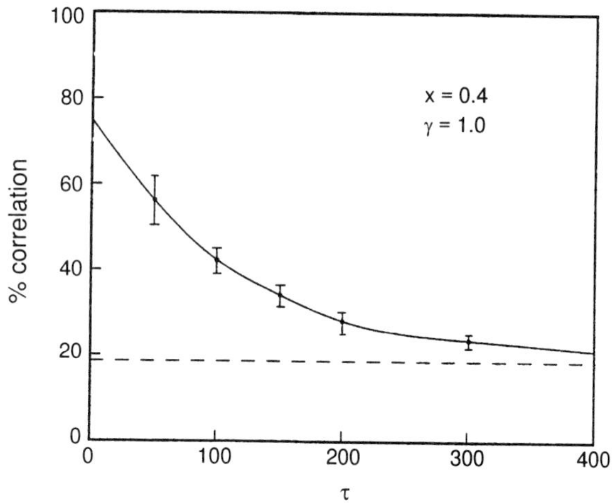

Figure 10: Plot of % correlation as a function of the time lag, τ, between first-level clusters ($x = 0.4$, $\gamma = 1.0$). There is a monotonic decrease in second-level memory performance with increasing τ. The decrease is asymptotic towards 19% correct recall (dashed lines). After Sutton and Trainor (1991).

trade-off between local autonomy at the first level of cluster organization and the global performance of the network with respect to second-level memory recall.

5 Learning and Development

5.1 Outline

Learning is a change in a system that results from reinforced experiences or interactions with the environment. In section 2, learning was discussed in the context of synaptic plasticity. Connecting synaptic plasticity to changes in the behavior of the brain as a whole was one of the great theoretical leaps in cognitive neuroscience. The link between plasticity at the level of individual neuron connections and memory at the scale of large neural networks was evident in the discussions in sections 3 and 4 concerning the HCM. Memory and learning are interdependent processes, and in this section, *learning* by selection is discussed as a mechanism of stabilizing higher-level *memory* patterns.

5.2 Perspectives on Learning

There are three basic views concerning the organization of information (Marler and Terrace 1984). It is generally agreed that no one view accounts for all systems. The first is known as the preformist view, wherein there is complete prior specification of information. The determination of synaptic connections according to equation (1) is an example of how information can be completely specified in a preformed manner. Learning occurs before there are interactions with the environment.

The second view of information organization is the empiricist view, in which the initial state of a system is considered to be a *tabula rasa* or blank slate. The entire internal organization of information results from direct instructive learning by means of environmental interactions. This is clearly not the case in the HCM, which is endowed with considerable structure based on its attractor properties within first-level clusters and on the clustering of networks formed from linkages among the first-level clusters.

In the third view, selection mechanisms increase the internal order of information by means of indirect environmental experiences. Very little specification of information needs to be encoded a priori. Interactions with the environment serve to selectively stabilize some of the variations in organization that have occurred spontaneously.

Historically, selectionist theories of learning and adaptation have played preeminent roles in biology. Darwin's theory of species evolution (Darwin 1859) and the network theory of antibody formation in immunology (Jerne 1955) are based on selectionist principles. Selectionist theories have also been proposed for neurobiological processes (Young 1964, Edelman 1978, Marshall 1980), including specific learning mechanisms (Changeux and Danchin 1976; Heidmann, Heidmann and Changeux 1984; Toulouse, Dehaene and Changeux 1986; Sutton 1993).

Two features are necessary in the selectionist approach. First, a generator of internal diversity that utilizes a combinatorial process must be present. Second, a mechanism for reinforcing some combinations and (or) rejecting others must exist in a way that allows for the exchange of signals with the environment. It has been suggested that the structure of the HCM provides a natural setting in which selective processes operate (Sutton 1988). The low-level memory patterns in the model are responses determined by their input and output connections to and from the neocortex. These are established early in development. With continued reinforcement, the patterns at this level become stabilized after a prespecified time. Within the cortex, combinations of low-level memory patterns provide internal diversity to establish the posited higher-level memory patterns.

What is required is a plausible mechanism for selecting and stabilizing combinations of first-level memory patterns. In accord with neurobiological principles, this mechanism should be directed at the time dependencies of neural interactions. Long term potentiation at the synaptic level is a key mechanism for learning, and disruptions in this process are associated with impaired learning (Sakimura et al. 1995). Interestingly, alterations in the synchronous activity between large scale brain regions are also linked with some disturbances of learning (Sutton, Whitton, Topa and Moldofsky 1986).

5.3 Learning by Selection in the HCM

5.3.1 Memory stabilization

To illustrate how learning by selection can be readily implemented within the HCM, consider the example network discussed in section 4. The network consists of three first-level clusters linked to form a second-level cluster, which is the entire network (Figure 5). Instead of each first-level cluster having three memory patterns as discussed in section 4, suppose that each cluster has only *two* first-level memory patterns. At the network level, 2^3 possible second-level memories $\rho = 1, ..., 8$ exist. Relaxation of the system is assumed to result in one of these eight memory patterns.

Initially, the probability ϵ_ρ that the system relaxes to any particular memory, ρ', is $\epsilon_{\rho'} = .125$, $\rho' = \rho$. Suppose that the network is subjected to t encounters with the environment, whereby each encounter elicits a response of one of the ρ patterns. In selectionist learning, not all of these second-level patterns are equally favorable. Selective reinforcement of favorable memory patterns can be represented by the updating rule

$$\epsilon_\rho(t+1) = M^+\Big[\epsilon_\rho(t)\Big] > \epsilon_\rho(t). \tag{6}$$

Here M^+ is a monotonically increasing function that may or may not be time-dependent. Selective inhibition occurs via a monotonically decreasing function M^-, such that

$$\epsilon_\rho(t+1) = M^-\Big[\epsilon_\rho(t)\Big] < \epsilon_\rho(t). \tag{7}$$

Global patterns are modified by M^+ and M^- accordingly, where these functions represent the integrated result of widespread plastic changes occurring at the synaptic level.

Different inputs are preferentially associated with different second-level memories. The preferences are based on (1) the particular input, because of the associative memory features of the model, (2) the favorable or unfavorable nature of the memory pattern, (3) the form of the learning functions M^+ and M^- and (4) the previous inputs or experiences. To illustrate this last point, consider the effect of previous inputs on two otherwise equally favorable second-level memories. Suppose the events are such that one of the two memory patterns is selected by chance more frequently than the other pattern early in the learning process. As the more frequent pattern is reinforced, its probability of being chosen again increases according to equation (6). On each subsequent trial, it becomes more and more difficult for the other favorable pattern to compete. Two memory patterns that were initially identified as equally favorable may end up with markedly different outcomes.

5.3.2 Numerical simulation

For illustrative purposes, consider a simple network with two reinforced and six inhibited second-level memory patterns. Beginning with $\epsilon_\rho(t = 0) = .125$, $\rho = 1, ..., 8$, the system evolves in discrete time steps. When a favorable memory pattern is obtained, its probability of occurrence is updated according to equation (6), where

$$M^+(\epsilon_\rho) = (\epsilon_\rho)^{1-\delta} \tag{8}$$

and $\delta = 0.2$. The probability of an unfavorable response is changed by

$$M^-(\epsilon_\mu) = (\epsilon_\mu)^{1+\delta}. \tag{9}$$

The learning trials consist of 50 discrete iterations. On any given iteration, a memory pattern is randomly chosen and weighted by the fraction of probability space that it instantaneously occupies. A small amount of biasing is given by δ in equations (8) and (9), and the data (which are not normalized) are shown in Figure 11. Normalization does not significantly alter the results. As seen in Figure 11, by the 10th iteration, the $\rho = 1$ memory pattern has increased its stability almost threefold. After 15 iterations, both $\rho = 1, 2$ show evidence of becoming stabilized, although a considerable difference exists in their rates over the first 30 iterations. By $t = 50$, both patterns approach maximal stabilization ($\epsilon \rightarrow 1$), with considerable background noise persisting among the unselected responses.

During the early iterations, a degree of randomness is present in learning. The randomness is partially due to the memory and learning mechanisms inherent in the model and partially due to the sequence and types of responses the system encounters with the environment. The result is that only a fraction of all possible favorable patterns are stabilized. The stabilized patterns are not

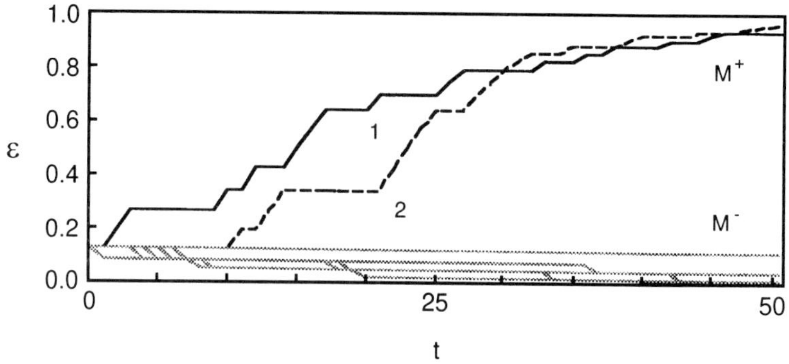

Figure 11: Probability of stabilization, ϵ, as a function of time, t, for eight memory patterns. All eight memories have equal a priori probabilities for stabilization. Memory 1 (solid line) and memory 2 (dashed line) become stabilized, albeit at different rates. The remaining six memories (hashed lines) remain unstabilized. At $t = 50$, some of the unstabilized memories have not had a significant reduction in ϵ, while the stability of others has approached zero (i.e., totally unstable). After Sutton (1993).

necessarily the best possible ones. However, the system does spend relatively little time inhibiting unfavorable responses.

Learning by selection is only one of several possible means of rapid learning in multi-level systems (Sutton 1993). Some of the details of how stabilization occurs lie in the effects of modulation on synaptic plasticity (Selverston 1985, Kandel et al. 1987). Neuromodulation plays an important role in regulating learning and in coordinating behavioral changes, such as those that occur during sleep and wake (Hobson 1988). This is relevant to brain development, since sleep accounts for a significant proportion of time during ontogeny when learning is maximal. Some preliminary research indicates that learning within neural networks in simulated sleep stages has extremely rich information properties (Sutton et al. 1992). This line of investigation has not yet been extended to the HCM.

6 Perturbation and Disease

6.1 Neuropathology as System Perturbation

Pathology, the study of disease, explores perturbed physiology. Studying the response to perturbation generally gives insight into the normal workings of a system, as well as into the possible mechanisms that underlie disease. This is related to the fact that normal homeostatic processes attempting to drive a system back to equilibrium are often the very same mechanisms involved in the development of disease.

In the central nervous system, synaptic loss is a progressive process that begins before birth (Changeux 1985). Since many pathways are overspecified and require pruning, it is sometimes difficult to distinguish what is normal development and aging and what is pathology. A characteristic feature of some neurological disorders is that they are selective for certain pathways: pyramidal tract and other motor neurons in amyotrophic lateral sclerosis; nigrostriatal neurons in Parkinson's disease; posterior and lateral column spinal cord neurons in Friedreich's ataxia (for a review of selective neuron disorders, see Bock and O'Connor 1987). Other disorders are selective for particular constituents of the brain. Multiple sclerosis involves lesions that affect white matter tracts in multiple places at multiple times (Figure 1C). Creutzfeldt-Jakob disease is a spongiform degeneration of the gray cells in the cerebral and cerebellar cortices. Strokes are vascular accidents that affect any part of the brain. They disrupt gray and white matter in brain regions supplied by particular arteries.

6.2 The HCM and Illness

It is surprising that given the amount of effort devoted to the study of artificial neural networks in the past one hundred years, there has been very little work done on neural modeling of cognitive and brain disorders. It turns out that modeling neuropathological processes is rather straightforward once an appropriate framework has been cast on the problem. For instance, by using perturbation techniques, it is possible to artificially lesion a network and observe its behavior. The more accurate the model is neuroanatomically and neurophysiologically, the more likely the simulated disturbances will give rise to biologically applicable predictions. Moreover, it is conceivable that modeling may provide new insights into brain disorders themselves, and even into treatment.

It is possible to examine the effects of simulated lesions on a network comprised of nested neural clusters. While the HCM lacks detailed information about the neocortex, to first order the correspondence between simulated neural lesions in the model and clinical data is strikingly high. One does not have to fine tune the parameters in the model to generate preconceived results.

Qualitatively, several features can be characterized. When neurons are randomly destroyed throughout the HCM, the performance of the model suffers at all levels simultaneously. There is a rapid deterioration of the entire network analogous to Creutzfeldt-Jacob disease.

Local lesions in the model that affect connections between clusters dissociate selected regions. Simulated lesions destroy all but the lowest level responses associated with the affected clusters, which is not unlike the events occurring in multiple sclerosis. Similarly, when local destruction affects neurons within clusters, all levels of functioning are impaired. Distant regions are altered slightly as well, due to connections emanating to and from the affected regions. This scenario captures the neurocognitive picture seen in stroke victims.

Finally, when there is random but selective destruction of large neurons, a progressive memory loss occurs. The loss proceeds sequentially from high level associations to low level memory recall. This phenomenon is described in more detail in the remainder of this section which focuses on Alzheimer's disease.

6.3 Alzheimer's Disease

Alzheimer's disease is a degenerative neuropsychiatric condition afflicting more than four million persons in North America, with approximately 300,000 new individuals diagnosed with the condition each year. It is the fourth leading cause of death, and the prevalence increases dramatically with age, with 20% of individuals over the age of 85 suffering from its dementia. This age group is the fastest growing segment of the population. The annual cost of providing care to its victims in the United States currently exceeds $20 billion (Ernst and Hay 1994).

Evidence of neuropathological changes in the disease dates back to a landmark case report early in this century which described the autopsy findings of a 55 year old woman who died with a history of progressive dementia (Alzheimer 1907). Newly available silver stains at the time demonstrated the presence of abnormal cortical neurons containing tangles of fibers, known as neurofibrillary tangles. Collections of degenerating nerve endings, or neuritic plaques, were also observed. To this day, a definitive diagnosis requires histopathological evidence obtained from a biopsy or autopsy.

While the neuropathological mechanisms leading to Alzheimer's disease are complex and not fully worked out, it is known that neurons degenerate and that there are neurochemical disturbances. These are correlated with disruptions in the pathways between the hippocampus and the neocortex, β-amyloid deposits and deficits of acetylcholine emanating from neurons in the basal forebrain (Lamour 1994). The neocortical degeneration is somewhat selective for layer III cells larger than 90 μm^2 residing in the association cortices (Terry et al. 1981). These large cells comprise roughly 20% of all neurons in the association neocortex, and they are important in providing cortico-cortical fibers (see section 2). In contrast to these large cells, small localized neurons remain relatively unaffected. Nevertheless, overall synaptic loss is a key factor in determining the severity of the disease.

One of the intriguing things about Alzheimer's disease is that despite its severe disruption of neural function, only about 10% of all neocortical neurons are actually lost in even the worst clinical presentations (Katzman 1986). The clinical manifestations are complicated, and they encompass a progressive dementia characterized by intellectual deterioration severe enough to interfere with occupational or social performance. The cognitive changes include not only disturbances in memory, but also disruptions of language, perception, praxis, learning, problem solving, abstract thought and judgment (Katzman 1986).

6.4 Hierarchical Memory Loss

One approach to modeling the selective neuron loss in disorders such as Alzheimer's disease is to construct a neural network containing heterogeneous populations of neurons. Then the premise of *selective* neuron destruction potentially has meaning. Topographically, the neurons are arranged such that the selective loss of a small fraction of key neurons has a significant, and possibly catastrophic, effect on the behavior of the entire network. Although a hierarchical framework is not necessary for this to occur, it provides a useful means of describing some disease processes affecting the neocortex.

It has been suggested that the HCM simulates selective neuron loss and its associated effects on higher brain operations, such as memory recall (Sutton et al. 1988a,b). The selective loss of neurons projecting between different levels of cluster organization destroys memory correlations between the clusters, while preserving memory correlations within the clusters. Although speculative, the projection neurons are like large layer III cells that establish cortico-cortical connections. Destruction of these neurons linking clusters at the highest levels can have profound negative effects on the hierarchical memory, since these neurons are strategically positioned and relatively few in number (Figure 12).

The loss of neurons in the HCM may be performed in a partial or complete manner. Partial destruction corresponds to the formation of neurofibrillary tangles and neuritic plaques. It can be simulated by decreasing the numbers and relative strengths of the non-zero connections formed by neurons projecting between clusters. Complete destruction of these neurons, if done randomly, would correspond mathematically to a site percolation problem.[d]

To demonstrate how the loss of neurons with inter-cluster connections affects memory correlations, Sutton et al. (1988a) considered a selective site

[d]See Essam (1980) and Stauffer (1987) for a review of percolation techniques.

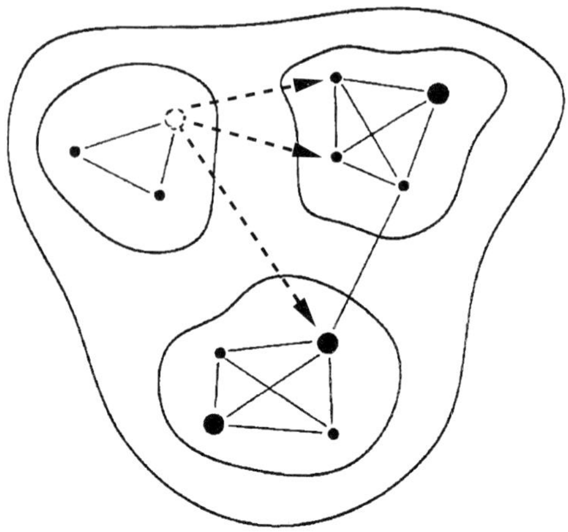

Figure 12: Nested network similar to Figure 5 showing the effect of the selected loss of a large neuron (dotted line). Only a few representative neurons and connections are drawn.

percolation on the model discussed in section 4 (Figure 12). They examined the case where there was a 50% selective loss of neurons with inter-cluster projections, which reduced the fraction, x, of such neurons from 0.2 to 0.1. Furthermore, they assumed that there was a reduction in the relative connection strength, γ, from 3.0 to 1.0 on the remaining inter-cluster connections. These alterations were not found to significantly alter the first-level memory performance of the clusters. However, they dramatically reduced the memory recall at the second level from greater than 70% to approximately 10% (Figure 13). The changes in x and γ effectively dissociated the first-level clusters with respect to inter-cluster correlations, and they eliminated second-level memory capabilities. Without clustering, a 10% random loss of neurons had minimal effects on memory loss at either level.

Although only suggestive of some of the neuropathological processes present in Alzheimer's disease, this simple computer simulation indicated that Alzheimer's disease may be, in part, a disconnection syndrome par excellence. Syndromes of this type have been postulated by Wernicke (1874) and Geschwind (1965), and they provide a link between neuropathology and psychopathology. Neural models, such as the HCM, have the capability to simulate how the disconnection may occur and what the consequences might be at multiple levels ranging from the neuron to the entire brain.

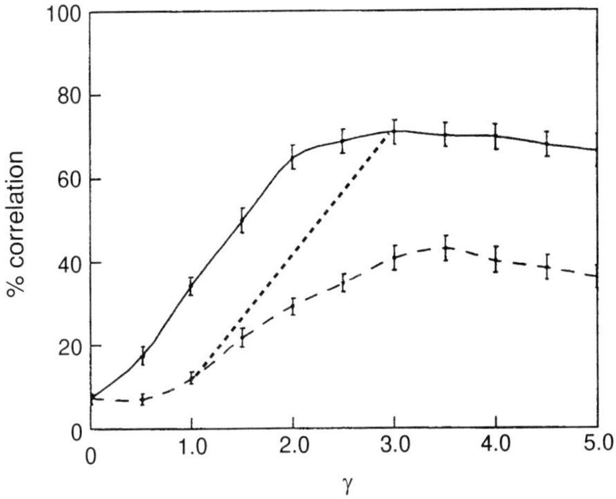

Figure 13: % correlation as a function of the inter-cluster connectivity strength γ for a selective percolation process. The solid line plots the curve for $x = 0.2$, and the dashed line gives the plot for $x = 0.1$. A transition from the solid line to the dashed line, for a given value of γ, represents a 50% reduction in the number of neurons that project between first-level clusters. The loss of such neurons corresponds to neuritic plaque formation in Alzheimer's disease. A reduction in γ along either curve corresponds to partial (as opposed to complete) neuron destruction. The associated biological process is neurofibrillary tangle formation. The dotted line shows the effect of selectively losing 10% of all neurons (20% of all neurons project between first-level clusters, and the percolation process selectively destroys half of them in random order) and reducing γ from 3.0 to 1.0. These conditions abolish the second-level memory properties of the network. The % correlation function drops from 70% to 10%. However, the first-level memory properties of the network are preserved despite the selective percolation. After Sutton et al. (1988a).

7 Integrative Neural Modeling

7.1 Perspectives

A central feature of the HCM that sets it apart from other models of neural computation is that it specifically addresses a mechanism for traversing multiple levels of neural organization in the neocortex. The relevance of this task to understanding the integration of brain and mind has been eloquently expressed by Minsky (1985). The HCM suggests a new approach to modeling neocortical function. It posits rather simple computational rules for an immensely complicated structure, and it suggests that these rules are plausible because they scale over many orders of magnitude.

Currently, the HCM is in an early stage of development. The model is speculative, but it is potentially valuable as a means of investigating brain organization and function. It presents a view of hierarchy within neuroscience that breaks from two other correct and commonly held notions about neural strata. One concept of hierarchy refers to disjoint levels of structure ranging from molecular constituents, to neurons, to networks and to the brain as a whole organ (Morris, Kandel and Squire 1988). In this scheme, there are no linkages between levels aside from the relationship that the brain is comprised of its constituent parts. The second view concerns brain regions that are arranged according to levels of computational sophistication. For example, in visual hierarchies, regions performing early vision tasks feed into regions performing more complex tasks, such as the directional selectivity of motion, and so on (Van Essen et al. 1992).

Even though the HCM implements hierarchical and parallel distributed principles in a novel and cohesive way, considerable work must be accomplished before a HCM approach to brain organization can be viewed as a brain theory. The model must apply to specific brain systems and it must make experimentally testable predictions. Some of these issues are discussed in this section. The topics touch on complex systems, neural computation and applications to technology and neuroscience.

7.2 Complex Systems

In the opening quotation of this chapter, Cajal conjectured that the complex nature of connections among neurons in the cortex is the basis for high level cerebral functions. For more than a century, in spite of ever increasing knowledge about structure and function, the brain continues to be seen as a complex organ whose organizational principles remain largely undiscovered. Understanding one phenomenon raises more questions than it provides answers for. One of the interesting aspects of studying models such as the HCM is that although they grossly simplify neuroscience, they possess rich and sometimes unexpected properties.

The HCM is a complicated model despite enormous neurobiological simplifications. It has a nested architecture with biological and physical complexities that are not well understood. Fortunately, there is a growing body of mathematical work on self-similar structures that may have relevance to the HCM, since the model possesses some fractal-like features (Ingber 1995; Martinez, Paredes, Borgani and Coles 1995). There is also an excellent understanding of the physics and computational properties of spin glass systems, which share associative memory capabilities of the non-hierarchical HCM or Hopfield model (Amit 1989). However, the effects of nested clustering on information storage and retrieval have not been well investigated.

Perhaps the most intriguing aspects of structures such as the HCM are their temporal characteristics. Clusters appear to form over very short time scales (Figure 4B). They also emerge over longer time scales as the system interacts with the environment and adapts to it through learning and development. Consolidation of lower-level memory patterns gives rise to higher-level memory patterns. If the HCM is even partially correct, an understanding of its spatiotemporal complexity may provide significant insight into how our own brains operate.

7.3 Neural Computation

The hierarchical memory and learning properties of the model have been discussed throughout this chapter. Two areas where the HCM may be useful in understanding the neural basis of psychological phenomena are in language and in categorical memory. The two topics are related. The nested structure of the model lends itself to categorizing information in a hierarchical manner. When the nesting pertains to components of language, some interesting properties may emerge. For instance, suppose that first-level clusters store letters, second-level clusters encode words, third-level clusters associate phrases, and so on. While simulations have not yet been performed, the nested architecture provides a means of testing hypotheses about language and its alterations, such as dyslexia.

From a neurobiological view, it is necessary to add physiological details to the HCM in order to corroborate or refute specific computational aspects of the model. For example, data concerning neural firing rates, synaptic connectivities and field potentials are required to test the model at the level of small neural circuits. Recordings from many individual neurons may provide the necessary data. At another level of analysis, new functional neuroimaging techniques, such as echo-planar magnetic resonance imaging, may be useful in examining spatiotemporal data about how brain regions interact with one another.

Even though more detail must be built into the model to elevate it from an impressionistic stage to an explanation stage, several predictions are possible. These include:

1. Clusters of neurons form networks at multiple levels of scale and obey similar rules of activation within and between levels.

2. Clusters at a single level perform heterogeneous functions. Thus, clusters representing brain regions such as V1 or V5 (MT) may be at the

same level of organization but are associated with relatively more or less sophisticated tasks. Similarly, individual neurons (trivial clusters) may be correlated with psychophysical phenomena, such as face recognition, while others perform low level processes. Heterogeneity exists at all levels.

3. Cluster boundaries are fluid. They can be modified by synaptic plasticity and by spatiotemporal patterns of activity among neurons.

4. Information transfer and learning occur at all levels simultaneously.

5. Memory patterns or responses are manifested by particular dynamic patterns distributed among different clusters. This property is ubiquitous and may challenge well accepted maps of cortical localization.

6. Distributed memory patterns that are formed among clusters are stable on the time scale of cluster activation. However, they are subject to change over longer time scales.

7. The memory capacity at a given level is finite. The limit is imposed by the attractor properties of the clusters at that level, but the capacity may vary from cluster to cluster. In principle, information storage scales exponentially with the number of levels.

8. Alterations in memory or learning at one level induce changes at all levels. The nature of the changes are similar across levels, in accord with the postulate of neural scale invariance (Sutton and Breiter 1994). Other predictions are also possible based on the implications listed in section 3.3.

7.4 Applications to Technology and Neuroscience

The architecture of the HCM has some unusual features pertinent to software and hardware design. First, computation is performed by similar clusters whose processing power is a function of the number of clusters. Second, the connectivity and processing speed scale linearly with network size. These relationships are advantageous since the storage capacity increases far more rapidly (exponentially) with the number of levels. Third, complex associations aid rather than hinder performance. This may account for relatively rapid convergence times.

Guan (1994) has applied the HCM to a variety of optimization problems, including image regularization and restoration. The algorithms are rapid, and the processes appear to be superior to several standard neural network models.

In particular, the HCM design works more efficiently than pyramid image analysis, wherein a scene is analyzed at a number of spatial scales simultaneously. The principal benefit of the HCM over other approaches lies in the unaveraged vector relationships linking the clusters in the HCM. Most multi-scale algorithms employ scalar averages (Anderson and Van Essen 1987). An exception is the network of networks model (Anderson, Gately, Penz and Collins 1990), which bears a strong resemblance to the HCM and has been used for sophisticated radar classification.

There are several attractions for very large scale integration (VLSI) implementations of the HCM. The modularity and scaling relationships are the main advantages. The disadvantages are that (1) the computation has serial components, including the activity spread and multi-level inputs, (2) the time constants are analog and, as always with new computer designs, (3) the software must be fitted to the architecture. Several parallel computers, including the Hypercube and Connection Machine, bear some resemblance in architecture to the HCM (Hillis 1985).

Finally, it should not be overlooked that technological advances inspired by the HCM may be used to improve the design of computational devices employed in brain studies. New computational techniques may also be required to analyze complex spatiotemporal data simultaneously recorded from many different brain regions and from different levels of neural organization. Finally, an overwhelming challenge facing the brain sciences today is handling the enormous amounts of data being generated in many laboratories. Solutions to this growing problem need to foster methods in data sharing, standardization and modification. Structures with multi-level and multi-tasking capabilities, such as the HCM, may play a role.

8 Concluding Remarks

As we move into the information age, the brain and computational sciences appear both inviting and daunting. It has taken millions of years for the neocortex, the most advanced of all computational devices, to evolve. It is the generator of our ideas and our technology, and its own computational principles may be partially accessible. The development of the HCM is one of several serious attempts to tease out some of the fundamental principles employed by this amazing structure. Even in its most rudimentary form, the HCM possesses spatiotemporal characteristics with unusual properties. These should be investigated because they are interesting and because they may yield further insight into the mystery of how our brains function.

Acknowledgements

The material presented in this chapter was adapted, in part, from a 1988 Ph.D. thesis entitled *Hierarchical Organization and Disordered Neural Systems* in the Department of Physics at the University of Toronto, Canada (Sutton 1988). The author is indebted to Lynn E. H. Trainor and Charles J. Lumsden who supervised the thesis and to the other members of the Physical Theory in Biology Group at the University of Toronto. This work was supported by the Medical Research Council of Canada, the McDonnell-Pew Program in Cognitive Neuroscience and the National Institutes of Health, Scientist Development Award MH01080.

References

Adams, R.D. and Victor, M. (1989) *Principles of Neurology.* Fourth edition. New York: McGraw-Hill.

Alzheimer, A. (1907) Uber eine eigenartige Erkrangkung der Hienrinde. *Allgemeine Zeitschrift für die Psychiatrie* **64**, 146–148.

Amit, D.J. (1989) *Modeling Brain Function.* Cambridge: Cambridge University Press.

Amit, D.J., Gutfreund, H. and Sompolinsky, H. (1985) Spin-glass models of neural networks. *Physical Review A* **32**, 1007–1018.

Anderson, C.H. and Van Essen, D.C. (1987) Shifter circuits: A computational strategy for dynamic aspects of visual processing. *Proceedings of the U.S. National Academy of Sciences* **84**, 6297–6301.

Anderson, J.A., Gately, M.T., Penz, P.A. and Collins, D.R. (1990) Radar signal categorization using a neural network. *IEEE Proceedings*, **78**, 1646–1657.

Anderson, J.A., Pellionisz, A. and Rosenfeld, E., eds. (1990) *Neurocomputing 2.* Cambridge, MA: MIT Press.

Anderson, J.A. and Rosenfeld, E., eds. (1988) *Neurocomputing.* Cambridge, MA: MIT Press.

Anderson, J.A. and Sutton, J.P. (1995) A network of networks: Computation and neurobiology. *World Congress of Neural Networks* **1**, 561–568.

Bock, G. and O'Connor, M. (1987) *Selective Neuronal Death.* New York: John Wiley.

Brodal, A. (1981) *Neurological Anatomy.* Third Edition. New York: Oxford University Press.

Brodmann, K. (1909) *Vergleichende Lokalisationslehre der Grosshirnirinde.* Leipzig: Barth.

Brown, T.H., Kairiss, E.W. and Keenan, C.L. (1990) Hebbian synapses: biophysical mechanisms and algorithms. *Annual Review of Neuroscience* **13**, 475–511.

Cajal, S.R.y. (1909) *Histologie du Systeme Nerveux de l'Homme et des Vertebres.* Paris: Maloine. (See also the recently translated Cajal classic: Cajal, S.R.y. (1891; 1990) *New Ideas on the Structure of the Nervous System in Man and Vertebrates.* Translated by N. Swanson and L.W. Swanson. Foreword by L.W. Swanson. Cambridge, MA: MIT Press.)

Cajal, S.R.y. (1937; translated 1991) *Recollections of My Life.* Translated by E.H. Craigie and J. Cano. Cambridge, MA: MIT Press.

Changeux, J. (1985) *Neuronal Man.* New York: Pantheon.

Changeux, J. and Danchin, A. (1976) Selective stabilisation of developing synapses as a mechanism for the specification of neuronal networks. *Nature* **264**, 705–712.

Churchland, P.S. (1986) *Neurophilosophy.* Cambridge, MA: MIT Press.

Darwin, C. (1859) *The Origin of the Species.* London: John Murray.

Donoghue, J.P., Leibovic, S. and Sanes, J.N. (1992) Organization of the forelimb area in squirrel monkey motor cortex: Representation of digit, wrist and elbow muscles. *Experimental Brain Research* **89**, 1–19.

Eckhorn, R., Bauer, R., Jordan, W., Brosch, W., Kruse, W., Munk, M. and Reitboeck, H.J. (1988) Coherent oscillations: A mechanism of feature linking in the visual cortex? *Biological Cybernetics* **60**, 121–130.

Edelman, G.M. (1978) Group selection and phasic reentrant signaling: A theory of higher brain functions. In G.M. Edelman and V.B. Mountcastle, eds., *The Mindful Brain.* Cambridge, MA: MIT Press, pp. 51–100.

Ernst, R.L. and Hay, J.W. (1994) The US economic and social costs of Alzheimer's disease revisited. *American Journal of Public Health* **84**, 1261–1264.

Essam, J.W. (1980) Percolation theory. *Reports on Progress in Physics* **43**, 833–912.

Freeman, W.J. (1975) *Mass Action in the Nervous System.* New York: Academic Press.

Freud, S. (1897; 1953–1966) Project for a scientific psychology. In *Standard Edition of the Complete Works of Sigmund Freud*, Volume 1. London: Hogarth, pp. 283–297.

Geschwind, N.G. (1965) Disconnexion syndromes in animals and man. *Brain* **88**, 237–294, 585–644.

Goldman-Rakic, P.S. (1988) Topography of cognition: Parallel distributed networks in primate association cortex. *Annual Review of Neuroscience* **11**, 137–156.

Guan, L. (1994) Image restoration by a neural network with hierarchical cluster architecture. *Journal of Electronic Imaging* **3**, 154–163.

Gutfreund, H. and Mezard, M. (1988) Processing of temporal sequences in neural networks. *Physical Review Letters* **61**, 235–238.

Hebb, D.O. (1949) *The Organization of Behavior: A Neuropsychological Theory.* New York: John Wiley and Sons.

Heidmann, A., Heidmann, T. and Changeux, J. (1984) Stabilisation selective de representations neuronales par resonance entre "prerepresentations" spontanees du researu cerebral et "precepts" evoques pazar interaction avec le monde exterieur. *Comptes Rendus des Seances de l'Academie des Sciences (Paris)* **299**, 839–844.

Hillis, W.D. (1985) *The Connection Machine.* Cambridge, MA: MIT Press.

Hobson, J.A. (1988) *The Dreaming Brain.* New York: Basic Books.

Hopfield, J.J. (1982) Neural networks and physical systems with emergent collective computational abilities. *Proceedings of the U.S. National Academy of Sciences* **79**, 2554–2558.

Horn, D. and Usher, M. (1989) Neural networks with dynamical thresholds. *Physical Review A* **40**, 1036–1044.

Ingber, L. (1995) Statistical mechanics of multiple scales of neocortical interactions. In P.L. Nunez, ed., *Neocortical Dynamics and Human EEG Rhythms*. Oxford: Oxford University Press, pp. 628–681.

Jackson, J.H. (1884; 1958) Evolution and dissolution of the nervous system. In J. Taylor, ed., *Selected Writings of John Hughlings Jackson*, Volume 2. New York: Basic Books, pp. 3–118.

Jerne, N.K. (1955) The natural-selection theory of antibody formation. *Proceedings of the U.S. National Academy of Sciences* **41**, 849–857.

Kandel, E.R., Klein, M., Hochner, B., Shuster, M., Siegelbaum, S.A., Hawkins, R.D., Glanzman, D.L., Castellucci, Y.F. and Abrams, T.W. (1987) Synaptic modulation and learning: New insights into synaptic transmission from the study of behavior. In G.M. Edelman, W.E. Gall and W.M. Cowan, eds., *Synaptic Function*. New York: John Wiley & Sons, pp. 471–518.

Kandel, E.R., Schwartz, J. and Jessell, T.M., eds. (1991) *Principles of Neural Science*. Third Edition. New York: Elsevier.

Kaplan, H.I. and Sadock, B.J., eds. (1989) *Comprehensive Textbook of Psychiatry*. Fifth Edition. New York: Williams and Wilkins.

Karni, A., Meyer, G., Jezzard, P., Adams, M.M., Turner, R. and Ungerleider, L.G. (1995) Functional MRI evidence for adult motor cortex plasticity during motor skill learning. *Nature* **377**, 155–158.

Katzman, R. (1986) Alzheimer's disease. *New England Journal of Medicine* **314**, 964–973.

Kohonen, T. (1987) *Content-Addressable Memories*. Second Edition. New York: Springer-Verlag.

Lamour, Y. (1994) Alzheimer's disease: a review of recent findings. *Biomedicine & Pharmacotherapy* **48**, 312–318.

Marler, P. and Terrace, H.S., eds. (1984) *The Biology of Learning*. New York: Springer-Verlag.

Marshall, J.C. (1980) On the biology of language acquisition. In D. Caplan, ed., *Biological Studies of Mental Processes*. Cambridge, MA: MIT Press, pp. 106–148.

Martinez, S., Paredes, S., Borgani, S. and Coles, P. (1995) Multiscaling properties of large-scale structure in the universe. *Science* **269**, 1245–1247.

Maunsell, J.H.R. and Van Essen, D.C. (1983) The connections of the middle temporal visual area (MT) and their relationship to a cortical hierarchy in the macaque monkey. *Journal of Neuroscience* **3**, 2563–2586.

McCulloch, W.S. and Pitts, W. (1943) A logical calculus of the ideas immanent in nervous activity. *Bulletin of Mathematical Biophysics* **9**, 115–133.

McKenna, T., Davis, J. and Zornetzer, S., eds. (1992) *Single Neuron Computation.* Cambridge: Academic Press.

Merzenich, M.M., Kaas, J.H., Wall, J., Nelson, R.J. and Sur, M. (1983) Progression of change following median nerve section in the cortical representation of the hand in areas 3b and 1 in adult owl and squirrel monkeys. *Neuroscience* **10**, 639–665.

Minsky, M. (1985) *The Society of Mind.* New York: Simon and Schuster.

Morris, R.G.M., Kandel, E.R. and Squire, L.R. (1988) The neuroscience of learning and memory: cells, neural circuits and behavior. *Trends in Neurosciences* **11**, 125–127.

Mountcastle, V.B. (1978) An organizing principle for cerebral function: The unit module and the distributed system. In G. Edelman and V.B. Mountcastle, eds., *The Mindful Brain.* Cambridge, MA: MIT Press, pp. 7–50.

Newsome, W.T., Britten, K. and Movshon, J.A. (1989) Neuronal correlates of a perceptual decision. *Nature* **341**, 52–54.

Peretto, P. (1992) *An Introduction to the Modeling of Neural Networks.* Cambridge: Cambridge University Press.

Phillips, C.G. and Porter, R. (1977) *Corticospinal Neurones: Their Role in Movement.* London: Academic Press.

Rakic, P. (1981) Developmental events leading to laminar and areal organization of the neocortex. In F.O. Schmitt, F.G. Worden, G. Adelman and S.G. Dennis, eds., *The Organization of the Cerebral Cortex.* Cambridge: MIT Press, pp. 7–28.

Rockel, A.J., Hiorns, R.W. and Powell, T.P.S. (1980) The basic uniformity of the neocortex. *Brain* **103**, 221–244.

Sakimura, K., Kutsuwada, T., Ito, I., Manabe, T., Takayama, C., Kushiya, E., Yagi, T., Aizawa, S., Inoue, Y., Sugiyama, H. and Mishina, M. (1995) Reduced hippocampal LTP and spatial learning in mice lacking NMDA receptor $\epsilon 1$ subunit. *Nature* **373**, 151–155.

Salthe, S.N. (1985) *Evolving Hierarchical Systems.* New York: Columbia University Press.

Selverston, A.I., ed. (1985) *Model Neural Networks and Behavior.* New York: Plenum Press.

Shepherd, G.M., ed. (1990) *The Synaptic Organization of the Brain.* Third Edition. New York: Oxford University Press.

Singer, W. (1994) Putative functions of temporal correlations in neocortical processing. In C. Koch and J.L. Davis, eds., *Large-scale Neuronal Theories of the Brain.* Cambridge, MA: MIT Press, pp. 201–237.

Sompolinsky, H. and Kanter, I. (1986) Temporal association in asymmetric neural networks. *Physical Review Letters* **57**, 2861–2864.

Stauffer, D. (1987) *Introduction to Percolation Theory.* London: Taylor and Francis.

Sutton, J.P. (1988) *Hierarchical Organization and Disordered Neural Systems.* PhD. Thesis, Department of Physics, University of Toronto, Toronto, CAnada.

Sutton, J.P. (1991) Mean field theory of nested neural clusters. *International AMSE Conference on Neural Networks* **1**, 47–58.

Sutton, J.P. (1993) Learning by selection in multi-layered networks. *World Congress on Neural Networks* **2**, 536–539.

Sutton, J.P. (1995) Neuroscience and computing algorithms. *Information Sciences* **84**, 199–208.

Sutton, J.P. and Anderson, J.A. (1995) Computational and neurobiological features of a network of networks. In J.M. Bower, ed., *Neurobiology of Computation.* Boston: Kluwer Academic, pp. 317–322.

Sutton, J.P., Beis, J.S. and Trainor, L.E.H. (1988a) Hierarchical model of memory and memory loss. *Journal of Physics A* **21**, 4443–4454.

Sutton, J.P., Beis, J.S. and Trainor, L.E.H. (1988b) A hierarchical model of neocortical synaptic organization. *Mathematical and Computer Modelling* **11**, 346–350.

Sutton, J.P. and Breiter, H.C. (1994) Neural scale invariance: An integrative model with implications for neuropathology. *World Congress on Neural Networks* **4**, 667–672.

Sutton, J.P. and Hobson, J.A. (1994) State-dependent sequencing and learning. In F.H. Eeckman, ed., *Computation in Neurons and Neural Systems.* Boston: Kluwer Academic, pp. 275–280.

Sutton, J.P., Mamelak, A.N. and Hobson, J.A. (1992) Network model of state-dependent sequencing. In J.E. Moody, S.J. Hanson and R.P. Lippmann, eds., *Advances in Neural Information Processing Systems* 4. San Mateo: Morgan Kaufmann, pp. 283–290.

Sutton, J.P. and Trainor, L.E.H. (1990) Real and artificial neural hierarchies. In C.N. Manikopoulos, ed., *Cybernetics and Systems*, Volume 1. Newark: New Jersey Institute of Technology, pp. 229–236.

Sutton, J.P. and Trainor, L.E.H. (1991) Information processing in multi-levelled neural architectures. *International AMSE Conference on Neural Networks* **1**, 59–66.

Sutton, J.P., Whitton, J.L., Topa, M. and Moldofsky, H. (1986) Evoked potential maps in learning disabled children. *Electroencephalography and Clinical Neurophysiology* **65**, 399–404.

Szentágothai, J. (1977) The neuron network of the cerebral cortex. *Proceedings of the Royal Society of London, Series B* **201**, 219–248.

Terry, R.D., Peck, A., DeTeresa, R., Schechter, R., and Horoupian, D.S. (1981) Some morphometric aspects of the brain in senile dementia of the Alzheimer type. *Annals of Neurology* **10**, 184–192.

Toulouse, G., Dehaene, S. and Changeux, J. (1986) Spin glass model of learning by selection. *Proceedings of the U.S. National Academy of Sciences* **83**, 1695–1698.

T'so, D.Y., Gilbert, C.D. and Wiesel, T.N. (1986) Relationship between horizontal interactions and functional architecture in cat striate cortex as revealed by cross-correlational analysis. *Journal of Neuroscience* **6**, 1160–1170.

Vaadia, E., Haalman, I., Abeles, M., Bergman, H., Prut, Y., Slovin, H. and Aertsen, A. (1995) Dynamics of neuronal interactions in monkey cortex in relation to behavioural events. *Nature* **373**, 515–518.

Van Essen, D.C., Anderson, C.H. and Felleman, D. (1992) Information processing in the primate visual system: An integrated systems perspective. *Science* **255**, 419–423.

Wernicke, C. (1874) *Der aphasische Symptomencomplex. Eine psychologische Studie auf anatomischer Basis.* Breslau: Cohn & Weigert. (A modern translation appears as "The aphasia symptom-complex: a psychological study on an anatomic basis." in G.H. Eggert (1977) *Wernicke's Works on Aphasia: A Sourcebook and Review.* Early Sources in Aphasia and Related Disorders, Volume 1. New York: Mouton, pp. 91–145.)

Young, J.Z. (1964) *A Model of the Brain.* Oxford: Clarendon Press.

Category Switching — A Neural Network Approach

L.E.H. Trainor, Wendy A.M. Brandts
and José-Leonel Torres

1 Introduction

In cases of social behavior (insects, people, whatever) among very similar individuals involved in a variety of activities, there are interesting problems as to how the group as a whole sorts itself out from an organizational or behavioral point of view. Let us suppose that there are a number of activities (activity categories) available to individuals in the group and that, in the course of their interactions with each other, individuals switch activity categories depending upon information coming to them from the rest of the group.

We are interested in whether and under what conditions the group settles down to a stable state, i.e. to a state in which the category populations no longer change. Our aim here is to develop an approach to this kind of problem using an appropriate model based on neural network theory.

In section 2, we review briefly the Hopfield (Hopfield 1982) formulation of a binary state neural network displaying emergent collective computational abilities. Although Hopfield's formulation was not the first attempt to draw an analogy between a network of on/off elements and neural activity (action potentials) in the brain, it has been the most instructive one to do so, particularly since it identifies an energy function whose minima correspond to stable attractors of the network.

In section 3, we introduce the concept of category switching and devise a model for instances where there are only two categories of activity available to individuals in the network (Brandts and Trainor 1996a). In section 4, we discuss the behavior of a species of harvester ants (*Pogonomyrmex barbatus*) studied extensively in the field by Gordon (Gordon 1986, 1987, 1989a, 1989b). In section 5 we introduce an eight-state binary switching model motivated by an attempt (Gordon, Goodwin and Trainor 1992) to understand the behavior of these colonies as information processing systems of a parallel, distributed type. In section 6, dynamics of the GTT model are altered to ensure that the system evolves to a steady state. The relative successes and limitation of the model are discussed in section 7.

Finally in section 8, we consider the possibility of extending this type of model to a wider class of applications of interest in the social sciences.

2 The Hopfield Model

The Hopfield model (Hopfield 1982), introduced as a model of associative memory, considers a set of identical elements, numbered say $i = 1, 2, \ldots, N$. These elements can exist in either one of two states, identified as "on" or "off," corresponding to values 1 and -1 of the state variables s_i, also referred to as "spins" by physical analogy to magnetic systems. In the Hopfield model, elements interact with strengths v_{ij} which are taken to be symmetric ($v_{ij} = v_{ji}$). The dynamics of the system proceed as follows. An element is chosen at random and its state up-dated according to the following algorithm. A local "field" defined by

$$F_i = \sum_{\substack{j=1 \\ j \neq i}}^{N} v_{ij} s_j \tag{1}$$

is evaluated for the chosen element i and the ith element is switched to the $i = 1$ state if $F_i \geq 0$, and to the $i = -1$ state if $F_i < 0$. We refer to this kind of dynamical rule as a "spin alignment rule".

Hopfield was able to show that this algorithm leads to stationary states of the system (candidates for memory states of the system in his application). It is straightforward to show that this dynamics guarantees a monotonic decrease in the quadratic "energy" function

$$E = -1/2 \sum_{\substack{i,j=1 \\ i \neq j}}^{N} v_{ij} s_i s_j \ , \tag{2}$$

which is a measure of how well the spins are aligned, on the average, with their local fields. In associative memory terminology, E is minimized when the overlap between the network configuration and the stored pattern ξ_i is largest (for details see, for example, Chapter 2 in Hertz, Krugh and Palmer 1991)).

Hopfield was then able to show that the v_{ij} could be selected beforehand to ensure that certain chosen "spin" configurations would correspond to stationary states (local minima) of the system, viz. if the number of memory states m is sufficiently small, the choice of interaction

$$v_{ij} = \sum_{\alpha=1}^{m} \xi_i^{\alpha} \xi_j^{\alpha} \tag{3}$$

ensures that the states $(\xi_1^{\alpha}, \xi_2^{\alpha} \ldots \xi_N^{\alpha})$, $\alpha = 1, 2, \ldots m$ will be memory states. For stable memory states, m must be less than about $0.15N$.

3 A Two Category Switching Model

We devise an alternative to Hopfield's model which incorporates the character of category-specific interactions as follows. Rather than taking the fundamental interaction between the ith and jth elements as fixed (once selected via equation (3)), as is done in the Hopfield model, we take the interactions to be category dependent. Since there are only two categories, there are 3 independent matrix elements, viz. those corresponding to the interaction between two elements in category one, to two elements in category two, and to one element in category one and the other in category two. In this latter case, we assume it does not matter which is which so that the interaction matrix is symmetric

$$V = \begin{pmatrix} v_{\mathrm{I\,I}} & v_{\mathrm{I\,II}} \\ \\ v_{\mathrm{II\,I}} & v_{\mathrm{II\,II}} \end{pmatrix} \tag{4}$$

with $v_{\mathrm{I\,II}} = v_{\mathrm{II\,I}}$.

In our model, all elements with $S_i = 1$ are defined to be in category I, all elements with $S_i = -1$ in category II. Let P_{I} and P_{II} represent the populations in categories I and II. One is then interested in whether these quantities eventually reach stable values starting from any given initial state.

Surprisingly, this association of interactions into categories leads to very different results from the those of the Hopfield model. In fact, using Hopfield's spin alignment dynamics, we find (Brandts and Trainor 1996a) that the quadratic energy function is no longer monotonic and stable populations are no longer necessarily reached by dynamic evolution. Of course, the nature of the interactions in the two models is fundamentally different. In Hopfield's model the interactions v_{ij} remain fixed throughout the dynamic evolution whereas in the category model the interaction between the $i^{'th}$ and $j^{'th}$ elements depends on which categories they chance to be in when one member is selected for updating.

We now adopt a dynamics (distinct from that of Hopfield) which is appropriate to the categories model. For every element i of the system we define a local field F_i acting upon it as follows:

$$F_i = \sum_{\substack{j \neq i \\ j=1}}^{N} v_{ij} S_j \ . \tag{5}$$

We then define an alignment function for the system as a whole

$$\mathcal{A} = \sum_{i=1}^{N} F_i S_i \;, \tag{6a}$$

which is a measure of how well the spins are aligned, on the average, with their local fields. We call $\mathcal{A}$ for convenience the "fitness function"[a] (Torres and Trainor 1993, Brandts and Trainor 1996a). Substituting (5) into (6), we obtain

$$\mathcal{A} = \sum_{i=1}^{N} \sum_{\substack{j \neq i \\ j=1}}^{N} v_{ij} S_j S_i \;. \tag{6b}$$

We now adopt the following dynamical procedure: choose at random an element, which may be in either category *I* or *II*. If the dynamical fitness is improved by switching the element to the other category, the switch is made; otherwise not. Another element is then selected at random and the procedure continued. From the form of $\mathcal{A}$ it is clear that in this procedure, the quadratic "energy" function

$$E = -\frac{1}{2} \sum_{\substack{i,j=1 \\ i \neq j}}^{N} v_{ij} S_i S_j = -\frac{1}{2} \mathcal{A} \tag{7a}$$

decreases monotonically, so that the system eventually settles down to a stable state corresponding to a local minimum of the "energy" function.

It is straightforward to show in this simple case that the quadratic "energy" function can be expressed directly in terms of the category populations:

$$E = -\frac{1}{2} \sum_{\substack{i,j=1 \\ i \neq j}}^{N} v_{ij} S_i S_j = -\frac{1}{2} v_{\mathrm{I\,I}} P_{\mathrm{I}}(P_{\mathrm{I}} - 1) - \frac{1}{2} v_{\mathrm{II\,II}} P_{\mathrm{II}}(P_{\mathrm{II}} - 1) + v_{\mathrm{I\,II}} P_{\mathrm{I}} P_{\mathrm{II}} \tag{7b}$$

An example of the dynamics in terms of category populations is shown in Figure 1A with the corresponding energy evolutions shown in Figure 1B.

4 The Challenge of *Pogonomyrmex barbatus*

Ants are fascinating social insects which occur in more than 8,000 species, each displaying distinctive features both at the level of individual ants and at the

[a]This is a dynamical fitness not to be confused with fitness for survival as occurs in evolutionary theory.

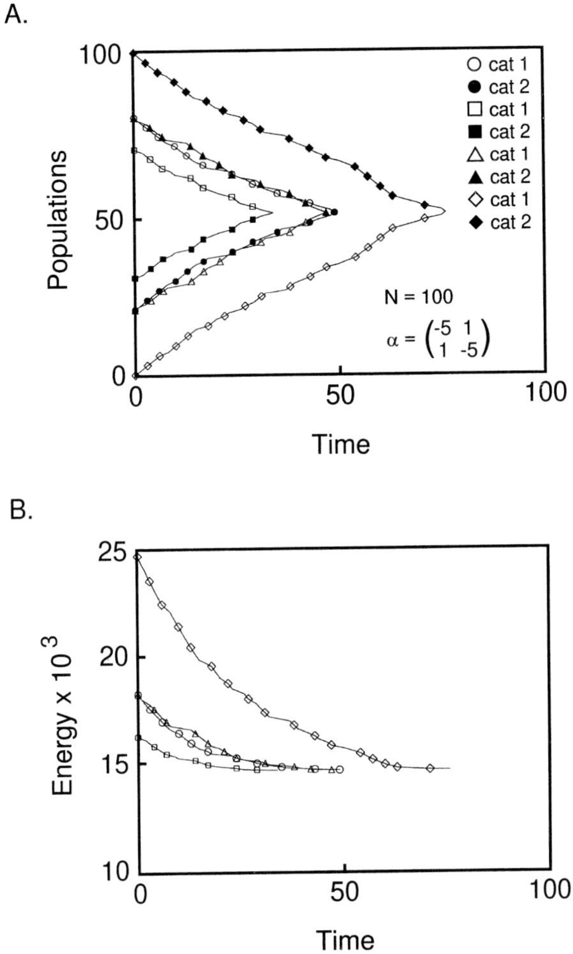

Figure 1: A. Evolution with a monotonic energy rule (MER) of populations in categories 1 and 2 for several initial distributions of 100 ants, viz. (100, 0), (80, 20) and (70, 30). Ants are selected sequentially, at random, and a switching of categories occurs only if the resultant energy does not increase. The interaction matrix α is such that the energy has a minimum corresponding to the state with equal populations in both categories. For the (80, 20) initial distribution two separate runs are shown to illustrate the moderate variations that occur in convergence to the attractor. B. Energy changes corresponding to the four sample cases in panel A.

Figure 2: *Pogonomyrmex barbatus* in combat. Photo courtesy Bert Hölldobler.

level of colony behavior (Hölldobler and Wilson 1990, 1994). The particular species of interest here, *Pogonomyrmex barbatus*, has been extensively studied in the field (Arizona) by Gordon (1986, 1987, 1989a, 1989b).

The primary parameters indicative of colony behavior are the numbers of ants engaging in various activities. Ants outside the nest engage in a variety of tasks such as foraging, collecting food, patrolling the nest mound and trails, maintaining the nest, clearing the trails, and midden work (sorting and upkeep of the refuse pile). To some extent these tasks are performed in a temporal sequence, referred to as the "daily round" by Gordon.

Most ant genera are monomorphic, i.e. ants are more or less identical despite their division at any time into different activity categories. The predominant view of social insect behavior has been that espoused by Wilson and collaborators (see e.g. Oster and Wilson 1978) who propose that the division of labor in a colony is primarily determined by the differentiation of its members into castes, based on such factors as age and physical characteristics. For polymorphic species, this view seems well established. For monomorphic species the specialization of activities is less clear. An alternative to the caste hypothesis has been expressed by Gordon (1986, 1987, 1989a, 1989b) on the basis of her observations that extensive task switching takes place under environmental or experimental perturbations. Without perturbations task fidelity tends to be high, even for ants that have returned to the nest for some period of time, but fidelity changes also with the age of an ant. Older ants tend to have greater task fidelity than do younger ants.

Generally speaking, decision making in the colony seems to be a collective affair arising out of frequent chemical communications (via pheromones presumably) between ant pairs, particularly near the entrance to the nest. The parallel between this type of organization and a parallel distributed network of model neurons is very suggestive. Discussions on this point led to development of the eight category model of Gordon, Goodwin and Trainor (1992) outlined in the next section.

5 Eight Category Model

The eight category model was developed (Gordon et. al., 1992, Trainor 1993, Torres and Trainor 1993) in an attempt to simulate the behavior of *Pogonomyrmex barbatus*. It was adapted from the neural network model of Hopfield (1982), reviewed in section 2 of this essay.

Extension of the Hopfield model to the *Pogonomyrmex* problem is not quite straightforward since in the latter, individual ants (elements) can be in any one of eight states (categories) rather than two as in the Hopfield model. The eight categories correspond to the activity states observed by Gordon in field studies. A useful characterization seems to be as follows for ants outside the nest: first a division between active and inactive ants; then in each case a division into four activity categories (for both active and inactive ants), viz. foraging, patrolling, nest maintenance and midden care. It is observed in field studies that ants switch activity categories under perturbations of the nest environment. The natural observables of the system are the numbers of ants in each category and how these change as the dynamics proceed. The behavior of the system as a whole is defined in terms of the category populations and their dynamical behaviors.

In establishing an eight category model the decision was made (Gordon et. al. 1992) to extend the binary Hopfield model states to an outer product of three such states.[b] The category for the kth ant is now described by a triad of spins (a_k, b_k, c_k) each corresponding to independent binary elements taking on the values ± 1. We use the letters P, F, N, M to refer to the patrollers, foragers, nest maintenance and midden categories for active ants, and their lower case counterparts p, f, n and m to the same categories for inactive ants. The eight categories are then specified by the triad values as follows:

[b] We refer to the individual spins as being in states ± 1 but a triad of spin values (a_k, b_k, c_k) defines an activity "category."

$$\begin{array}{rrrcl} (1, & 1, & 1) & \equiv & \text{active patroller } P \\ (-1, & 1, & 1) & \equiv & \text{inactive patroller } p \\ (1, & 1, & -1) & \equiv & \text{active forager } F \\ (-1, & 1, & -1) & \equiv & \text{inactive forager } f \\ (1, & -1, & 1) & \equiv & \text{active nest maintenance worker } N \\ (-1, & -1, & 1) & \equiv & \text{inactive nest maintenance worker } n \\ (1, & -1, & -1) & \equiv & \text{active midden worker } M \\ (-1, & -1, & -1) & \equiv & \text{inactive midden worker } m \end{array} \quad (8)$$

Thus a_k distinguishes active ants from inactive ants. Within these two broad categories, b_k distinguishes the patroller/forager subclass from the nest maintenance/midden worker subclass. Finally c_k distinguishes patrollers from foragers and maintenance workers from midden workers.

In the model, ants are assumed to interact pairwise (presumably by the exchange of pheromones) and that all pairs of ants interact (fully connected nest). This latter assumption is probably unrealistic and could be refined in a subsequent model which considered spatial distributions as well as time development of the colony. It is partially justified because there seems to be great activity among ants just outside the entrance to the nest.

We now assume that three separate and independent fields act locally on each ant. The three fields acting on ant i at any particular time step are taken to be

$$\begin{aligned} E_i &= \sum_{\substack{j=1 \\ j \neq i}}^{N} \alpha_{ij} a_j \\ L_i &= \sum_{\substack{j=1 \\ j \neq i}}^{N} \beta_{ij} b_j \\ K_i &= \sum_{\substack{j=1 \\ j \neq i}}^{N} \gamma_{ij} c_j \end{aligned} \quad (9)$$

with $\alpha_{ij} = \alpha_{ji}$, $\beta_{ij} = \beta_{ji}$ and $\gamma_{ij} = \gamma_{ji}$. Thus, the local fields acting on ant i depend at any time step on the states of all other ants $j \neq i$. There are three separate fields, E_i acting on state a_i, L_i acting on state b_i and K_i acting on state c_i.

The alignment dynamics then proceeds as follows: an ant, say the ith ant, is selected at random and the fields (9) evaluated. Switching for the state a_i

then occurs if a_i is not lined up with the field E_i. For example if $a_i = 1$ and $E_i \geq 0$, no switching of this state occurs, but if $E_i < 0$, switching occurs and the value of a_i changes to -1. A similar dynamic applies to the state b_i with respect to its field L_i and to c_i with respect to its field K_i.

Thus beginning with some particular initial set of category populations for the system, it evolves in successive time steps to a steady state (fixed populations), or to a limit cycle behavior of repeating cycles of populations. In the next section, we show (Torres and Trainor 1993) how the dynamical rule can be altered to ensure convergence to a steady state.

6 Modified Category Switching Model

The eight category model as formulated in section 5 has some serious limitations. Although in computer simulations the system tends to evolve to one or another steady state, it is not clear that a monotonically decreasing "energy" function exists for the 3-field case, analogous to the quadratic function of Hopfield (Brandts and Trainor 1996a). However, Torres and Trainor (1993) have shown that the quadratic 3-field energy function

$$E = -1/2 \sum_{\substack{i,j \\ i \neq j}} (\alpha_{ij} a_i a_j + \beta_{ij} b_i b_j + \gamma_{ij} c_i c_j) \tag{10}$$

is monotonically decreasing under a modified decision rule analogous to that introduced in section 3, namely that a randomly chosen ant changes category only if the spin alignments with their associated fields are collectively improved thereby, that is if

$$(a_F E_F + b_F L_F + c_F K_F) - (a_I E_I + b_I L_I + c_I K_I) \geq 0 \tag{11}$$

where the subscript F refers to the new category explored, while the subscript I refers to the original category before transition. In this dynamic, alignment of spins and fields plays the role of what we have termed a "dynamical fitness" parameter. A randomly chosen ant belonging to some category, say I, explores the possibility of making a transition to some randomly chosen category F. The ant makes the transition if its alignment is thereby improved, otherwise it remains in category I.

7 Discussion

Gordon (1996) has discussed the organization of work in social insect colonies with particular reference to task switching in the model of Gordon et. al. (1992)

based on interactions among individuals in colonies of *Pogonomyrmex barbatus*. In general there seems to be good accordance between actual colony behavior and a network model emphasizing parallel distributed dynamics of the type discussed here. However, more detailed work in the field is necessary to fix parameters and make refinements in the model. Field observations are time consuming and laborious and the turn round between theory and experiment for the model proposed here is a slow process. Nonetheless the correspondence is promising. In the first place, the model is in accord with the non-hierarchical organizational and collective decision making structure of the colony. There is sufficient flexibility in the choice of matrix elements, both as to magnitude and sign, to simulate detailed information on switching preferences between categories which constitute an important part of the observational parameters. The model also captures the nonlinear behavior of the colony when subjected to successive perturbations (the results are not additive). Starting from any particular initial state, the model (as the colony) evolves toward a stable state in which the activity populations are steady; and the stable states reached depend upon the initial distributions in the populations. One can envision that after some major perturbation has occurred, the colony populations relax toward a stable state by a series of category switching events. Finally, a test of the model can be made using the protocol suggested by Torres and Trainor (1993), in which original observations of the various category populations at an attractor state can be used to fix the matrix elements. Subsequent behavior following a perturbed initial state then puts the model to a severe test.

The three field binary model has a possible disadvantage in that the binary formulation (Kauffman 1969) implies equalities between some switching rates that might not correspond to observed rates, although this circumstance has not been investigated. An alternative approach to an eight category model which does not have a basis in binary representation is being investigated by Brandts and Trainor (Brandts and Trainor 1996b).

Finally, there is the question of how the model parameters, the matrix elements α_{ij}, β_{ij}, and γ_{ij} are to be set for a given colony. Presumably, one has to establish some correspondence between these parameters and the colony observables, which are the populations in the various categories. In principle one can proceed as follows: from some identifiable initial population, the colony evolves toward an observable set of stationary populations. The matrix elements are then adjusted so that the simulated stationary state predicts the observed populations. However, in practice there may be many choices of matrix elements which will give the correct populations so that a one-to-one correspondence does not exist. The analysis of Torres and Trainor (1993), indicates how this in principle could be done.

8 Extended Applications

There are many instances in the social sciences where a category switching model might be of analytical advantage. One example would be in the analysis of the job market, where workers possessing a broad spectrum of talents switch occupations on the basis of a variety of economic, social and political factors. Category switching models of the neural network type suggested here might provide a forum for analysis and discussion. What such models show is how the pairwise interaction between individuals in a group can lead to large changes in the group behavior as a whole. Such analysis might succeed in identifying what parameter change stabilizes good group behavior. In more complicated situations, one can consider higher order interactions where fundamental interactions involve groups of three or four individuals rather than two (see essay by Busch and Trainor, this collection). One can also conceive of organizational structures (Issac and Trainor 1995) in which attitudes shift from one category to another, possibly with hierarchical features included (see essay by Sutton, this collection).

References

Brandts, W.A.M. and Trainor, L.E.H. (1996a) Two-category model: spin dynamics and energy functions. In preparation.

Brandts, W.A.M. and Trainor, L.E.H. (1996b) Eight-category network model of an ant colony: evolution with monotonic energy functions. In preparation.

Gordon, D.M. (1986) The dynamics of the daily round of the harvester ant colony (*Pogonomyrmex barbatus*). *Animal Behaviour* **34**, 1402–1419.

Gordon, D.M. (1987) Group level dynamics in harvester ants; young colonies and the role of patrolling. *Animal Behaviour* **35**, 833–834.

Gordon, D.M. (1989a) Caste and change in social insects. In P. Harvey and L. Patridge, eds., *Oxford Surveys in Evolutionary Biology*, Volume 6. New York: Oxford University Press, pp. 53–72.

Gordon, D.M. (1989b) Dynamics of task switching in harvester ants. *Animal Behaviour* **38**, 194–204.

Gordon, D.M. (1996) The oganization of work in social insect co lonies. *Nature* **380**, 121–124.

Gordon, D.M., Goodwin, B.C. and Trainor, L.E.H. (1992) A parallel distributed model of behaviour of ant colonies. *Journal of Theoretical Biology* **156**, 293–307.

Hertz, J., Krugh, A. and Palmer, R.G. (1991) *Introduction to the Theory of Neural Computing.* Reading, MA: Addison-Wesley.

Hölldobler, B. and Wilson, E.O. (1990) *The Ants.* Cambridge, MA: Belknap Press of Harvard University Press.

Hölldobler, B. and Wilson, E.O. (1994) *Journey to the Ants.* Cambridge, MA: Belknap Press of Harvard University Press.

Hopfield, J.J. (1982) Neural networks and physical systems with emergent collective computational abilities. *Proceedings of the U.S. National Academy of Sciences* **79**, 2554–2558.

Kauffman, S.A. (1969) Metabolic stability and epigenesis in randomly constructed genetic nets. *Journal of Theoretical Biology* **22**, 437–467.

Isaac, M. and Trainor, L.E.H. (1996) *Application of Category Switching to Leadership and Organizational Effectiveness.* In preparation.

Oster, G.F. and Wilson, E.O. (1978) *Caste and Ecology in the Social Insects.* Princeton, NJ: Princeton University Press.

Torres, J.L. and Trainor, L.E.H. (1993) Energy function for a model of ant colonies. *Journal of Theoretical Biology* **165**, 177–184.

Trainor, L.E.H. (1993) Modelling the behavior of ant colonies as an emergent property of a system of ant-ant interactions. In W.D. Stein and F.J. Varella, eds., *Thinking About Biology: An Introduction to Current Theoretical Biology.* Santa Fe Institute Studies in the Science of Complexity, Volume III. Reading, MA: Addison-Wesley, pp. 303–313.

V

Evolution

A Model of Molecular Evolution Based on the Statistical Analysis of Nucleotide Sequences

Liaofu Luo

1 Statistical Analysis of Nucleotide Sequences

Nucleotide sequence data are increasing explosively. At present about 10^7 base pairs worth of sequences are stored in the main data banks all over the world, which provides mankind with the most important life information. However we know little about their theoretical implications. As Trifonov (1989) said, "there are well organized libraries of sequences, but the 'reading rooms' are empty. Of course, readers are not interested in just looking at sequences when they have no idea how to read them." Or as a Chinese proverb says, "this is a heaven book, indeed." We know almost nothing about how nature writes it. We are reminded of the last years of the nineteenth century. At that time, the rules governing the line spectra of atoms and molecules were so curious for physicists, just as if atoms were sending telegrams to mankind but the receivers could not understand them at all. However, as soon as they understood them, quantum mechanics was born. Similarly today, many quantitative relations exist behind the nucleotide sequences, but we hardly have any idea how to read them. It is reasonable to hope that when the meaning of the book of life has been grasped, great progress will be made in the natural sciences.

About the relation between physics and biology, I would like to cite some words of a noted physical scientist (Volkenstein 1982, p. 8): "We may say with some exaggeration that if physics gave biology the microscope, biology gave physics the law of conservation of energy." Of course, there is a great opportunity today for physicists to learn from biology once again.

The usual method of analysing nucleotide sequences is by computer comparison. Any two sequences can be examined for regions of significant similarity. Alternatively, in a set of similar sequences, we can find an optimal homologous alignment and discover what positions are important to their common function. Scientists have obtained a great deal of knowledge from sequences. However, there are some restrictions in studying the evolutionary problem by use of these methods alone. As is well known, the formation and stabilization of nucleic acid sequences is a very complex process which is influenced by natural selection as well as by random genetic drift. It is also constrained by the principles of quantum biochemistry and by statistical thermodynamics as well as the requirements of various biological functions. There are many accidental factors in the historical formation of any given sequence. Therefore, in addition to the commonly used method of sequence determination, a statistical approach is useful which considers a large amount of sequence data at the same time and attempts to find the statistical relations among them. These relations may turn out to be the rules dominating the molecular evolutionary process. In any case they provide a starting point to establish a reasonable model to describe molecular evolution. In the first section we shall sketch the main results from the statistical analysis of nucleotide sequences (protein-coding sequences mainly) (Luo 1991).

For each sequence, we define an informational entropy and a Markovian entropy

$$H = -\sum_i p(i) \log_2 p(i) \tag{1}$$

$$H_M^{(1)} = -\sum_{ij} p(i)p(j|i) \log_2 p(j|i) \tag{2}$$

and the corresponding redundancies

$$D_1 = H_{\max} - H \tag{3}$$

$$D_2 = H - H_M^{(1)} \tag{4}$$

Here, $p(i)$ describes the statistical distribution of bases in a given sequence. When $p(i) = \frac{1}{4}$ for each i, H attains its maximum $H_{\max} = 2$. $p(i|j)$ is the conditional probability of base i occurring after base j, which relates to base correlations in the sequence. D_1 describes the divergence of base composition from equiprobability while D_2 describes the deviation of neighboring bases from independent occurrence. Gatlin first tried to investigate nucleotide sequences using D_1 and D_2 (Gatlin 1972). However, only a few data were available at that time, so her results were far from conclusive. In addition to D_1 and D_2, one may use another parameter X

$$X = D_2/(D_1 + D_2) \tag{5}$$

and F, (G+C content), to analyse sequences. X has a smaller deviation than D_1 and D_2 for sequences within a given category.

In Luo, Tsai and Zhou (1988) and later Luo and Shengli (1990), we calculated D_1, D_2, X and F for many sequences and averaged them for each category. The average will be denoted by $\langle \quad \rangle$. We discovered meaningful statistical correlations between these parameters and molecular evolution, namely $\langle D_2 \rangle$ varies from category to category, e.g., from 0.03 for mitochondria to $0.05 - 0.06$

for phages and bacteria to 0.11 for mammals; $\langle X \rangle$ varies from 0.2 to 0.8 correspondingly. $\langle D_1 \rangle$ changes with evolution too, though the correlation is poorer than for $\langle D_2 \rangle$. For lower organisms, the sequence has a comparatively large $\langle D_1 \rangle (0.2 - 0.1)$, while for higher species the value is lower (0.02). The decrease of D_1 means $p(i)$ is changing statistically to the value 1/4 in the course of evolution. Instead of D_1 one may use $(p(i) - 1/4)^2$ to describe evolution. Related to the change of $\langle D_1 \rangle$, $\langle F \rangle$ increases from near 0.3 (lower organisms for example, mitochondria, chloroplasts) to near 0.5 (higher species). The increase of $(G + C)$ content in evolution has been pointed out in earlier work (Heijne 1987).

In the statistical analysis we should notice that in order to diminish the standard deviation of averaged parameters a sufficiently large number of sequences in a given category is necessary. Otherwise, the statistical correlation of the parameter with evolution will be concealed by deviations. On the other hand, the sequence length effect also needs to be taken into account. For a short sequence, the stochastic fluctuation may give contributions to D_1 and D_2 and thus to $\langle X \rangle$. Through stochastic simulations we find that the contributions become negligible only if the length is greater than 600.

We may look at the nucleic acid sequence as a genetic language. The informational parameter $\langle D_1 \rangle$ describes the vocabulary constitution for the language and $\langle D_2 \rangle$ its grammatical construction. The smaller $\langle D_1 \rangle$, the higher is the randomness of base composition. The smallness of $\langle D_1 \rangle$ correlates with a large vocabulary. The larger $\langle D_2 \rangle$, the stronger is the correlation of neighboring bases. The largeness of $\langle D_2 \rangle$ implies the degree of clarity of the grammatical rules. So the trends of evolution, roughly towards large $\langle D_2 \rangle$ and small $\langle D_1 \rangle$ — which results in large $\langle X \rangle$ — means that the genetic language evolves along the direction of the clarification of grammatical construction but complexifying of vocabulary composition.

Since the nucleotide sequence is nonMarkovian (Granero-Porati, Porati and Zani 1980; Luo, Tsai and Zhou 1988), one should investigate the joint probabilities $p(i\ j\ k)$, $p(i\ j\ k\ l)$, etc. (Sibbald, Banerjee and Maze 1989). However, the length of the protein-coding sequence is generally 1000–2000 nucleotides. In order to improve statistics, we may introduce the non-neighboring conditional probabilities $p(j|i*)$ $p(j|i**)$, etc. (the sequence denoted by $p(j|i*1)$, ..., $p(j|i*k)$, *etc.*) instead of joint probabilities. Here $*$ means any base. Correspondingly, instead of (2) and (4) we define the Markovian entropy with lag $(n-1)$ (Luo and Li 1991)

$$H_M^{(n)} = -\sum_{ij} p(i) p(j|i*n-1) \log_2 p(j|i*n-1) \tag{6}$$

and the $(n+1)th$-order redundancy

$$D_{n+1} = H - H_M^{(n)} \ . \tag{7}$$

By use of periodic boundary conditions to guarantee $D_{n+1} \geq 0$ one can show that

$$D_k = D_{N+2-k} \ . \tag{8}$$

The set of Markovian entropies $\{H_M^{(k)}\}$ describes all correlation characteristics of bases — neighboring, short-range and long-range. For example, for an independent sequence, since there is no statistical correlation between any pair of bases, one has

$$D_2 = D_3 = \ldots = 0 \ . \tag{9}$$

For a stochastic sequence, in addition to (9) one has

$$D_1 = 0 \ . \tag{9'}$$

However for a finite sequence of length N, equalities (9) and (9′) are broken due to fluctuation.

It can be shown that (Lee 1989)

$$D_1(\text{fluctuation}) = 8.15/N \tag{10}$$

$$D_{k\geq 2}(\text{fluctuation}) = 15.65/N \ . \tag{11}$$

So if D_k is lower than its fluctuation bound then one may consider it ~ 0. $D_k \neq 0$ means it is larger than its fluctuation bound. Define the correlation length of a sequence to be s if

$$D_{s+1} \neq 0, \quad D_{s+2} \sim 0, \quad D_{s+3} \sim 0, \quad \ldots, \quad D_{N-s} \sim 0 \ . \tag{12}$$

Evidently, the correlation length describes the range of the statistical correlations between bases in a sequence. It can be proved that for a sequence of correlation length s one has

$$p(j|i * k) = p(j) \quad (k \geq s) \ . \tag{13}$$

We calculated $\{D_k\}$ for 2,341 sequences and obtained the correlation length of each sequence. The results are shown in Table 1 (Luo and Li 1991).

Table 1 shows that the correlation length is smaller than three for two-thirds of the sequences. In this respect Rowe and Trainor (1983a) first investigated a viral DNA and demonstrated that only strong triplet correlation exists. Their analysis is consistent with our general result. Of course, one

Table 1: The Correlation Character of Nucleotide Sequences. Pri means primates, Rod — rodents, Mam — other mammals, Vrt — other vertebrates, Inv — invertebrates, Pln — plants, Fun — fungi, Bct — bacteria, Vru — viruses, Phg — phages, Chl — chloroplasts, Mit— mitochondria.

	Pri	Rod	Mam	Vrt	Inv	Pln	Fun	Bct	Vru	Phg	Chl	Mit	Total
s=0	0.25%	1.5%	0.7%	1.6%	11%	3%	19%	9.2%	9%	31%	33%	21%	6.7%
$1 \leq s \leq 2$	67%	71%	50%	56%	21%	48%	64%	61%	65%	50%	42%	60%	59%
3-period	10%	5.4%	25%	14%	24%	22%	0	2.7%	11%	0	0	2%	9.7%
9-period	0.5%	0.5%	0	4%	3.6%	0	0	0	0	0	0	0	0.64%
aperiodic long-range	22%	21%	25%	24%	40%	27%	17%	27%	15%	19%	25%	17%	24%
Number of Sequences	397	389	129	122	168	169	141	477	234	32	36	47	2,341

notes that the range of correlation introduced here is meant only in the sense of statistical correlation expressed by probabilities of bases. Thus the short-range correlation of a sequence does not mean that any given nucleotides at a distance could not pair to form a definite structure, nor that the full sequence data does not contain any further information.

Since the correlation of most sequences is of short range we can use D_1, D_2 and X to describe their statistical properties approximately. From Table 1 one finds that the proportion of sequences with $s = 0$ decreases with evolution. Since many sequences in lower organisms have $D_2 \sim 0$ this would lead to a small $\langle D_2 \rangle$ for these categories. The proportion of $s = 0$ sequences decreasing in the course of evolution would lead to a larger $\langle D_2 \rangle$ and $\langle X \rangle$ for higher species.

Furthermore, the statistical analysis can also be done with subsequences (Tavare and Song 1989). According to the position of a base in the codon (i.e. whether in the first, second or third position), we break the data of coding regions into three constituent subsequences and calculate their informational parameters (Luo and Shengli 1990). For example, for subsequences 1, which comprise all bases that occur in the first codon position, $D_1^{(1)}$ describes the base composition of the first codon position in the sequence, $D_2^{(1)}$ describes the correlation of a base in the first codon position with its nearest neighboring base to the left in the sequence, and $X^{(1)} = D_2^{(1)}/(D_1^{(1)} + D_2^{(1)})$, etc. We find (Luo and Shengli 1990) that only for the first subsequence does there exist a good statistical correlation between $\langle D_2^{(1)} \rangle$ or $\langle X^{(1)} \rangle$ with molecular evolution. On the other hand $D_1^{(3)}$ of the third subsequence is larger than $D_2^{(1)}$ and $D_1^{(1)}$ (by a factor of $\sim$ 1.5–2.5 for most sequences), which is related to the nonrandom usage of synonymous codons. In connection with this $\langle F^{(3)} \rangle$ changes with evolution remarkably.

About the usage of synonymous codons, the diversity in GC content at the third position of codons in vertebrates was reported in earlier work (Aota and Ikemura 1986, Bernard and Bernard 1986). By statistical analysis of the frequencies of codon usage for each codon multiplet, and for each category of species, we found that it exhibits a definite relation with evolution. For example, the C/U ratio and G/A ratio of the third bases in a degenerate multiplet change with evolution. The C/U ratio in Phe, Cys, Tyr, Asp, His and Asn changes from 0.4/0.6 for chloroplasts and phages to 0.6/0.4 for eukaryotes. In degenerate quartets a similar relation holds, too.

2 Reconstruction of the Evolutionary Tree

The results of statistical analysis provide a clue to reconstructing the evolutionary relationships among species. The popular methods to extract phylogenetic information from sequences are distance matrix treeing and maximum parsimony analysis (Woese 1987). The distance matrix method is based on the knowledge of the sequence distances between pairs of sequences, i.e., the fraction of positions in which the two sequences differ. It assumes that the "correct" tree is the one that best fits the data by a least-squares analysis. The maximum parsimony analysis does not reduce the differences among sequences to a single number, a distance. It treats the positions individually and assumes that the correct phylogenetic tree is the most parsimonious one, the one for which the smallest overall number of mutational changes have to be postulated to arrange the set of considered sequences upon it. Both methods require correct sequence alignment first which is as yet an unsolved problem.

Another general method for analyzing sequence data is cluster analysis. It groups sequences on the basis of how similar they are to one another and it does not definitely require the preliminary alignment (Blaisdell 1986). For sequences with correlation length smaller than three, the knowledge of $p(i)$, $p(i|j)$ and $p(i|j*)$ is the most important factor. Based on these data we define similarity of any two sequences under consideration through these probability components, then we use the concept of fuzzy classification proposed by Zadeh (1971) to reconstruct the evolutionary relationship among these sequences. The method of cluster analysis suggested by us can be stated as follows (Luo, Ji and Li 1995):

1. Find the probability vectors

$$\begin{aligned} E &= (p(A) - a,\ p(G) - a,\ p(C) - a,\ p(U) - a) \\ F &= (p(A|A) - a,\ \ldots,\ p(U|U) - a) \\ G &= (p(A|A\ *) - a,\ \ldots,\ p(U|U\ *) - a) \end{aligned}$$

   for each sequence under investigation. Here the additive constant $-a$ is optionally introduced to make the difference between different probability components more obvious.

2. Deduce the fuzzy similarity relation R. For n sequences to be classified, R is a $n \times n$ matrix, the elements of which are denoted by r_{ij}, $i, j = 1 \ldots n$:

$$r_{ij} = aE_i \cdot E_j / \max(E_i^2, E_j^2) + \beta F_i \cdot F_j / \max(F_i^2, F_j^2) + \gamma G_i \cdot G_j / \max(G_i^2, G_j^2) \qquad (14)$$

Here, α, β, γ are chosen to be positive and satisfy

$$\alpha + \beta + \gamma = 1 \ .$$

For realistic nucleotide sequences we find

$$0 < r_{ij} < 1 \ .$$

The fuzzy similarity relation R is symmetric ($r_{ij} = r_{ji}$) and reflexive ($r_{ii} = 1$). It is a measure of the similarity of sequences.

3. Deduce the fuzzy equivalence relation T

$$T = R^{n-1} \tag{15}$$

where $R^k = R \cdot R^{k-1}$, and in the multiplication of matrices the sum is replaced by operation max and product by operation min. It can be shown that the matrix T is not only symmetric ($T_{ij} = T_{ji}$) and reflexive ($T_{ii} = 1$) but also transitive ($(T \cdot T)_{ij} < T_{ij}$). The transitivity of the matrix T guarantees a correct classification of sequences.

4. Construct the partition tree from the fuzzy equivalence relation by use of the α-cut (α-level set) method. The method can be demonstrated through an example (see Zadeh 1971):

$$\begin{array}{cc} & \begin{array}{cccccc} a & b & c & d & e & f \end{array} \\ T = & \left[\begin{array}{cccccc} 1 & 0.2 & 1 & 0.6 & 0.2 & 0.6 \\ 0.2 & 1 & 0.2 & 0.2 & 0.8 & 0.2 \\ 1 & 0.2 & 1 & 0.6 & 0.2 & 0.6 \\ 0.6 & 0.2 & 0.6 & 1 & 0.2 & 0.8 \\ 0.2 & 0.8 & 0.2 & 0.2 & 1 & 0.2 \\ 0.6 & 0.2 & 0.6 & 0.8 & 0.2 & 1 \end{array}\right] \begin{array}{c} a \\ b \\ c \\ d \\ e \\ f \end{array} \end{array}$$

The equivalence relation is represented by a "similarity-tree". In this tree each level represents an α-cut of the equivalence relation. The sets of elements on specific α-levels can be considered as similarity classes of α-level. For the above equivalence relation the similarity tree is shown below. S_α is called a (weak) α cut set or α level set — the elements in the set have "similarity" $\geq \alpha$. The tree is rooted at the point $[abcdef]$ where the similarity takes its minimum:

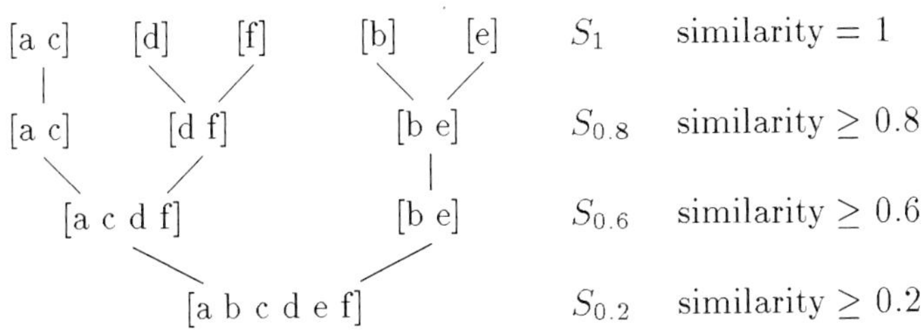

Of course, there exist many methods of cluster analysis apart from fuzzy clustering. We prefer the fuzzy cluster since it is simpler in logic. In other cluster analyses, after grouping the two most similar sequences into one category one must redefine the distance between the new category and other samples. Additionally, in comparison to other kinds of cluster analysis, e.g., the distance matrix method and maximum parsimony analysis, the method of fuzzy clustering requires the least computation, though it is not as accurate.

We have used the fuzzy clusters to analyze ribosomal RNA sequence data, and to reconstruct the evolutionary relationship among bacteria (Luo, Ji and Li 1995). The universal phylogenetic tree obtained by this method is consistent with other trees deduced by the maximum parsimony analysis and the distance matrix method. The success of the method supports the concept that the base composition and its short-range correlation are the most important factors for characterizing molecular evolution. In fact, we can use a series of $\{D_k\}$ to define the similarity, namely, instead of equation (14), we define

$$r_{ij} = \vec{D_i} \cdot \vec{D_j} / \max\{\vec{D_i}^2 , \vec{D_j}^2\} \tag{16}$$

in which $\vec{D_i}$ means $(D_1, D_2, \ldots D_L)$ of the i-th sequence. By fuzzy clustering of 1569 protein-coding sequences from 13 species, we obtain a basically correct evolutionary tree if $3 \leq L \leq 6$. However, for large L the result is not very consistent with the true phylogenetic tree (Ji and Luo 1992).

3 Maximum Information Principle — A Formulation on Molecular Evolution

From the statistical analysis we know that the evolution of nucleotide sequences is dominated statistically by two major factors. The first is the random mutation of bases which maximizes the entropy of base composition. The second is natural selection (including biochemical constraints) which strengthens the statistical correlation between bases, especially the correlation between neighboring bases, and results in the decrease of Markovian entropy. This hypothesis has been verified by the reconstruction of evolutionary trees in section 2. In this section we shall use the Maximum Information Principle (MIP) as a formulation of molecular evolution to examine the evolutionary mechanism further.

The formulation was suggested several years ago as a macroscopic approach to synergetics (Haken 1988), but originated in 1877 with Boltzmann. He proposed that a molecular system has the maximum disorder consistent with the conservation of energy and momentum that constrains the system. This approach has come to be called the maximum entropy formalism in the literature. In the MIP, the probability that maximizes the entropy is to be determined, subject to a set of auxiliary conditions handled by use of Lagrange multipliers.

Now we shall use MIP to discuss the statistical distribution of nucleotides in nucleic acid sequences. Since the molecular evolution of nucleotide sequences is a process of random mutation of bases subject to the constraints of natural selection, the problem is maximization of entropy under some constraints and hence the application of MIP to sequence evolution is direct.

Evidently, the first constraint incorporates the Markovian entropy $H_M^{(1)}$ which describes the nearest-neighbor correlation of bases. Another constraint we could use is the correlation of next-to-nearest neighbor bases or alternatively the $G+C$ content. Thus the MIP formulation can be stated as follows (Luo 1989, Luo and Bai 1995):

1. We maximize the entropy H under constraints

$$N = -\sum_i p(i) = 1 \tag{17}$$

and fixed $H_M^{(1)}$

$$\text{where} \quad H_M^{(1)} = -\sum_i p(i) F_i$$

$$\text{with} \quad F_i = \sum_j p(j|i) \log p(j|i) \ , \tag{18}$$

and fixed $G+C$ content

$$C = p(C) + p(G) \ . \tag{19}$$

Thus, one has

$$\delta H - \lambda_0 \delta N - \lambda \delta H_M^{(1)} - \lambda' \delta C = 0 \ . \tag{20}$$

2. If alternatively, we maximize the entropy H under constraints (17), fixed $H_M^{(1)}$ and $H_M^{(2)}$

$$\begin{aligned} \text{where} \quad H_M^{(2)} &= -\sum_i p(i) G_i \\ \text{with} \quad G_i &= \sum_j p(j|i*) \log p(j|i*) \end{aligned} \tag{21}$$

then instead of (20) one has

$$\delta H - \lambda_0 \delta N - \lambda_1 \delta H_M^{(1)} - \lambda_1' \delta H_M^{(2)} = 0 \ . \tag{22}$$

The solutions of (20) and (22) are

$$\begin{aligned} p(i) &= 2^{-(\ell+\lambda_0)} \prod_j p(j|i)^{\lambda p(j|i)} \quad (i = A, U) \\ p(i) &= 2^{(-\ell+\lambda_0+\lambda')} \prod_j p(j|i)^{\lambda p(j|i)} \quad (i = G, C) \\ & \qquad \text{(case A)} \end{aligned} \tag{23}$$

$$\begin{aligned} p(i) &= 2^{-(\ell+\lambda_0)} \prod_j p(j|i)^{\lambda_1 p(j|i)} \cdot p(j|i*)^{\lambda_1' p(j|i*)} \\ & \qquad \text{(case B)} \end{aligned} \tag{24}$$

($\ell = 1/\ell n2$) respectively. In (23) and (24) λ_0 can be eliminated by the normalization condition (17). Hence there are only two free parameters λ and λ' (or λ_1 and λ_1') in (23) (or (24)) and therefore one has a testable relation between probabilities $p(i)$. If $\lambda' = 0$, $\lambda_1' = 0$ then case A is equivalent to case B and one has

$$p(i) = 2^{-(\ell+\lambda_0)} \prod_j p(j|i)^{\lambda p(j|i)} \ . \tag{25}$$

This leads to

$$\lambda H_M^{(1)} = H - (\ell + \lambda_0) \ . \tag{26}$$

For a sequence with correlation length $s = 0$ (stochastic sequence) $H \sim H_M^{(1)}$. In this case

$$\lambda = 1 \ , \quad (\lambda_0 = -1/\ell n2) \ . \tag{27}$$

For coding regions of nucleic acid sequences one always has $H_M^{(1)}$ smaller than but near to H, so that

$$\lambda = \lambda_1 = 1, \quad \lambda' = \lambda_1' = 0 \; . \tag{28}$$

This is a first approximation to cases A and B, and will be called case C.

Now we will use the MIP formulation to analyze the nucleotide sequences and see if the principle can give an adequate description of base composition. For given λ and λ', or λ_1 and λ_1', one can calculate the base probabilities according to equations (23) or (24). The theoretical results are denoted by p_i^*, while the corresponding observational values are denoted by p_i. Define an error parameter

$$YY = \sum_i (p_i - p_i^*)^2 \; . \tag{29}$$

By minimizing YY with respect to λ and λ' or λ_1 and λ_1', around the first approximation point $\lambda = \lambda_1 = 1$, $\lambda' = \lambda_1' = 0$, the final result of probability distribution p_i^* can be obtained for each sequence. We have chosen 1,469 protein-coding DNA sequences and calculated p_i^*. The results show that the deviation of the theoretical probability from the experimental value is generally lower than 10% for case A and B and lower than 20% for case C. The agreement of MIP analysis with observational data is satisfactory. So the formation of nucleotide sequence (at least for coding regions) can be explained under the MIP principle in spite of the complexity inherent in the evolutionary process of nucleic acids. Furthermore, from the calculation of Lagrange multipliers of each sequence we find that the multipliers λ, λ' and λ_1 change systematically from species to species. The averaged Lagrange multipliers $\langle\lambda\rangle$, $\langle\lambda'\rangle$ and $\langle\lambda_1\rangle$ decrease roughly with evolution. The change of $\langle\lambda\rangle$ and $\langle\lambda_1\rangle$ means the Markovian entropy is decreasing or base correlation is increasing in the course of evolution which reflects the constraints of various biological functions due to natural selection. The change of $\langle\lambda'\rangle$ is related to $G+C$ content, growing with evolution which is essentially a biochemical constraint since the more $G+C$ bases contained, the more stable the double helix will be. Since there are also many accidental factors in the historical formation of any given sequence, the correlation of Lagrange multipliers with evolution can only be understood in the sense of a statistical average over many sequences.

4 A Model of Molecular Evolution

Nucleic acid sequences incorporate enormous amounts of information about living systems. Understanding the evolution of nucleotide sequences is central to the problem of the evolution of life. For example, how does order emerge from an evolutionary system? How does the diversity of species occur, given a variety of environmental conditions? etc. On the other hand, a nucleic acid is a well-defined one-dimensional informational system composed of only four letters, so the problem of molecular evolution is also of great interest in physics. As Feynman said: "Physicists always have the habit of taking the simplest example of any phenomenon and calling it 'physics', leaving the more complicated examples to become the concern of other fields." Nucleotide and amino acid sequences provide a simple and well-defined example of complex systems which may belong to a "universal class" of all such systems. It is expected that thorough analysis of the physics behind the evolution of sequences should given an important clue to the future development of a general nonequilibrium statistical theory, at least in some respects. Anderson and Stein (1984) have reported an attempt to model the initial stages of the evolutionary process in which the first information-containing molecule appears. They emphasized the role of diversity as well as selection and stability. For a long-N string written by two letters, if no diversity occurs, then the system will have an information quantity of the order of $O(1)$, but not $O(N)$. Likewise, if there is no selection and stability, then 2^N possibilities take place equally well, and the information quantity is still $O(1)$. However, if both diversity and stability are present, then we have a "broken symmetry in information space" and we gain a lot of information.

On the other hand, Kauffman and Levin (1987) proposed a general theory of adaptive walks on rugged landscapes. They pointed out that order arises as a result of self-organization of the complex systems and the selective forces play a role in further molding. Following the approach suggested by Stein and Anderson (1984), combining the idea proposed by Kauffman and Levin (1987), but emphasizing the results obtained from the statistical analysis of nucleotide sequences, we proposed a stochastic evolutionary model of molecular sequences (Luo and Trainor 1992). The work started from a simplified model of molecular sequences which comprise an ensemble of sequences composed of just two letters, A and B. As stated by the statistical analysis of nucleotide sequences, the trends of evolution are roughly towards increasing entropy and decreasing Markovian entropy. The two aspects, complexification of vocabulary and clarification of grammatical construction of genetic language, appear to correspond respectively, to two evolutionary paradigms: random drift (Kimura 1968, King and Jukes 1969) and natural selection. To incorporate these results, our model (Luo, Tsai and Lee 1990) assumes:

1. Mutation via $A \rightleftarrows B$ with probability λ in one time step. This rule models base mutation. It represents the mechanism of mutation and random drift in molecular evolution.

2. Selection. The unit of selection is called the "codon." The simplest case is a codon composed of two neighboring bases. There are four types of bases, namely, AA, AB, BA, and BB. Introduce a stochastic fitness function
$$S = S(n_1, n_2, n_3, n_4) \qquad (30)$$
where n_i's denote the numbers of codons AA, AB, BA and BB in a sequence.

We assume that if the mutation in one step leads to

$$\delta S \geq T_+ \qquad (31a)$$

for a given sequence (the advantageous mutation) it reproduces a new identical one with a probability β. On the other hand, for

$$\delta S \leq T_- \qquad (31b)$$

(the lethal mutation) it will be eliminated from the ensemble with probability β. T_+ and T_- are some threshold values which reflect the selectivity strength. These rules model Darwin's natural selection at the molecular level. As is well known in genetics, a single mutation in nucleic acid sequence could cause a correspondingly different amino acid to be coded for. The selection pressure on the amino acid would influence the final outcome of mutation. The fittest survives while any lethal mutations would be eliminated. For example, the mutations which change the amino acid class (hydrophilic to hydrophobic, etc.) can generally be so dangerous for protein function that they are eliminated by natural selection.

According to the above two rules ($31a$, b) we can deduce the evolutionary equation for an ensemble of strings. Suppose the ensemble is composed of M strings of equal length L. There are $ML/2$ codons in total. The probability p_i of the i-th codon in the ensemble obeys the following equations:

$$dp_i/dt = \left(1 + \frac{1}{M} dM/dt\right)^{-1} \{\lambda \sum p_j \tau_{ji} - \beta\lambda \sum p_k \theta_{kj} p_i + \beta\lambda \sum \theta_{ji} p_j\}$$
$$(i = 1, 2, 3, 4)\,. \qquad (32)$$

The first term on the right-hand side (RHS) of equation (32) represents the effect of a mutation in which τ_{ji} can be put in a matrix form, that is, τ_{ji} is an element of the j-th row and i-th column of the matrix

$$\begin{array}{c} \\ AA \\ AB \\ BA \\ BB \end{array} \begin{array}{c} \begin{array}{cccc} AA & AB & BA & BB \end{array} \\ \begin{pmatrix} -2 & 1 & 1 & 0 \\ 1 & -2 & 0 & 1 \\ 1 & 0 & -2 & 1 \\ 0 & 1 & 1 & -2 \end{pmatrix} \end{array} \tag{33}$$

The last two terms on the RHS of equation (32) represent the effect of selection in which θ_{ij} can also be put into the matrix form

$$\begin{array}{c} \\ AA \\ AB \\ BA \\ BB \end{array} \begin{array}{c} \begin{array}{cccc} AA & AB & BA & BB \end{array} \\ \begin{pmatrix} 0 & x & y & 0 \\ -x & 0 & 0 & z \\ -y & 0 & 0 & w \\ 0 & -z & -w & 0 \end{pmatrix} \end{array} \tag{34}$$

The elements of one-mutant neighbors have been denoted by x, y, z and w which take values $+1, -1$ or 0 stochastically. $+1$ means the corresponding transition is selectively advantageous, -1, the transition is selectively lethal and 0 the corresponding neutral transition. For example, $\theta_{12} = x = +1$ means $\delta S \geq T_+$ due to $\delta n_1 = -1$ and $\delta n_2 = +1$ (corresponding to the transition $AA \to AB$).

The selection is a stochastic force. If it is a rapidly-changing fluctuation we can approximate the set of equations (32) as equations of Langevin's type. They can be put into a form of Fokker-Planck equations. It can then be shown that the fast-varying part of stochastic selection only produces a diffusion effect on the motion along the flow line. If the time scale of stochastic selection is long enough, then the set of equations (32) can be regarded as the usual deterministic equations. We will confine discussion to the latter case.

The equations (32) can be solved numerically or by perturbation methods if β is small. The solution was obtained under a definite initial condition, for example $p_1 = 1, p_2 = p_3 = p_4 = 0$, which represents the case of M identical strings $AA \ldots A$ in the initial state. Then the informational parameters D_1, D_2 and X are calculated. We find that D_1 decreases and X increases monotonically. D_2 increases from 0 to a stable value across a maximum (for large β) or monotonically (for small β). The stable value of X for $t \to \infty$ is (for small β)

$$X = [1 + 4(x+y+z+w)^2/(x+y-z-w)^2]^{-1} . \tag{35}$$

This result shows how order emerges from a system under an appropriate selective pressure since D_2 and X can be viewed as order parameters of the system. It also explains the statistical relation between the evolutionary level of nucleotide sequences with its informational parameters. In addition, we can prove $D_k = 0$ if $k \geq 3$. So the correlation length is 1 in the simple model. With the same procedure one expects the correlation lengths to be smaller than 3 for most real nucleic acid sequences (Luo, Tsai and Lee 1990).

In summary, when the selective force is stochastic, different selection rules lead to different evolutionary trajectories of nucleotide sequences. The selection pressure reflects the influence of the environment on the evolution. Different environmental conditions lead to different selection forces, which in turn results in the diversity of species which develop with different evolutionary trajectories and rates and arrive at different levels at the present stage of evolution. It is interesting to note that the stable value of the order parameter X is invariant under some transformations in selective parameter space and the order emerges under some conditions, for example,

$$x + y \neq z + w$$

as seen from equation (35). This means that the order is related to the breakdown of the full symmetry between x, y, z and w in stochastic selection.

Our model is somewhat similar to Kauffman-Levin's concept of adaptive walk (Kauffman and Levin 1987). In K-L's model, an adaptive walk begins at an arbitrary sequence and on each trial, samples a one-mutant variant of the current sequence, and moves to that sequence only if it is fitter. The number of sequences is fixed. However, in our model the system may move to either a fitter sequence or a more non-fit sequence through one mutation. In the former case the fitter sequence is reproduced and in latter case the non-fit sequence eliminated. So the sequence number in the ensemble may be increased or decreased by 1 after one mutation. The system may get trapped on local optima (maxima) in K-L's theory. But this difficulty can be avoided in our model since the sequence can either go up or go down in the rugged landscape. Apart from these differences, in both models the stochastic selective force dominates the process and each site in a sequence of length N accepts K inputs, that is, the fitness contribution of each site depends on itself and K other sites among the N. The $K+1$ sites compose a selection unit which we call a codon. We have discussed the $K = 1$ case in detail. The discussion can easily be generalized to $K > 1$. The larger K is, the more rugged the fitness landscape will be. The number of independent stochastic elements grows rapidly with ruggedness of K. However, as discussed in previous sections, for most real nucleotide sequences the correlation length is smaller than three. So the cases of low K are more important for the purpose of modelling nucleic acids.

How shall we go from the simplified model to a more realistic evolutionary model of nucleic acids? The first step is to use four letters instead of two letters A and B. The mutation rates between different pairs of bases are, of course, not the same. For example, the transitional mutation rate (the mutation $A \leftrightarrow G$ or $C \leftrightarrow T$, in which the purine or pyrimidine base remains unchanged) is larger than that of transversional rate (the mutation between purine and pyrimidine bases). The mutation of a base in the third position of a codon takes place more frequently than in either of the other two. The second step is to formulate a more realistic selection rule, which accounts for not only the statistical correlation in a codon of three bases, but also the correlation between different codons. The latter will be able to explain the statistical correlation between $D_2^{(1)}$ and $X^{(1)}$ with evolution. The third step is to investigate the coding regions and the noncoding regions simultaneously and to introduce two kinds of sequences and the interaction between them which models the interaction between proteins (the functional molecules) and nucleic acids (the informational molecules).

5 Further Discussions on Molecular Evolution of Nucleic Acid Sequences

We have studied molecular evolution of nucleic acid sequences from the view point of information theory.

In mathematics, both the Hausdorff dimension (or fractal dimension) and the informational dimension are related to the order-q Renyi entropies (Glassberger 1983), so we can investigate sequence evolution in the framework of fractal dimension (FD). First we generalize the concept of FD to base space. The four bases A, T, C, G will correspond to positive X, negative X, positive Y and negative Y axes respectively. A nucleic acid sequence is related to walks with correlated jumps in a two-dimensional base space (Gates 1986). Consider a sequence consisting of N bases. The mean-square separation of the end points of a segment containing n bases is denoted by $\langle R_n^2 \rangle_N$. The local FD is defined as

$$D_N(n) = \ell n(n + 1/n)/\ell n\{\langle R_{n+1}^2\rangle / \langle R_n^2\rangle_N\}^{1/2} \qquad (36)$$

We find that $D_N(n)$ changes with n smoothly only for $n < N/2$.

So we introduce an average $FD(AFD)$ denoted as D_f which is defined by the average value of $D_N(n)$ in the range of $n < N/2$. We calculate D_f for each sequence and take the mean value in a given category. The standard deviation is smaller than 20% of its mean generally. The result shows that $\langle D_f \rangle$ grows gradually accompanying the increasing evolutionary level. It takes a value ranging from 1.1 to 1.2 for mitochondria, and to 1.7 for higher organisms. There exists a good statistical correlation between $\langle D_f \rangle$ and $\langle X \rangle$ although the two parameters are defined very differently (Luo and Tsai 1988).

Why can $\langle D_f \rangle$ play the role of an evolutionary parameter? For the case of a random walk, it follows from equation (36) that $D_N(n) = 2$ for n large. So the increase of D_f means the randomization of the base composition, which is a result of the evolutionary paradigm of random drift. On the other hand, from computer simulation, one can demonstrate that the correlation of neighboring bases in sequences also increases D_f. Thus the growth of $\langle D_f \rangle$ in evolution also means the clarification of grammatical construction of the genetic language which may be a result of the evolutionary paradigm of natural selection. Therefore, $\langle D_f \rangle$ describes the two factors simultaneously. This is the very reason why there exists a good statistical relation between the fractal dimension as defined above and evolution of nucleotide sequences.

So far we have discussed the statistical characteristics of protein–coding sequences and established a simple model to explain these features. We have emphasized the statistical distribution and correlation of bases. Now we will discuss a new aspect of sequences, namely, the *repeated fragments* in sequences and their theoretical implication. The problem of repeated fragments in amino acid sequences from the point of complexity of genetic language has been studied (Ebeling and Jimenez-Montano 1980). We investigated the repeated fragments of nucleic acid sequences in coding regions of length $N < 4000$. To minimize stochastic effects only segments with length ≥ 6 were taken into account. Evidently, for a sequence of $N < 4^6 = 4096$ the repetition of a fragment of length ≥ 6 would not likely be due to random recurrence. Using 875 sequences of different species, we calculated that (Zhou and Luo 1992) the ratio of total length of repeated fragments to sequence length N is

$$A = 0.55 \sim 0.63 \ . \tag{37a}$$

The ratio of total number of repeated fragments (including second-order repeated fragments, i.e., the repeated modes occurring in the repeated fragments) to N is

$$B = 0.083 \sim 0.090 \ . \tag{37b}$$

The parameters A and B take nearly the same values for different categories of species (primate, rodent, etc.) and show their independence of evolution. We call it the frozen ratio of repeated fragments.

Furthermore, we investigated the structure of repeated segments (Zhou and Luo 1992). We define the first-order informational redundancy D_1' and GC content F', of an artificial sequence which is obtained by joining together all repeated segments. We discovered that for most sequences, D'_1 in repeated fragments is larger than D_1 in the corresponding real sequence by a factor of magnitude $2 \sim 3$ to 5. The total length of repeated fragments is about half N as seen from the value of (37a). The fluctuation due to a short length would make D'_1 larger. However, detailed analysis shows that it could not explain such a large D'_1. In fact, it is related to the biased GC content in repeated fragments. We found that if $F > 0.5$ then the corresponding $F' > F$ and if $F < 0.5$ then $F' < F$, the change of which (namely, $(F' - F)/F$) is several percent to ten percent or more. The above results mean that the vocabulary usage of genetic language in repeated segments is more biased.

How do we understand the characteristics of repeated fragments? It seems that the two mechanisms, mutation and selection are not enough to explain the frozen ratio and related properties. We should extend our model of molecular evolution to the very early epoch of sequence formation to include an inflation stage (Luo 1991). Since the observed repetition is not due to statistical recurrence and since its biased GC content implies that these fragments could hardly be formed in evolution, we suppose that these repeated fragments are therefore remnants of prebiotic evolution of the sequence. There may have existed a prebiotic epoch in which the sequence was enlarged rapidly through repeating and linkage of bases and small fragments. The process could be understood from thermodynamic considerations as is pointed out in Rowe and Trainor (1983b). The primitive DNA fragments first formed in this way were strongly codon-biased in general. To ensure the superiority of a definite bias in a long segment one should assume a homogeneous condition for the sequence growth, which requires the process to proceed rapidly and sharply. In fact, "life had to arise very quickly", "all the prebiotic processes we know about are fast" (Waldrop 1990). As for why these repeated fragments, resisting molecular noise, have remained a frozen ratio in the long years of evolution is a difficult problem.

The repetitious DNA in noncoding regions has been suggested to "exist for no other purpose than its own perpetuation" (Grant 1981). Of course, the reason for repetition in coding regions may be different. A possible answer is as follows. The repeated fragments with length ≥ 6 in DNA correspond to the repeating of 2 or 3 residues in the amino acid sequence. We have studied the statistical correlation between peptide frequency in amino acid sequences and the secondary structure of protein, and obtained $r_1 = 0.098$ for a single peptide, $r_2 = 0.42$ for neighboring dipeptide, $r_8 = 0.33$ for next-to-neighboring dipeptide, etc. (Luo and Dong 1988, Luo 1993). Here r_k has a similar definition as D_k for nucleic acid sequence, namely, it is defined by {(entropy − Markovian entropy with lag) /entropy }. So the neighboring dipeptide correlation is a major factor in determining the secondary structure of protein. The observed repetition in sequence may be necessary for sustaining some secondary structure, even though the fragment itself may not participate in any active site of the protein.

In the prebiotic epoch, free from the constraints of function, mutation proceeded randomly and frequently. Accompanying the enlargement of sequences, the frequent mutations broke a lot of repeated fragments and made A and B decrease gradually. After a comparatively short time the length N of sequences and the amount of information possibly stored (roughly proportional to $\log N$) would have largely accumulated. In the meantime, the ratios A and B decreased by a large amount. As a definite length of sequence and a definite value of A and B were reached, then a new stage of evolution began. The above-mentioned first prebiotic epoch is what we called the *inflation stage* of sequence evolution.

According to modern cosmology, to overcome a series of difficulties such as why the monopole has not been found, why the Universe is so homogeneous, why the mass of the Universe is so near its critical value, etc., an inflation theory has been proposed which suggests an inflation stage in the very early epoch of the Universe's evolution. Likewise, to explain the statistical characteristics of nucleotide sequences it seems desirable to assume an inflation stage in the early evolution of sequences, too. The great similarity between life and the Universe reminds us of the idea of ancient Chinese philosophers: "the unity of Universe and Man."

6 Future Perspectives

The evolution of nucleic acid sequences is a very complicated process. Many parts of the genome frequently interact with each other. Moreover, the nucleotide sequence is not an isolated system. Its evolution is closely related to the evolution of protein sequences. On the other hand, the competition between species largely influences the evolution of the nucleic acid sequence of a particular species. The model developed in this article emphasizes the point and shows how order emerges from a molecular sequence in the process of evolution. Evidently, this is an important aspect of evolution. However, the model is too simple as compared with real nucleotide sequences. To incorporate other important aspects of molecular evolution in the model, the following points should first be noticed:

1. The evolution of noncoding sequences in a genome takes place parallel to the evolution of coding sequences. But the former has many new peculiarities. For example, there are a lot of repeated segments, especially in

eukaryotic genomes, which occupy a large portion of the noncoding sequences. What does it mean in terms of evolutionary mechanism? There exist long-range correlations between bases in the genome. After large-scale averaging the power spectrum shows the $1/f$ noise (noise inversely proportional to frequency) (Voss 1992). What is the mechanism responsible for the large-scale, long-range correlations? So the first issue in the development of a model is to account for these peculiarities of noncoding sequences (introns, $5'$-caps, $3'$-tails, spacers etc.) in the whole genome and to give a picture of the associated evolution of coding and noncoding parts of the genome.

2. In our model, two genetic operations on the sequence — mutation and selection — are considered. In fact, there exists another important operation — recombination of several fragments to produce new sequences. For eukaryotes, recombination is widely used by nature and it is the source of genetic diversity. Recombination is also of special importance in biotechnology which permits us to generate vast numbers of novel sequences. So the second issue in the development of the model is to include recombination as a third element of genetic operation.

3. The interactions of proteins and nucleic acids are the key rhythms of "the music of life" at the molecular level. A DNA sequence not only contains the information about the primary structure of a protein, but also programs its synthesis. On the other hand, proteins interact with nucleic acids and regulate the expression of genetic information from the requirement of biological function. Hofstadter (1979) described the interaction between enzymes and nucleic acids vividly in his famous monograph *Gödel, Escher, Bach: an Eternal Golden Braid.* So a deeper theory should include the modeling of the two kinds of sequences and investigate their coevolution.

4. The competition between species plays an important role in biological evolution. The competitive exclusion principle, based on Gause's experiments and Volterra's and Lotka's simple model, stated that two species cannot occupy exactly the same niche and coexist (Toft and Mangel 1991). Niche is a multidimensional space, each dimension of which is a variable of some resource or other quantity of importance to individuals in that population. Evidently, for a definite species, a niche is related to its selection. The fourth important issue is to discuss the competition of species from the level of molecular genetics by relating the selective parameters to niche variables. This would enable us to understand how

species are eliminated and how a new species is produced from the picture of interspecific competition of nucleic acid sequences.

References

Aota, S. and Ikemura, T. (1986) Diversity in $G+C$ content at 3rd position of codons in vertebrate gene and its cause. *Nucleic Acid Research* **14**, 6345–6355.

Bernard, G., Bernard, G. (1986) Compositional constraints and genome evolution. *Journal of Molecular Evolution* **24**, 1–11.

Blaisdell, B.E. (1986) A measure of the similarity of sets of sequences not requiring sequence alignment. *Proceedings of the U.S. National Academy of Sciences* **83**, 5155–5159.

Ebeling, W. and Jimenez-Montano, M.A. (1980) On grammars, complexity and information measures of biological macromolecules. *Mathematical Biosciences* **52**, 53–71.

Gates, M.A. (1986) A simple way to look at DNA. *Journal of Theoretical Biology* **119**, 19–28.

Gatlin, L.L. (1972) *Information Theory and the Living System.* New York: Columbia University Press.

Glassberger, P. (1983) Generalized dimension of strange attractors. *Physics Letters* **97A**, 227–230.

Granero-Porati M.I., Porati, A. and Zani, L. (1980) Informational parameters of an exact DNA base sequence. *Journal of Theoretical Biology* **86**, 401–403.

Grant, B. (1981) The safe-neighborhood hypothesis of junk DNA. *Journal of Theoretical Biology* **90**, 149–150.

Haken, H. (1988) *Information and Self-Organization. A Macroscopic Approach to Complex Systems.* New York: Springer-Verlag.

Heijne, G. (1987) *Sequence Analysis in Molecular Biology.* New York: Academic Press.

Hofstadter, D.R. (1979) *Gödel, Escher, Bach: an Eternal Golden Braid.* New York: Basic Books.

Hu, S.Z. and Luo, L.F. (1989) The frequencies of codon usage and its relation to evolution. *Acta Scientarium Naturalium Universitatis Intramongolicae* **20**, 483–490.

Ji, F.M. and Luo, L.F. (1992) Fuzzy classifying of nucleotide sequences and molecular evolution. *Acta Scientarium Naturalium Universitatis Intramongolicae* **23**, 194–199.

Kauffman S.A. and Levin, S. (1987) Toward a general theory of adaptive walks on rugged landscapes. *Journal of Theoretical Biology* **128**, 11–45.

Kimura, M. (1968) Evolutionary rate at the molecular level. *Nature* **217**, 624–626.

King, J.L. and Jukes, T.H. (1969) Non-Darwinian evolution. *Science* **164**, 788–798.

Lee, W.J. (1989) The distribution of informational redundancies of random sequences. *Acta Scientarium Naturalium Universitatis Intramongolicae* **20**, 490–494.

Luo, L.F. (1989) MIP and informational parameters of nucleic acids. *Acta Scientarium Naturalium Universitatis Intramongolicae* **20**, 184–189.

Luo, L.F. (1991) An evolutionary model of nucleic acid sequence. *Evolutionary Theory* **10**, 75–81.

Luo, L.F. (1993) A statistical mechanics approach to prediction of secondary structure of protein. *Evolutionary Theory*, Special Issue, pp. 6–8.

Luo, L.F. and Bai, G.Y. (1995) Maximum information principle and the evolution of nucleotide sequence. *Journal of Theoretical Biology* **174**, 131–136.

Luo, L.F. and Dong, Y.Z. (1988) Statistical analysis of peptide correlation and prediction of protein conformation. *Chinese Biochemical Journal* **4**, 173–183.

Luo, L.F., Ji, F.M. and Li, H. (1995) Fuzzy classification of nucleotide sequences and bacterial evolution. *Bulletin of Mathematical Biology* **57**, 527–537.

Luo, L.F. and Li, H. (1991) The statistical correlation of nucleotides in protein-coding DNA sequences. *Bulletin of Mathematical Biology* **53**, 345–353.

Luo, L.F. and Shengli (1990) The information-theoretic investigation of heterogeneous DNA sequences split into 3 subsequences. *Acta Scientarium Naturalium Universitatis Intramongolicae* **21**, 229–234.

Luo, L.F. and Trainor, L.E.H. (1992) A stochastic evolutionary model of molecular sequences. *Journal of Theoretical Biology* **157**, 83–94.

Luo, L.F. and Tsai, L. (1988) Fractal dimension of nucleic acid sequences and its relation to evolutionary level. *Chinese Physics Letters* **5**, 421–424.

Luo, L.F., Tsai, L. and Lee, W.J. (1990) Model of evolution of molecular sequence. *Physical Review A* **41**, 5441–5450.

Luo, L.F., Tsai, L and Zhou, Y.M. (1988) Information parameters of nucleic acid and molecular evolution. *Journal of Theoretical Biology* **130**, 351–361.

Luo, L.F., Zhou, Y.M. and Tsai, L. (1988) Statistical distribution and homological and non-Markovian analysis of nucleic acid sequence. *Journal of Biomathematics* **3**, 10–17.

Rowe, G.W. and Trainor, L.E.H. (1983a) On the informational content of viral DNA. *Journal of Theoretical Biology* **101**, 151–170.

Rowe, G.W. and Trainor, L.E.H. (1983b) A thermodynamic theory of codon bias in viral genes. *Journal of Theoretical Biology* **101**, 171–203.

Sibbald, P.R., Banerjee, S. and Maze, J. (1989) Calculating higher order DNA sequence information measures. *Journal of Theoretical Biology* **136**, 475–483.

Stein, D.L. and Anderson, P.W. (1984) A model for the origin of biological catalysis. *Proceedings of the U.S. National Academy of Sciences* **81**, 1751–1753.

Tavare, S. and Song, B. (1989) Codon preference and primary sequence structure in protein-coding region. *Bulletin of Mathematical Biology* **51**, 95–115.

Toft, C.A. and Mangel, M. (1991) From individuals to ecosystems; the papers of Skellam, Lindeman and Hutchinson. *Bulletin of Mathematical Biology* **53**, 121–134.

Trifonov, E.N. (1989) Searching for codes in the sequences. In R.R. Colwell, ed., *Biomolecular Data*. New York: Oxford University Press, pp. 199–210.

Volkenstein, M.V. (1982) *Physics and Biology.* New York: Academic Press.

Voss, R.F. (1992) Evolution of long-range fractal correlations and $1/f$ noise in DNA base sequences. *Physical Review Letters* **68**, 3805–3808.

Waldrop, M. (1990) Goodbye to the warm little pond? *Science* **250**, 1078–1080.

Woese, C.R. (1987) Bacterial evolution. *Microbiological Reviews* **51**, 221–271.

Zadeh, L.A. (1971) Similarity relation and fuzzy orderings. *Information Sciences* **3**, 177–206.

Zhou, Y.M. and Luo, L.F. (1992) The statistical analysis of repeated fragments in nucleic acid sequences. *Acta Biophysica Sinica* **8**, 581–584.

Codon Space: Exploring the Origins and Evolution of the Genetic Code

L.E.H. Trainor, G.W. Rowe and G.J. Nelson

1 Introduction

The (almost universal) genetic code is a language of 3-letter words (called "codons") based on the 4-letter the alphabet A,C,G and T (standing for the four bases adenine, cytosine, guanine and thymine, respectively), with A and T, and C and G pairing in the famous Watson-Crick base pair cross-bridges between the two strands of the DNA double helix. The code is the degenerate translation between the 64 possible codons in the language and the 20 amino acids which occur in biological proteins. The translations between codons and their amino acid counterparts are reproduced in Table 1 for the convenience of the reader. The degeneracy is by no means uniformly distributed across the amino acids; serine, for example, is coded for by no less than 6 different codons from the set of 64.

As DNA or RNA is believed to contain the information required to orchestrate the development of a complete organism from a single cell (or particle, in the case of viruses and bacteriophages), much effort has been made to extract information from the sequence of bases that makes up the genome. Most of this effort has, understandably, concentrated on the relation between the base sequences of DNA and RNA molecules and their expression as protein molecules or as regulators for their production. Among these investigations are several comprehensive computer analyses which attempt to classify DNA sequences as coding or non-coding regions, or to determine the type of source organism given a fragment of nucleic acid (Gonnet, Cohen and Brenner 1992; Guigo et al. 1992). More recently, it has become evident that each polynucleotide base sequence itself possesses a complex system of internal patterning, the functional significance of which remains unclear (Lewin 1980).

One major goal of these various projects is to provide a method whereby newly sequenced DNA (such as that provided by the Human Genome Project) may be quickly and accurately analysed by computer. Much emphasis has therefore been placed on identifying key features in the sequences and correlating these with the known properties of specific sequences.

A principal impediment to sequence analysis work is the infamous difficulty of representing the alphabetic patterning of polynucleotide strings in a way that can be quickly and intuitively grasped. We have shown that a

Table 1: The Genetic Code. This presentation of the universal code for DNA is patterned after the format used in Stryer (1975). The letters A, T, C, G correspond to the nucleotide bases adenine, thymine, cytosine and guanine respectively. (For RNA sequences, thymine is replaced by uracil). Codons correspond to triplets of bases, with individual bases occupying slots in the codon referring to the first, second and third codon positions, respectively. For example from the table we see that ACT codes for the amino acid threonine. The abbreviations for all 20 amino acids are standard.

First position (5′ end)	Second position				Third position (3′ end)
	T	**C**	**A**	**G**	
T	Phe	Ser	Tyr	Cys	**T**
	Phe	Ser	Tyr	Cys	**C**
	Leu	Ser	Stop	Stop	**A**
	Leu	Ser	Stop	Trp	**G**
C	Leu	Pro	His	Arg	**T**
	Leu	Pro	His	Arg	**C**
	Leu	Pro	Gln	Arg	**A**
	Leu	Pro	Gln	Arg	**G**
A	Ile	Thr	Asn	Ser	**T**
	Ile	Thr	Asn	Ser	**C**
	Ile	Thr	Lys	Arg	**A**
	Met	Thr	Lys	Arg	**G**
G	Val	Ala	Asp	Gly	**T**
	Val	Ala	Asp	Gly	**C**
	Val	Ala	Glu	Gly	**A**
	Val	Ala	Glu	Gly	**G**

mathematical structure we call *codon space*, based on the frequencies of nucleotides occurring in each of the three codon positions, provides a natural arena in which polynucleotide strings may be represented by single points, and thereby readily visualized and compared (Rowe 1981; Rowe and Trainor 1983a,b; Rowe, Szabo and Trainor 1984; Trainor, Rowe and Szabo 1984). Our previous study of 332 DNA sequences in codon space searched for clustering with respect to the broad biological classes of eukaryotes, prokaryotes, viruses, bacteriophages, and organelles (mitochondria, etc.). We found that certain areas in codon space were favored by sequences from one or two of the five biological classes considered.

The world database of sequence information has expanded enormously in the past several years, providing an unprecedented opportunity to map codon space much more comprehensively than heretofore possible. At the same time, the very complexity of the potential organization of inter-sequence relationships calls into question the value of standard cluster-mapping methods.

In this essay we survey the properties of codon space and the organization of terrestrial life forms within it. We then summarize the results of a new mapping effort in which we describe the relational clustering of 16,432 protein-coding DNA sequences as represented in the genome library of the European Molecular Biology Laboratory (EMBL hereafter). We "look" at the codon space organization of this large data set with an artificial neural network, and apply an algorithm of unsupervised learning which allows the neural network to assess the degree to which the codon space is organized into distinguishable regimes or "clouds" of sequence points. The cluster patterns formed by living systems in codon space may, as we discuss further below, provide exciting insights about the origins and early evolution of the genetic code. Furthermore, the exploration of codon space motivates the search for additional novel mathematical structures suited to further clarifying the early history of the genome. We will suggest that one such representation, based on the symmetry group of the tetrahedon, is particularly suited to this task.

2 Codon Space

Codon space is based on the idea that a protein-coding DNA sequence can be classed according to the frequencies of the four bases in each of the three positions in a codon. For example, the codon CGT has a cytosine (C) in position 1, a guanine (G) in position 2, and a thymine (T) in position 3. For a complete DNA sequence, we count up the numbers of each of the four bases present in each of the three positions to obtain the 12 quantities $A_1, G_1, T_1, C_1, A_2, G_2, T_2, C_2, A_3, G_3, T_3, C_3$, where A_1 is the number of adenines in the first codon position, T_2 is the number of thymines in position two, and so on. Given the total length N of the sequence, we see that not all of these 12 quantities are independent, since we must have

$$A_i + G_i + T_i + C_i = N/3 \ , \quad i = 1, 2, 3 \tag{1}$$

We may define $a_i \equiv A_i/N$, $g_i \equiv G_i/N$, $t_i \equiv T_i/N$, $c_i \equiv C_i/N$ and rewrite equation (1) as

$$a_i + g_i + t_i + c_i = 1/3 \tag{2}$$

Clearly, only three of the four parameters describing each codon position are independent, so we may eliminate one base from each set. We choose to eliminate c_i. We may then characterize a DNA sequence by a set of nine quantities, which may be conveniently arranged in a 3×3 matrix M:

$$M \equiv \begin{bmatrix} a_1 & t_1 & g_1 \\ a_2 & t_2 & g_2 \\ a_3 & t_3 & g_3 \end{bmatrix} \tag{3}$$

These nine numbers may be thought of as coordinates in a 9-dimensional Euclidean space which we have previously dubbed *codon space* (Rowe et al. 1984), although a more appropriate name might have been *sequence space*.

Having represented DNA sequences by points in a codon space, we can ask if the points can be grouped into clusters in any meaningful fashion. Indeed, they can. When we considered an initial data set consisting of 332 DNA sequences drawn from a broad range of taxonomic categories and applied standard methods (hierachical cluster anaysis; Anderberg 1973), rather well separated clusters corresponding to phages, viruses, bacteria, and various kinds of higher order organisms tended to form. A 9-dimensional space is, admittedly, difficult to visualize, so that it is useful to make various 3-dimensional cuts through the 9-dimensional space, such as the one shown in Figure 1. In general, it seems that genes from higher organisms in the phylogenetic chain have higher GC content at the expense of AT, while projection onto the AT plane of cluster centroids shows that one tends to find higher-order organisms closer to the origin in codon space.

Can these qualitative impressions of codon space, based on the visual inspection of selected 3-dimensional cuts through the 9-dimensional hypervolume, be made quantitative? To some extent they can. The technique of multidimensional scaling, for example, (abbreviated MDS; Kruskal 1964a,b) is designed to find such lower-dimensional representations of sets of points originally plotted in spaces of higher numbers of dimensions. Using this technique,

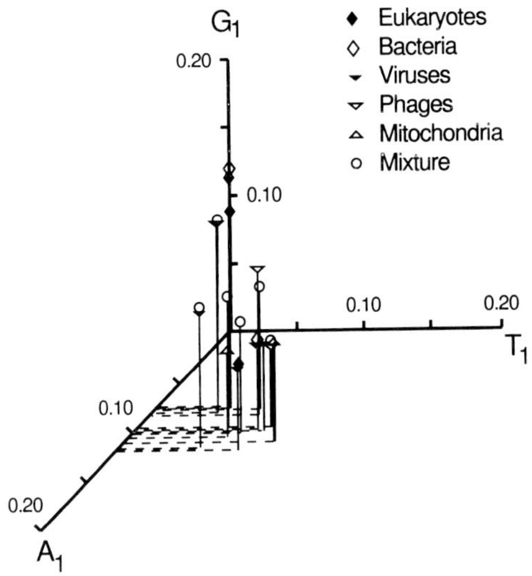

Figure 1: Three-dimensional transect through 9-dimensional codon space, showing taxonomic clustering. Adapted from G.W. Rowe, V.L. Szabo and L.E.H. Trainor (1984) *Journal of Molecular Evolution* **20**, 167–174, Springer-Verlag, with permission.

we found that the DNA sequences used in our analysis could be meaningfully represented in a space of only three dimensions (Rowe 1985). We also found that one of the two-dimensional projections of this three-dimensional representation split the collection of DNA sequences into five broad groups: genes from organelles, viruses, bacteriophages, eukaryotes and prokaryotes.

Although the MDS analysis seems to reveal some sort of underlying order in the base composition of DNA sequences, it suffers from several problems. First, the algorithm for reducing the dimensionality of the space used to represent the sequences does not scale well to large numbers of sequences. In order to construct a lower-dimensional representation for a set of points, the distances between all possible pairs of points must be calculated and compared. With a set of n points, we must calculate $n(n-1)/2$ distances, which means that the algorithm is $O(n^2)$: the execution time increases as the square of the number of points! Even allowing for the increased speed and memory capacity of computers that has been achieved over the past decade, a routine MDS analysis of, say, 15–30,000 sequences is out of the question.

A second problem with the MDS approach is the method by which sequences are classified. To determine how a new sequence fits into the MDS representation requires the entire MDS analysis to be rerun to include the

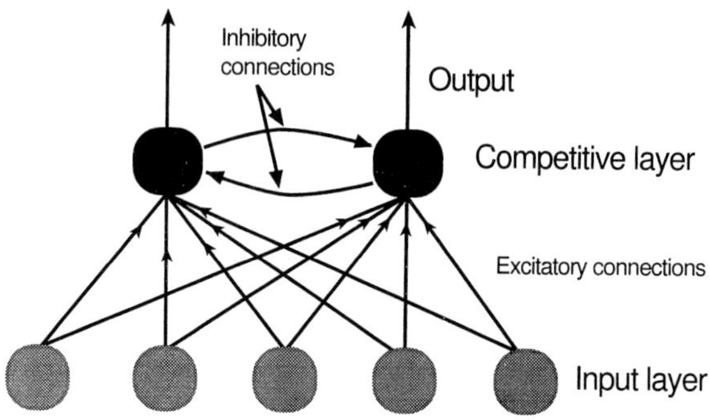

Figure 2: A simple competitive learning network.

extra point. Since the number of known DNA sequences is constantly increasing, it would be preferable to have a scheme in which sequences are treated individually.

The final problem is one of interpretation. In MDS analysis it is often not at all obvious what the final coordinates in the lower-dimensional space mean, if anything. Although we do find a three-dimensional representation for our original nine-element matrix, how to relate the three coordinates to the base compositions of each codon position remains obscure. Ideally, one would like to define a measurement procedure on codon space that gives simple low-dimensional representations in which the biological and physical meaning of the coordinates are clear. We will suggest here that one can do this by creating a class of idealized codon space "observers," in the form of artificial neural networks that detect sequence points distributed in codon space and, without external "guidance," construct concise representations of the sequence distribution pattern. Unlike standard "divide and conquer" methods such as MDS, the work load faced by such nets scales only linearly with the number of sequence points n, rather than as n^2. Very comprehensive analyses of sequence clustering in codon space can therefore be undertaken. Let us consider these idealized observers and their action on codon space.

3 Competitive Learners in Codon Space

A competitive learning network (Rumelhart and Zipser 1986) consists of several layers of information processing units conventionally called neurons (see Figure 2; also the chapters by Busch and Trainor and by Sutton, this volume). In our codon space nets, the first layer consists of cells which act as a "retina" — they "see" the input pattern which is presented to the network. A pattern, in this context, is a set of input cells which can be "turned on," in the sense that they can be stimulated to send signals to neurons in the second layer. A stimulated cell is said to *fire* when it sends a signal. Different patterns will stimulate different sets of cells in the input layer.

Each cell in the input layer is connected to each cell in the second layer, which is the first layer in which processing of the input image occurs. Each of these connections has a *weight* associated with it. We shall call the weight between cell i in the first layer and cell j in the second layer w_{ij}. This weight determines the strength of the signal received by cell j if cell i should fire. All these weights are positive numbers, indicating that any signal from the first to the second layer is *excitatory* — it will tend to stimulate the cell that receives it.

The cells in the second layer are divided into one or more distinct clusters. Each cluster receives weighted input from all cells in the input layer. Within each cluster, each cell sums up the input it receives and sends an *inhibitory* signal of a size equal to its total input from the first layer to all other cells in the same cluster. The only cell in that cluster to fire will be the one that has received the most input from the cells in the input layer, since that cell is the only one that will have a net positive input after all the inhibitory signals have been processed. This is known as a *winner-takes-all* network, since the cell with the largest input wins the competition to see which cell will fire.

The simplest form of a competitive learning network is one with only two layers: an input layer and a second layer. The output of the network is the pattern produced by those cells (one cell per cluster) in the second layer that fire when a certain input pattern is presented. The simplest output layer is one consisting of a single cluster containing only two cells. Such a network is capable of making a "yes-no" decision on an input pattern: "yes" if cell 1 fires and "no" if cell 2 fires.

More complex networks are possible. We may increase the number of cells in each cluster, increase the number of clusters, or even add more layers before we read the output. If a third layer is added to the network, it too would be divided into one or more winner-takes-all clusters, each of which receives input from all the cells in the second layer.

There are many possibilities for the architecture of the higher layer of a network suited to codon space mapping. Since our main goal is to see if the network will train itself to classify the input sequences into the five groups observed in the MDS analysis, we therefore need an output layer which allows at least five different categories. The simplest such architecture is to have a single upper layer consisting of a single cluster with five neurons. However, such a simple network appears incapable of classifying the input data meaningfully.

We have examined several architectures consisting of two or more clusters, each of five mutually inhibitory neurons. The number of possible classifications produced by a network with m clusters in its output layer, each containing five neurons, is 5^m. If $m > l$, we are faced with a large number of possible classifications, relative to the five biological classes. We shall return to this point below.

It may seem unusual that we are going to use unsupervised learning for codon space analysis, where the network forms its own classifications of the input patterns without the aid of a teacher, instead of supervised learning, where an input pattern is presented to the network and the difference between the observed output and the desired output is used to train the network. It is true that if our goal were to design a neural network that could be used to classify unknown or newly sequenced DNA fragments, a supervised learning algorithm would make more sense. However, we are trying to show that there are fundamental differences between the base usage patterns in the genes of different taxonomic categories. If these differences are fundamental and general, an unsupervised network should be able to identify them without any prompting. The specific formal steps in competitive learning algorithms implemented by our networks are discussed further in the Appendix.

The EMBL database separates sequences into several large groups, such as viruses, prokaryotes, phages, organelles (mitchondria, etc.), rodents, primates, plants and so on. Since we found that sequences seemed to split naturally into five groups in MDS analysis, we divided the EMBL files into five groups corresponding to viruses, prokaryotes, bacteriophages, organelles and eukaryotes. The final category included the EMBL files in the directories labelled fungi, invertebrates, mammals, plants, primates, rodents, and vertebrates. (We did not examine the synthetic DNA sequences listed on the tape.) All the sequences (Table 2) were protein-coding genes containing 51 or more bases (we would have set the lower limit at 50, but all sequences must have a length that is a multiple of 3).

Having extracted the DNA sequences, it is straightforward to calculate the entries of the matrix M (equation (3)) for each one. The most obvious way of presenting the data to the neural network is to use an input layer with 9 neurons, each of which corresponds to one of the entries in the matrix M and fires with a strength proportional to that entry. However, we found that such a network does not have enough degrees of freedom in the weights to allow a meaningful classification of the various input patterns. We need a larger input layer in order to capture the details of the data.

Table 2: Codon Space DNA. Numbers of protein-coding DNA sequences used in the analysis. For example, 751 of the sequences used came from phages.

Source	Number
eukaryotes	6955
organelles	862
phages	751
prokaryotes	5302
viruses	2612

We therefore need a binary representation for the nine entries of the matrix M. This representation should preserve the relative distances between the numbers in the matrix, so that two sequences with similar base compositions lie close to each other in the binary image. How this might be done may be found in one of the examples given by Rumelhart and Zipser (1986). They consider the case of a 4×4 array of dots where a single pattern consists of two nearest-neighbor dots being "on" with the remaining 14 dots being "off." When a competitive learning network with a single binary output (allowing classification of the input into two patterns) is trained with a collection of such *dipole patterns*, the square of points is divided in half, with the dipoles in the upper half of the square being classified in one group and those in the lower half in the other. Two dipoles are "similar" to each other if they share a point. In other words, two patterns are similar, and more likely to be classified in the same group, if parts of their patterns overlap.

We implemented this *dipole representation* in codon space. For each of the nine numbers in the matrix M, we defined 101 neurons in the input layer, making an input layer containing 909 neurons. The set of 101 neurons corresponding to each entry in M may be pictured as a linear array of points numbered from 0 to 100. Then, if the value of a matrix entry is x, neuron numbers $[100x]$ and $[100x]+1$ are fired, with all other neurons remaining dormant. (Here the notation $[100x]$ means the largest integer less than or equal to $100x$.) Consider the entry a_1 as an example. If the value of a_1 is 0.342, neuron numbers 34 and 35 would be turned on, since these two neurons bracket the interval in which the value of a_1 lies.

Thus each DNA sequence is presented as a pattern of nine dipoles in the input layer. In practice, very few of the sequences analyzed gave rise to any matrix elements with values below 0.05 or above 0.55 so, to avoid wasting half of the input array, we scaled the elements of M so that 0.05 maps into 0.0 and 0.55 maps into 1.0 according to the formula

$$M'_{ij} = 2(M_{ij} - 0.05) \tag{4}$$

where M_{ij} is an entry from the matrix M and M'_{ij} is the scaled entry which is used in training the network. If any matrix element M_{ij} is mapped to values less than 0 (greater than 1), as will happen for sequences with base frequencies less than 0.05 (greater than 0.55), it is assigned the scaled value $M'_{ij} = 0.0$ (1.0).

We noted above that a competitive learning network is designed to seek its own classification of the input patterns with which it is trained. The number of possible classifications in one of our networks with m output clusters is 5^m. For the simplest network, containing only a single cluster of five neurons in its output layer, classification into five groups is automatic. For larger networks, the sequences are spread over many classifications.

To interpret the results, we kept a tally of which sequences were classified as each output pattern. To store the classifications efficiently, a code number was calculated for the output pattern produced by sequence. For each output pattern, the following data for the entire test set were recorded:

1. The number of sequences of each biological class that have been classified to this output pattern.
2. The average of each of the nine base frequencies for all sequences that have been classified to this output pattern.

After the entire test set had been presented to the network, a summary of the results can be obtained by listing all the different classifications, together with the number of sequences derived from each biological class, and the averages of each of the nine base frequencies that have been classified to each pattern.

The network does its classifications using only the data on base frequencies in each sequence. It has no information concerning the biological class to which each sequence belongs. We are interested in seeing whether a classification done in this fashion is able to distinguish between biological classes based solely on an analysis of the base frequencies. If such a classification has indeed taken place, we should find that most of the output patterns are dominated by one particular biological class. We would therefore like to sort the patterns so that those with a predominance of one biological class are in one group. We can do this by applying the non-hierarchical cluster analysis technique (Anderberg 1973) to the data on biological classes stored at each pattern.[a]

[a]Unfortunately, the introduction of cluster analysis at this point leads to two different meanings for the word "cluster" in what follows: a cluster of neurons in the output layer of the neural network, and a cluster of data points in the cluster analysis. Hopefully, the correct interpretation will be clear from the context, but we have tried to refer to neuron clusters as "output clusters" and data clusters as simply "clusters".

The cluster analysis algorithm is as follows:

1. Determine how many clusters are required (we take 5 — one for each biological class).
2. Choose random locations for the center of each cluster. (The center of a cluster is determined by taking the arithmetic average of each coordinate for each member of the cluster, and is known as a *centroid*.)
3. Assign each data point to that cluster whose centroid is closest to it.
4. Recalculate the centroid positions by averaging the coordinates of all points that have been assigned to that cluster.
5. Repeat from step 3 until no data point is assigned to a different cluster from the last iteration.

The clustering procedure does not move any of the data points — it merely moves the centroids until a consistent assignment of data to clusters is obtained.

We applied the clustering technique to the collection of patterns as follows. The cluster analysis operates on the array of counts of biological classes. We define the array for one particular output pattern to be

$$[E, M, F, P, V] \tag{5}$$

where E is the number of eukaryotic sequences, M = organelles (mitochondria), F = phage, P = prokaryote and V = virus. We define $B = E + M + F + P + V$ to be the total number of sequences assigned to that pattern in the tree. The procedure is then to

1. Normalize each set of counts of biological classes to obtain an array of fractions $[e, m, f, p, v]$ where $e \equiv E/B, m = M/B$, etc., and $e + m + f + p + v = 1$.
2. Apply the clustering algorithm above to the normalized data, with the exception that in calculating the new centroid positions in step 4, each data point is weighted by B, the number of sequences assigned to that pattern.

The result of this cluster analysis is a set of 5 centroids $X_i, j = 1 \ldots 5$, where each X_i is an array of 5 numbers giving the average biological composition of the patterns from which it is formed. We may therefore write a centroid as follows:

$$X_i = [X_{ie}, X_{im}, X_{if}, X_{ip}, X_{iv}] \tag{6}$$

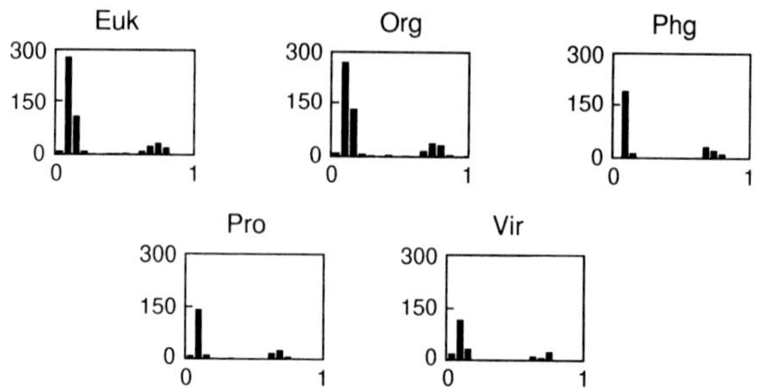

Figure 3: Number of cluster centroids as a function of the fraction sequences of each biological class making up the centroid. Euk= Eukaryote; Org = Organelle (mitochondria, etc.); Phg = Phage; Pro = Prokaryote; Vir = Virus. Data for a network with 4 output clusters.

where X_{ie} is the fraction of centroid i composed of eukaryotes, and so on for the other four components.

If the network has been successful in classifying sequences according to their biological origin, we should see, for each centroid, components such that one of the five components is fairly large (approaching 1.0) and the others are small (approaching 0.0) with each centroid representing a different biological class.

4 What They Saw

We chose a different training set and different testing set for each of 100 repeat runs for each network, classified the testing set, and obtained a set of five cluster centroids for each run. We ran the program for networks with 2, 3, 4 and 5 output clusters. The number of cluster centroids as a function of the fraction in which each biological class was examined. Since each run produced five centroids, the 100 runs produced 500 cluster centroids. In order to eliminate bias from clusters containing relatively few sequences, only those clusters containing more than 100 sequences were used in the histograms. Figure 3 shows the results for networks with four output clusters.

Networks with two output clusters were not particularly efficient at classifying the sequences according to their biological origin. The situation improves dramatically with 3 output clusters however, and appears to improve even more with 4 or 5 clusters. Because the 4-cluster network includes about 82% of the testing set in its results, and illustrates the points most clearly, we will use it in what follows.

The question of exactly what a neural network is using to form its classifications has vexed users of neural networks ever since they first became popular tools in analyzing patterns. In many applications, it doesn't matter what the network "sees," as long as it produces the correct classification. In our case, we would like some insight into the source of the distinctions.

We can get some idea by considering the correlation between the cluster centroids and the average base frequencies of those sequences making up each cluster. For a cluster centroid $i, j = 1 \ldots 5$, we therefore have an array Y_i of nine real numbers which we may write as follows:

$$Y_i = [Y_{iA_1}, Y_{iG_1}, Y_{iT_1}, Y_{iA_2}, Y_{iG_2}, Y_{iT_2}, Y_{iA_3}, Y_{iG_3}, Y_{iT_3}] \qquad (7)$$

Here, Y_{iA_1} is the average A_1 content of all those sequences comprising cluster i, and so on for the other eight components of Y_i.

We may now attempt to correlate the nine average base frequencies in equation (7) with the five biological frequencies in equation (6). The easiest way to see any correlations that may exist is to draw a scatter plot for each of the 45 possible pairings of an average base frequency selected from the array (7) with a biological frequency selected from the array (6).

As mentioned above, the 100 runs produced a total of 500 cluster centroids, each of which may be characterized by the two arrays (6) and (7). On each of the 45 scatter plots, we plotted one point from each of the 500 centroids. If there is any correlation between the base frequencies and biological frequencies making up the various clusters, we should see the points in the scatter plots forming separate, compact groups, rather than spreading out uniformly over the plotting area.

Scatter plots for the 4-output cluster network are shown, for the EMBL eukaryotes and prokaryotes in Figures 4 and 5 respectively, where each Figure shows the nine plots corresponding to the correlations between one particular biological class and the nine base frequencies. The base frequencies used in these plots are the original, unscaled frequencies. The results show a strong correlation between base usage and biological origin in all nine base frequencies and all five biological classes. There were varying degrees of separation of the data into distinct groups of centroids. The least spread of values occurs for A_1 and G_2, while the greatest occurs for A_3 and G_3. Clear separation into five distinct groups can be seen on some of the plots for G_1 (eukaryotes). T_1 (phage group) and G_3 (phage group), with less clear separation for the other base frequencies. If we examine a table of maximum and minimum values for each base frequency (Table 3), we see some interesting patterns. In the first codon position, the extremes are shared by organelles and prokaryotes, in the second, by organelles and eukaryotes (with the single exception of the viral maximum for G_2), and in the third, by organelles and prokaryotes again. The organelle

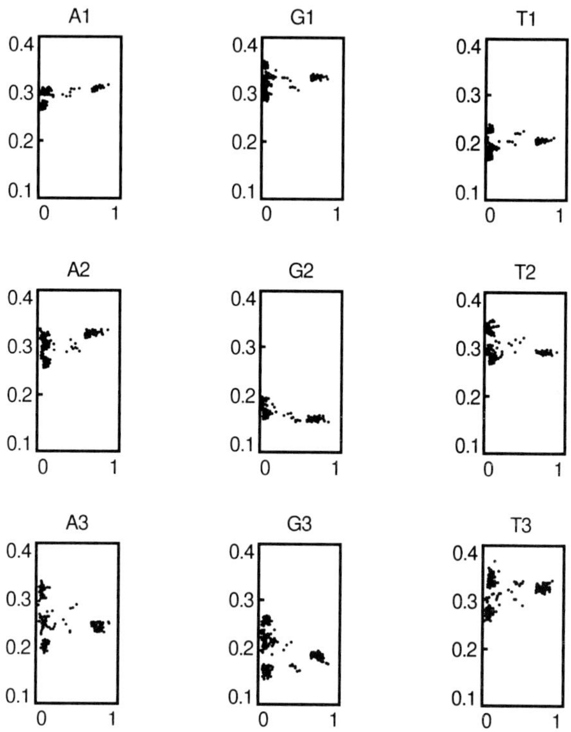

Figure 4: Average base frequencies of sequences comprising a cluster as functions of the fraction of those sequences that are from eukaryotes.

sequences consistently show either the highest or lowest base frequency for all bases in all codon positions (with one exception). Although mitochondria have their own genetic code, the codons there are not significantly different from the normal code, so it would not seem to be enough to warrant such extreme base usages.

The strongest correlation between biological class and base frequency appears in the third codon position, which is somewhat surprising since the third codon position is the least relevant to the choice of amino acid in the genetic code. Possibly this is related to the availability of various species of transfer RNA molecules (Kowald and Kirkwood 1993) or to some ancestral requirement for the stability of the DNA or RNA molecule (Rowe and Trainor 1983b; see further below).

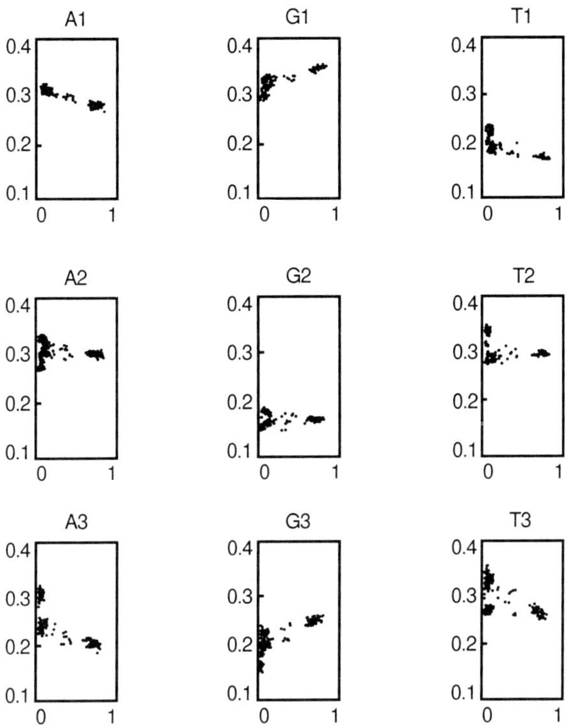

Figure 5: Average base frequencies of sequences comprising a cluster as functions of the fraction of those sequences that are from prokaryotes.

5 The Tetrahedral Representation and the Primordial Code

Thermodynamic calculations (Rowe and Trainor 1983b) suggest that under prebiotic conditions, certain repeating polymers of DNA would tend to be favored over others, in particular the polymers GATGATGAT... and GCT-GCTGCT..., on the basis that these polymers showed a smaller free energy of successive base pair/base pair interactions in DNA (or its equivalent in RNA). The sequence of steps and associated mechanisms bridging between polymer formation and the establishment of a meaningful genetic code are, of course, unknown, so that the present remarks have suggestive value only. There is even some controversy as to whether RNA or DNA played the primary role in the earlier stages of this event sequence. What seems likely, however, is that as translation and transcription mechanisms evolved, the choice of reading frame

Table 3: Base Frequency Extremes. Biological classes with extremes of base frequency for each base/codon position. For A_1, all classes except prokaryotes use it with about equal frequency.

Base	Maximum	Minimum
A_1	-	Pro
G_1	Pro	Org
T_1	Org	Pro
A_2	Euk	Org
G_2	Vir	Euk
T_2	Org	Euk
A_3	Org	· Pro
G_3	Pro	Org
T_3	Org	Pro/Vir

would have been a part of this evolution. One might even anticipate that in the early stages of the evolution of life, such possibilities as overlapping and confusion of reading frames and of transcription of DNA in both directions might well have to be considered. Indeed, some present forms of primitive organism like $\phi X174$ (Smith 1979) are observed to make use of overlapping reading frames in their DNA complement.

Our work suggests that a polymerized form of DNA with a repeating base sequence might well have been an early favored polymer (Rowe and Trainor 1983b). Combining this work with our observations on phages (Rowe and Trainor 1983a), namely, that there is a tendency for G to occur in the first codon position, A or C in the second, and T in the third, suggests that we may be observing a remnant of prebiotic or early biotic conditions corresponding to the predominance of the polymers GATGATGAT... and GCTGCTGCT.... Moreover, one might anticipate in early life forms, the appearance of the phenomena both of overlapping reading frames and of reverse strand transcription. The possible relevance of these to our work will appear shortly.

Starting with the strand GATGATGAT... and its DNA complement CTACTACTA..., the above considerations would suggest exploration of six different three-letter codons, viz. GAT, ATG, TGA, CTA, TAC and ACT, depending upon choice of strand and choice of reading frame. According to the modern genetic code (Table 1) these six codons correspond to a STOP and a START signal and four amino acids covering three of the five functional groups identified by Dayhoff (Dayhoff, Eck and Park 1972). Combined with a similar analysis on the other favored strand GCTGCTGCT..., this yields 9 of the 20 permissible amino acids plus one STOP signal. Additional codons such as TTG would readily arise from strand breaking and reforming in a primordial soup containing GAT and GCT repeating polymers. The possibility of an important role for repeating polymers in early evolution of the genetic code suggests that deep within codon space lies a novel, fascinating and suggestive representation of the codons and their associated amino acids in the present (almost universal) code (Trainor, Rowe and Szabo 1984). We call this structure the *tetrahedral representation*.

In order to obtain the tetrahedral representation we first observe that the 64 codons which are possible as three-letter words using a four-letter alphabet A,C,T and G can be uniquely arranged into 20 sets, each set being composed of codons which can be obtained from one another by a permutation of the three symbols in the codon. For example, the codon AAA is the single member of one set, while the codon AAT belongs to a three-member set comprised of the codons AAT, ATA and TAA. The codon ACG, on the other hand, belongs to the set of six codons: ACG, AGC, CGA, CAG, GCA and GAC. One easily verifies that, defined in this way, there are exactly 20 sets, with a suggestive bridging from 64 codons to the 20 sets. We now show that the regular tetrahedron, which has four identical faces, each consisting of an equilateral triangle, provides an attractive and suggestive representation of the codon/amino acid correspondence.

The tetrahedron is labeled as follows. Each apex in turn, but in any order, is labeled by the single entry sets AAA, CCC, GGG and TTT. Consider now any one of the four triangular faces, say that labeled by apices AAA, TTT and CCC, as shown in Figure 6. As one moves down from AAA to TTT along the left hand edge, one replaces an A by a T at the one-third point to obtain the set {AAT,ATA,TAA}. At the two-thirds point along the AAA/TTT line, one replaces a second A by a T and labels the point with the set {ATT,TAT, TTA}. A similar process labels the points along the AAA/CCC line, and along the TTT/CCC line. All sets containing triplets composed of the letters A, C and T are now identified as points on the triangular face, except the set of six codons {ACT,ATC,CTA,CAT,TCA and TAC}. This set is identified with the centroid of the equilateral triangle ACT. Note that the diagram has three principal directions oriented at 60° to one another and that various points on the diagram can be reached by taking the appropriate number of steps in each of the three directions.

A similar analysis can be applied to the other three sides of the tetrahedron. One can visualize the whole process beginning with the intact tetrahedron as follows. Imagine a set of four parallel planes cutting through the tetrahedron, with the first plane coincident with one face of the tetrahedron, the second

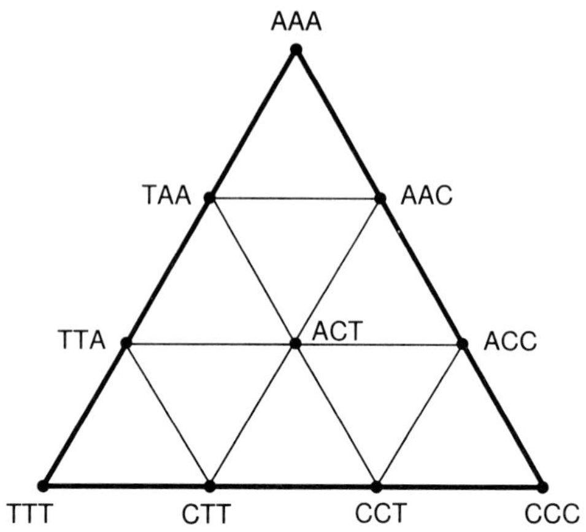

Figure 6: Exposed ACT face of the codon tetrahedron. The codon labeling interstices designates the set of codons obtained by permutations of the one labeled. Slightly modified from L.E.H. Trainor, G.W. Rowe and V.L. Szabo (1984) *Journal of Theoretical Biology* **108**, 459–468, Academic Press, with permission

cutting at a "height" 1/3 above it, the third at a height 2/3 above it and the fourth at a height of 3/3 just passing through the apex opposite to the chosen face. We visualize this process repeated for all four faces. Figure 7 shows how to construct a tetrahedron from a flat surface consisting of four contiguous equilateral triangles by appropriate folding (along the heavy lines AAA/CCC, CCC/GGG and GGG/TTT). The labeling of points in Figure 7 is such that *sets* of codons are implied, e.g. TAA implies the set of TAA and its permutations ATA and AAT. As can be seen, there are exactly 20 unique points on the tetrahedron corresponding to the 20 sets of permutable codons defined above.

The tetrahedral representation takes on an added interest with the following observation and construction. Consider any particular one of the four faces of the labeled tetrahedron, such as that shown in Figure 6. We note from the diagram that for points like AAA corresponding to a single codon and coding for a single amino acid, there is but one 60° wedge with apex at that point, whereas for a point like TAA which represents a set of three codons coding for three distinct amino acids, there are three 60° wedges with apex at that point. Moreover, for the point ACT at the centroid, corresponding to the set of six codons specifying six distinct amino acids, there are six 60° wedges with apex at that point. Thus we can enter in each distinct wedge in the diagram an amino acid corresponding to one of the codons represented at each point.

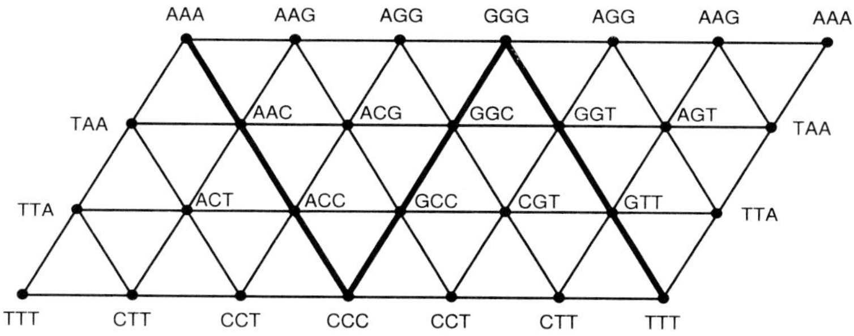

Figure 7: Constructing the codon tetrahedron by folding from an appropriate flat surface. Folding lines are marked in bold. Slightly modified from L.E.H. Trainor, G.W. Rowe and V.L. Szabo (1984) *Journal of Theoretical Biology* **108**, 459–468, Academic Press, with permission.

The result is shown in Figure 8. Moreover, we observe that amino acids can be associated with wedges in such a way that a given amino acid forms an island of contiguous wedges on the surface of the tetrahedron, as is shown in Figure 9 for the flattened-out tetrahedron. For any permutation set of codons, the amino acids specified thereby are all distinct. For example, serine, which is six-fold degenerate, is specified by six codons, one from each of six different codon sets.

The tetrahedral representation organizes and displays the 64 codons and their associated 20 amino acids (plus START and STOP signals) in a manner suggestive of the importance of the permutation grouping of codons, which itself relates back to the possible importance of repeating polymers, thermodynamically selected, in the early stages of the development of the genetic code and, over evolutionary time, to the ordering of our planet's life forms within codon space.

References

Anderberg, M.R. (1973) *Cluster Analysis for Applications.* New York: Academic Press.

Dayhoff, M.O., Eck, R.V. and Park, C.M. (1972) *Atlas of Protein Sequences and Structure.* Volume 5. Georgetown University Medical Center, Washington, D.C.: National Biochemical Research Foundation, p. 89.

Gonnet, G.H., Cohen, M.A. and Brenner, S.A. (1992) Exhaustive

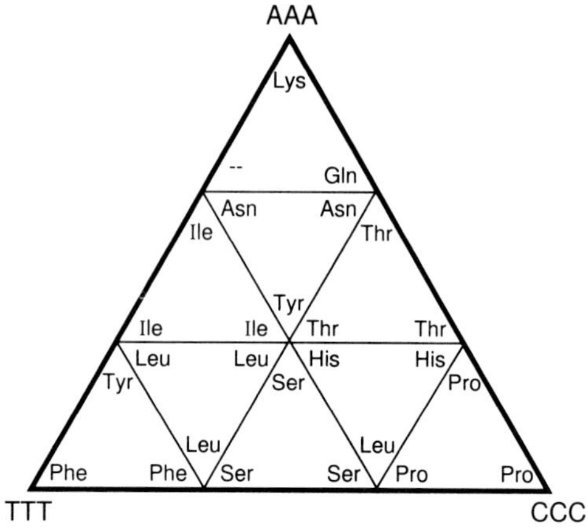

Figure 8: Filling in the ACT face of the tetrahedron with associated amino acids. Slightly modified from L.E.H. Trainor, G.W. Rowe and V.L. Szabo (1984) *Journal of Theoretical Biology* **108**, 459–468, Academic Press, with permission.

matching of the entire protein database. *Science* **256**, 1443–1445.

Guigo, R., Knudsen, S., Drake, N. and Smith, T. (1992) Prediction of gene structure. *Journal of Molecular Biology* **226**, 141–157.

Kowald, A. and Kirwood, T.B.L. (1993) Accuracy of RNA charging and codon:anticodon recognition: Relative importance for cellular stability. *Journal of Theoretical Biology* **160**, 493–508.

Kruskal, J.B. (1964a) Multidimensional scaling by optimizing goodness of fit to a nonmetric hypothesis. *Psychometrika* **29**, 1–27.

Kruskal, J.B. (1964b) Nonmetric multidimensional scaling: A numerical method. *Psychometrika* **29**, 115–129.

Lewin, B. (1980) *Gene Expression 2. Eucaryotic Chromosomes.* Second Edition. New York: Wiley-Interscience.

Oliver, J.L., Bernaola-Galvan, P., Guerrero-Garcia, J. and Roman-Roldan, R. (1993) Entropic profiles of DNA sequences through chaos-game derived images *Journal of Theoretical Biology* **160**, 457–470.

Rowe, G.W. (1981) *Information Content, Thermodynamics and Codon Bias in Viral DNAs.* PhD Thesis, Department of Physics, University of Toronto, Canada.

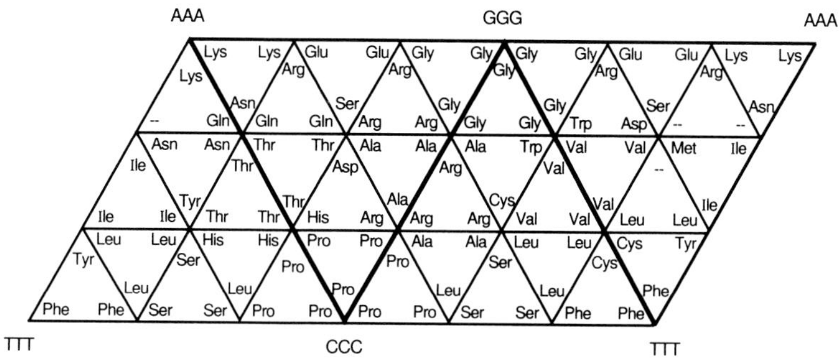

Figure 9: Islands of amino acids formed by filling in all four faces of the tetrahedron. Slightly modified from L.E.H. Trainor, G.W. Rowe and V.L. Szabo (1984) *Journal of Theoretical Biology* **108**, 459–468, Academic Press, with permission.

Rowe, G.W. (1985) A three-dimensional representation for base composition of protein-coding DNA sequences. *Journal of Theoretical Biology* **112**, 433–444.

Rowe, G.W., Szabo, V.L. and Trainor, L.E.H. (1984) Cluster analysis of genes in codon space. *Journal of Molecular Evolution* **20**, 167–174.

Rowe, G.W. and Trainor, L.E.H. (1983a) On the informational content of DNA. *Journal of Theoretical Biology* **101**, 151–170.

Rowe, G.W. and Trainor, L.E.H. (1983b) A thermodynamic theory of codon bias in viral genes. *Journal of Theoretical Biology* **101**, 171–203.

Rumelhart, D.E. and Zipser, D. (1986) Feature discovery by competitive learning. In D.E. Rumelhart and J.R. McClelland, eds., *Parallel Distributed Processing: Explorations in the Microstructure of Cognition*, Volume 1: *Foundations.* Cambridge, MA: MIT Press, pp. 151–193.

Smith, M. (1979) The first complete nucleotide sequence of an organism's DNA. *American Scientist* **67**, 57–67.

Stryer, L. (1975) *Biochemistry.* First Edition. New York: W.H. Freeman. Table 25-4, p. 636.

Trainor, L.E.H., Rowe, G.W. and Szabo, V.L. (1984) A tetrahedral representation of poly-codon sequences and a possible origin of codon degeneracy. *Journal of Theoretical Biology* **108**, 459–468.

Appendix. Learning Codon Space by Inter-Neuron Competition

When a competitive learning network is first set up, it usually contains randomly assigned weights between the various layers. Obviously, it would be highly unlikely for such a network to give a meaningful classification of a set of input patterns. The utility of the network comes from its ability to *learn* to classify a set of patterns after repeated exposure to them. Learning is achieved by varying the weights w_{ij} between the layers. The algorithm, as reported in Rumelhart and Zipser (1986), is as follows;

Each cell j has a weight w_{ij} on each incoming line from cell i in the previous layer. These weights are normalized so that their sum is 1:

$$\sum_i w_{ij} = 1 \tag{A.1}$$

When a pattern is presented to the input layer, a certain set of cells in that layer will fire, with the others remaining dormant. The signals from the input layer may be binary, meaning that each input neuron either fires at full strength or does not fire at all. Alternatively, the input neurons may fire with a certain strength, which can be any non-negative real number.

Signals will be sent from those cells that fire to the cells in the second layer. The competition described above is carried out to determine which cell in each second-layer cluster wins and hence fires. If cell j in the second layer wins the competition, all weights into that cell are changed according to the formula

$$\Delta w_{ij} = g\left(\frac{c_i}{A} - w_{ij}\right) \tag{A.2}$$

where c_i is equal to the strength with which cell i in the input layer fired, and A is the total activity in the input layer, so that

$$A = \sum_i c_i \ , \tag{A.3}$$

where the sum is taken over all neurons in the input layer. If we are using binary input values, this equation can be rewritten as follows:

$$\Delta w_{ij} = g\left(\frac{c_i}{n} - w_{ij}\right) \tag{A.4}$$

where c_i is 1 if cell j in the input layer fired that round and zero if it did not, and n is the total number of cells in the input layer that fired.

The parameter g is a scaling factor which determines the size of step taken by each weight. Note that equations (A.2) and (A.4) preserve the normalization condition (A.1) since, using that condition and the fact that

$$\sum_i c_i = A \ , \tag{A.5}$$

we have

$$\sum_i \Delta w_{ij} = 0 \ . \tag{A.6}$$

This procedure for altering the weights may be performed for a collection of patterns which serve as a *training set*. The patterns are presented to the net one at a time, in random order, either until none of the weights changes when a complete cycle of patterns is presented or, because there is no guarantee that the process will converge, for a fixed number of cycles. After a training session, the net has "decided" for itself a set of categories into which its input patterns must fall. The number of categories is determined by the number of distinct possible outputs in the final layer. In the case of a two-layer yes-no network, all input patterns are classified into two groups. If the final layer has C clusters, each with M neurons, there are M^C possible categories. Once a network has been trained, it can be used to classify patterns which were not part of the training set: in other words it can apply its training to novel situations.

In order to take advantage of the learning capability of the network, we can train the net on a subset of all the sequence data available and then test the trained net both on the original training set and the remainder of the data that was not used in training. Since we have five groups of sequences, we constructed the training set from equal numbers of sequences from each of these five groups. As we see from Table 3, the category with the fewest representatives is the phage group with 751 sequences so, if we wish to use the same number of sequences from each group, our training set cannot contain more than $5 \times 751 = 3755$ sequences. However, to avoid bias in the final results, we should also use an equal number of sequences from each of the five biological classes when we test the net on data that it has not previously seen. This means that we are limiting the size of the final testing set to 3755 sequences, so that the training set must be smaller than this. In practice we used 375 sequences randomly chosen from each of the five classes as a training set, for a total training set size of 1875 sequences. The testing set was twice the size, for a total training set size of 3750 sequences. We did explore the effect of using a training set with unequal numbers of sequences from each group. The network then shows a bias towards classifying most of the testing set as sequences from the most common biological group.

To implement training, a network of the required architecture was generated and the weight w_{ij} between the input and the output layers seeded with random numbers from a pseudo-random number generator in the open interval (0, 1) (i.e., no weights were allowed to be zero). The weights were then normalized to satisfy condition (A.1). The program was written in C^{++}. In order to train the net, we must specify a value for the learning rate g in equation (A.2). We found that values of g in the closed interval [0.05, 0.1] gave satisfactory performances, with $g = 0.1$ giving a slightly better performance, especially for the larger networks. Since the value of g, as long as it is kept within this range, does not affect the results very much, we will only consider results for nets trained with $g = 0.1$.

The 1,985 sequences in the training set were selected randomly from the total database of 16,432 sequences, such that 375 sequences were taken from each biological class. Once the training set was determined, the sequences in this set were presented to the input layer one at a time, in random order, until the complete set had been seen by the network. As the standard training set consists of 1875 sequences, the net has been trained quite thoroughly after only a single cycle through the set. We experimented with the amount of training by presenting the training set to the network up to 32 times, but found that no significant difference to the network's ability to classify sequences was obtained. It therefore appears that a single pass through the training set is sufficient.

After training, the net was used to classify a test set of 3,750 sequences, again chosen at random from the total database in such a way that 750 sequences came from each biological class. This means that the entire set of phage sequences is used in each test, with randomly chosen subsets of 750 sequences for each of the other classes.

Evolution of Development: The Shuffling of Ancient Modules by Ubiquitous Bureaucracies

Ellen W. Larsen

1 Introduction

The complex dynamical events transforming a fertilized egg into a multicellular organism are nothing short of miraculous. We want to understand how such miracles arise from the lawfulness we believe underlies the universe. As a biologist, I have been trained to consider two sorts of constraints which affect complicated biological processes like development. The first constraint is imposed by the physical laws governing the behavior of matter. The second constraint is the evolutionary history of the organism undergoing development. The evolutionary history of a species may be responsible for otherwise inexplicable aspects of a particular sequence of developmental changes. For example, during human embryogenesis, we make three different kidneys: first the pronephros, then a mesonephros and then a metanephros. The pronephros quickly degenerates while the mesonephros probably functions for a period before the metanephros, the permanent kidney, takes over. This sequence seems strange until we realize that the functional fish kidney is the pronephros and the functional amphibian kidney is the mesonephros. Reptiles, birds and mammals with functional metanephri carry their evolutionary history in the form of developmental sequences.

For the physicist, a preoccupation with evolution may seem beside the main point. Isn't the point to understand the physical mechanisms underlying biological structure and function? For a dyed-in-the-wool biologist, the answer would have to be a qualified "not entirely." The underlying quest driving biology is to understand both the nature and origin of biological diversity.

We want to comprehend how the myriad contemporary and extinct organisms could have evolved. For multicellular organisms, this means understanding the changes which have taken place during development in order to produce new kinds of organisms. How do we think about development and evolution? Any change of evolutionary significance must be encoded in the heredity of the organism. Understanding the evolution of development must of necessity include understanding the way the heredity of an organism influences development.

There are some like Brian Goodwin (essay, this volume), who believe that the universal laws underlying form generation will be missed if we study morphogenesis from historical (e.g., evolutionary) perspectives. Surely each approach will reveal different aspects of biological morphogenesis, which will together give a more complete understanding of this complex phenomenon.

In this essay I will present a perspective integrating development, heredity and evolution in terms of well-known ideas about hierarchies which are applicable to any complex system. The consequences of these ideas for morphogenesis will be seen to depend on the peculiarities of biological systems which differ from other physical systems in the pervasiveness of their hereditary component. Within the context of the perspective presented, instances where physical models have been developed will be pointed out. It is my hope that this discussion will trigger in readers new ideas for clarifying the interrelationship of hereditary and physical constraints on morphogenesis in biological organisms.

To begin, I suggest that two features of multicellular development[a] provide a framework for understanding developmental genetic strategies: (1) development results in hierarchically arranged structures (cells, tissues, organs, organ systems, individuals), which (2) require gene-based bureaucracies for their construction and maintenance. A key difference between the hierarchical nature of organisms and hierarchies in the institutional world (e.g., clergy in the Roman Catholic church) is that biological hierarchies can multiply using a hereditary mechanism while in institutions recruitment is used. After discussing developmentally relevant features of hierarchical organization, the term "bureaucracies" will be defined and it will be shown how these bureaucracies are an essential link between the hereditary components and the developing structures of an organism.

2 Modules As the Structural and Functional Basis of Biological Hierarchies

An essential aspect of hierarchical structures is their modular organization (Simon 1973). Modules can be defined as any level of biological organization from polymer building blocks (nucleotides for DNA, amino acids for proteins and monosaccharides for polysaccharides) to cellular building blocks of tissues and the diverse tissue types comprising organs. Indeed, some organisms are best viewed as comprising morphologically equivalent modules. Among multicellular animals, "iterated modules" are found in the hydrozoa, bryozoa and colonial ascidians while in higher plants (Harper and Bell 1979) the leaf and its axillary bud provide a unit of construction.

[a]I use the term development interchangeably with morphogenesis. Another aspect of development, namely cell differentiation, will not be addressed here.

In addition to such structural modules, what can be called "process" (or functional) modules are also found in a variety of biological contexts. For example, in mRNA synthesis the cohort of transcription factors, DNA binding sites, and RNA polymerases which are required can be viewed as comprising a functional module to transcribe a variety of genes, just as cells are used to make a variety of tissues. Process modules may also be defined in terms of developmental stages. For example, each stage of development from embryo to larva (where appropriate), to adult, may be considered as a different life history module.

The enormous number of easily discerned modules at every level begs the question: Why are "modules" so common in biological organization? The answer may lie with the important issue of stability. The general principle states that the difficulty of assembling or evolving a structure with unstable intermediates is overcome if instead stable subassemblies are joined together (Simon 1973, and the essays by Brandts and by Lumsden in this volume). H. A. Simon put forth these stimulating ideas over 20 years ago illustrating his points with the parable of two watchmakers, each assembling a 10,000-part watch and subject to an average frequency of telephone calls, corresponding to one call for every 150 pieces assembled, at which time the partially completed watch falls apart. The watchmaker who puts together 100-piece units which are stable and will not dissociate during telephone calls will complete the watch in about 11 calls, whereas the watchmaker who doesn't work with subassemblies will likely never complete the watch. Thus, linking stable subassemblies, or modules, speeds up the production of structures for which intermediate forms would be unstable. Simon also pointed out that the stable subassembly or modular method has interesting implications for the time required to evolve structures with large numbers of elements. Because the time for evolution is proportional to $\log_s k$, where k is the number of elements in the hierarchy and s is the average number of components at a given level in the hierarchy which forms a part of the next higher level, the time to evolve from a structure with 10 elements to one with 1,000 is the same as to evolve from one with 1,024 to one with 1,026 elements. Thus, according to Simon, modular construction overcomes two potential difficulties in development and evolution. It can speed up the development of structures with unstable intermediate forms and it may allow the rapid evolution of structures with very large numbers of elements.

Another feature of modules in a hierarchy is that individual modules are often only loosely coupled to other modules in the system through inputs and outputs. For example, so long as mitochondria produce requisite amounts of ATP, they are free to evolve without affecting cell survival. In this way, mitochondria and the rest of the cell are loosely coupled. More generally, it was suggested by Needham more than half a century ago that growth, organogenesis and metabolism may be partly uncoupled from each other during development (Needham 1933). This loose coupling or dissociability provides latitude for evolution in several ways. It allows for genetic variability, some of which may prove to be preadaptive (that is, a genetic variant is established before changing circumstances provide a new function for it). Dissociability provides a potential mechanism whereby genetic deletions or substitutions can occur without lethality. To give a dramatic example, the loose coupling of the larval stage (module) of development to the embryonic and adult stages has allowed the evolution of direct developers (omitting the larval stage) in lineages of several groups of animals (del Pino 1989, Jeffery and Swala 1992, Raff 1992) by suppressing the larval module.

It seems that modular biological organization is virtually a necessity for development and the evolution of development. The variety of combinations and permutations achievable with modular organization is essential for the flexibility of form and function required for organisms to have adapted to over a billion years of environmental change. The problem now is to understand the organization of a hereditary system which can meet the challenge of guiding modular development while allowing for evolutionary flexibility.

3 Biological Bureaucracies Coordinate Modular Development

Many mutations giving rise to dramatic morphological effects turn out to be changes in genes which directly or indirectly activate or repress other genes. Gene products (usually proteins) whose function is to regulate other genes are referred to here as "biological bureaucrats." The majority of "biological bureaucrats" studied to date fall into two functional groups, those involved in activating or repressing genes more or less directly as part of transcription activation and message processing, and those acting indirectly through signal transduction of external signals (e.g. light in plants) or internal signals (e.g. hormones or growth factors). What distinguishes bureaucrats from "worker" molecules (enzymes for energy metabolism, cytoskeleton, membrane structural elements) is that bureaucrats do not themselves influence cell behavior. They function to coordinate the activation of genes required for cell division, growth, shape change, movement, death and matrix secretion (Figure 1). Biological bureaucrats are part of the infrastructure for stimulating the workers to modify cells in the right place at the right time. Some molecules may function as both workers and bureaucrats. For example, a transmembrane molecule with a receptor at the cell surface may function in membrane integrity as well as in signal transduction. Just as we can separate the teaching (worker) functions

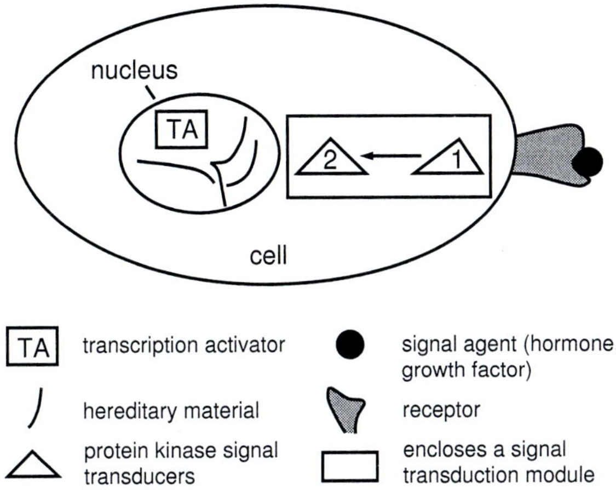

Figure 1: Signal transduction elements 1 and 2 form a module in that they operate in more than one cell type and carry signals between different receptors and transcription factors. The module is conserved in evolution.

of a department head from bureaucratic roles, there should be little ambiguity in categorizing the functional significance of a biological molecule.

As previously mentioned, among the well-studied bureaucrats are the transcription factors. These factors bind to DNA, usually as part of a transcription complex, and inhibit or activate transcription. Currently three groups of eukaryotic DNA binding domains are well characterized structurally: helix-loop-helix, zinc finger and leucine-zipper domains (Harrison 1991). The most famous of these for developmental biologists is the homeodomain, a helix-loop-helix motif with ancestral forms perhaps dating back to prokaryotic ancestors (Harrison 1991). Signal transduction factors include membrane receptors, second messengers, phosphatases and protein kinases (Hepler and Gilman 1992). As with transcription factors, there are families of structurally similar transducers. For example, members of the G protein family transduce signals between receptors and second-messenger systems (e.g. adenylate cyclase or phosphatidylinositol).

Molecular bureaucrats and the bureaucracies of which they are a part also appear to be modular in construction and function. A transgene (genetically engineered DNA which is then integrated into the genome), when truncated, is frequently able to rescue some but not all of the function of the complete gene. It was recently shown that *fushi tarazu*, known as a DNA binding protein with a homeodomain, modified *Drosophila* segmentation in unexpected ways when its homeodomain was removed. Thus in addition to its DNA binding module, *fushi tarazu* must have at least one other module, probably involved in protein-protein interactions (Fitzpatrick, Percival-Smith, Ingles and Krause 1992).

As mentioned above, molecular modules such as DNA binding regions are often evolutionarily conserved. Entire signal transduction pathways, (e.g., the phosphatidylinositol second-messenger pathway), have been conserved in a wide range of organisms and in all types of cells (Berridge 1993). This suggests that just a few types of transcription factors and a few signal transduction pathways have been used to coordinate the functions of virtually all nucleated cells. Presumably the reason evolution has been able to be so parsimonious at this level is that the same signal pathway or transcription factor can be used in many different ways in different cells and at different times, providing an enormous potential for flexibility.

We can see this flexibility by examining "bureaucratic pathways" in development. Bureaucratic pathways are found where bureaucrats activate one another in a prescribed sequence. In the segmentation of embryonic fruit flies, cascades of activation of transcription factors are required. (Here many bureaucrats activate or inactivate the transcription of other bureaucrats.) One molecule, shown from mutational studies to be important for defining parasegment boundaries, is the protein *engrailed* (itself a transcription factor). The bureaucrats involved in turning on the *engrailed* gene differ in the so-called even and odd parasegments. In the even parasegments *fushi tarazu* turns on *engrailed*, while in the odd parasegments, paired is the positive regulator (Manoukian and Krause 1993). Thus, in the evolution of development, not only may the same pathway be used to do many different things, but the same function can be accomplished in different ways.

4 Bureaucracies Influence Morphogenesis by Influencing Cell Behavior

According to the view of development advanced here, bureaucratic genes function to coordinate life processes. What are they coordinating in the context of morphogenesis? The author has previously suggested that morphogenesis results from basically six cell behaviors (cell division, growth, death, shape change, movement, matrix secretion) (Larsen and McLaughlin 1987). Bureaucratic genes, then, would be involved in the coordination of these six cell behaviors (Figure 2). Some evidence for this supposition is beginning to emerge. *Ultrabithorax* and *Antennapedia* genes encode homeodomain-containing transcription factors in fruit flies. Their mutations transform appendages from one

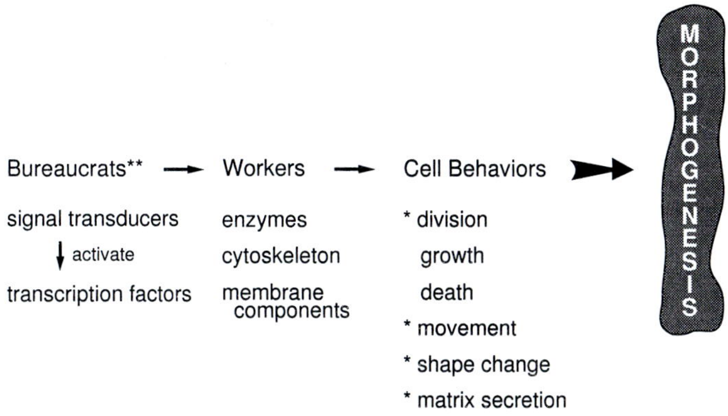

Figure 2: An eclectic view of factors contributing to morphogenesis. See text.

type to another (homeotic mutations). Among the direct downstream target genes of *Ultrabithorax*, is one putative "worker gene," *connectin* (Gould and White 1992). For *Antennapedia*, there is evidence correlating ectopic expression of the gene and changes in cell arrangements (Larsen and Zorn 1989), possibly driven by changes in cell division patterns (Mathi and Larsen 1988).

5 Mutant Phenotypes Suggest Developmental Genetic Strategies

Ultrabithorax and *Antennapedia* mutations transform one type of fruit fly appendage precursor into another type. Although each of these has structurally homologous genes in mammals (McGinnis and Krumlauf 1992), it is striking that no mutation turns a human hand into a foot. Why should there be such a difference in mutant expression? An explanation rests on the observation that the cell/tissue basis of appendage morphogenesis is very different in fruit flies and vertebrates (Larsen 1992). In flies, appendages are fashioned out of sacs one-cell thick called imaginal discs. Genes like *Antennapedia* can act autonomously in these discs, so that if the gene is expressed in only a few cells of the disc, those few cells will respond to the gene product. If *Antennapedia* is expressed in a region of the antenna disc, only that region will transform to leg-like structures. In contrast, the formation of vertebrate appendages depends on the interaction of cells in two distinct tissues, an ectodermal sheet and a clump of migrating mesenchymal cells. Experimental manipulations have demonstrated that two-way interactions between these tissues are necessary to form a normal limb. Thus, if a single mutation is to transform a hand into a foot, it would need to produce complementary effects in each tissue.

Such mutations apparently do not occur. Thus, fly and vertebrate appendage development differ in the degree of autonomy of the tissues involved in their formation and therefore in the degree of morphological change that results from single gene changes. It may be that, like flies and other contemporary arthropods, the earliest multicellular animals developed essentially from sheets in which even a few gene changes could have large effects. This could explain, in part, the diversity of animal body plans (Brusca and Brusca 1990) which evolved in the 10–20 million years of the so-called "Cambrian explosion." The cessation of major body plan evolution after this period may be related to the evolution of strategies employing interacting tissue which would require a greater amount of genetic change to produce new body plans.

6 The Evolution of Development — A Bureaucratic Shuffle?

For development to have evolved, a mechanism had to be found for incorporating the information coded in DNA into the development while ensuring the maintenance of the organism. The task for this temporal-spatial coordination problem fell to a vast bureaucracy. A biological bureaucracy can coordinate different types of processes, just as a management team in business can run an auto factory or a steel plant. For example, in fly development, the *engrailed* and *wingless* pathway is crucial in making both embryonic parasegments and adult appendages (Cohen and di Nardo 1993). Furthermore, bureaucratic structure in biology has the flexibility to change its function in evolution so as to make the same structure in different ways (e.g., in the parasegment boundaries discussed above). From this point of view, developmental genetics is now largely a study of developmental bureaucracies. Indeed, a perusal of the current literature leaves the distinct impression that most genes with notable morphological effects are bureaucrats. To test this impression a list was found of spontaneous fruit-fly mutations (Morgan, Bridges and Sturtevant 1988). These "pre-1925" mutations were identified because of their visibly abnormal structure, thus avoiding potential selection biases found in the current practice of inducing mutations and screening for those of interest. In the list were 14 morphological mutants which have recently been cloned or otherwise identified as to function. Of these, 11 are putative bureaucrats and only three are workers. In terms of understanding the genetic organization of complex systems it will be necessary to discover if indeed there are more bureaucrats than workers or if other explanations for the apparent inequality are required. Curiously, the two common explanations for the paucity of "workers" are somewhat contradictory; either workers are deemed to be redundant so that single mutations often do not affect the phenotype, or "worker mutations" are lethal and therefore not picked up phenotypically. These alternative explanations reveal how much we still have to learn about the organization of the genome before we can understand the evolution of "bureaucrat-worker" strategies. For example, one will want to know in what ways the "bureaucracies" are organized to control highly lineage-dependent morphogenesis in the nematode *Caenorhabditis elegans*, and how this organization might differ from those used to control epithelial-sheet morphogenesis in fruit flies and tissue-interaction morphogenesis in vertebrates.

Have we gained a better understanding of the issues in development by reducing all the richness of morphogenesis into a framework in which a few ancient cell behaviors are directed by a few ancient bureaucracies? I believe this point of view provides a general perspective for systematizing the exponentially growing information on molecular aspects of development. It lays to rest any lingering ideas that the genome may have organ morphology "leg genes" or "leaf genes;" bureaucrats are general-purpose tools, just as ATP is a general-purpose fuel. In a more specifically evolutionary context, this perspective can direct us to potentially valuable lines of inquiry. For example, a still-contentious issue is the phylogenetic conservation of segmentation. Molecular biology has given us the tools to determine which parts, if any, of the bureaucratic modules have been conserved (Slack, Holland and Graham 1993). Such studies will shed light both on questions of phylogeny (what happened) as well as on the nature of the evolution of process (how it happened).

This perspective underlines the desirability of looking beyond the evolution of structure (e.g. segments) to the evolution of process (evolution of the organization of bureaucracies). Can we, for example, learn how developmental organization allows a larval stage to be deleted from a life history? It is likely that larval bureaucracies are still largely intact in direct developers but key linkages have been repressed. Investigation of these cases may give us insight into the nature of "loose coupling" of modules. Understanding the conditions under which loose coupling can be achieved is a first step in more interventionist programs to devise ways, using transgenic organisms, to create predetermined morphologies (designer organisms). Such efforts would provide a test of our understanding of developmental/genetic strategies.

Hierarchical organization is a pervasive feature of complex structures. It provides a perspective for integrating the properties of gene, cell, morphogenesis and evolution; yet it figures little in physical models in development. It remains to be seen if the ideas presented here can be usefully incorporated in a broader view which considers both hereditary and physical constraints on morphogenesis.

References

Berridge, M.J. (1993) Inositol triphosphate and calcium signalling. *Nature* **361**, 315–325.

Brusca, R.C. and Brusca, G.J. (1990) *Invertebrates.* New York: Sinauer.

Cohen, S.M. and di Nardo, S. (1993) *wingless* from embryo to adult. *Trends in Genetics* **9**, 189–192.

del Pino, E.M. (1989) Modifications of oogenesis and development in marsupial frogs. *Development* **107**, 164–187.

Fitzpatrick, V.D., Percival-Smith, A., Ingles, S.C. and Krause, H.M. (1992) Homeodomain-independent activity of the *fushi tarazu* polypeptide in *Drosophila* embryos. *Nature* **356**, 610–612.

Gould, A.P. and White, R.A. (1992) *Connectin*, a target of homeotic gene control in *Drosophila. Development* **116**, 163–174.

Harper, J.L. and Bell, A.D. (1979) The population dynamics of growth form in organisms with modular construction. In R.M. Anderson, B.P. Turner and C.P Taylor, eds., *Population Dynamics.* Oxford: Blackwell, pp. 29–52.

Harrison, S.C. (1991) A structural taxonomy of DNA-binding domains. *Nature* **343**, 715–719.

Hepler, J.R. and Gilman, A.G. (1992) G Proteins. *Trends in the Biological Sciences* **17**, 383–387.

Jeffery, W.R. and Swala, B.J. (1992) Evolution of alternative modes of development in ascidians. *BioEssays* **14**, 219–225.

Larsen, E. (1992) Tissue strategies as developmental constraints: Implications for animal evolution. *Trends in Ecology and Evolution* **7**, 414–417.

Larsen, E. and McLaughlin, H.M. (1987) The morphogenetic alphabet: Lessons for simple-minded genes. *BioEssays* **7**, 130–132.

Larsen, E. and Zorn, A. (1989) Cell patterns associated with normal and mutant morphogenesis in silver-impregnated imaginal discs of *Drosophila. Transactions of the American Microscopical Society* **108**, 51–57.

Manoukian, A. and Krause, H. (1993) Control of segmental asymmetry in *Drosophila*. *Development* **118**, 785–796.

Mathi, S. and Larsen, E. (1988) Aspects of cell division in *Drosophila* imaginal discs. *Tissue and Cell* **20**, 461–472.

McGinnis, W. and Krumlauf, R. (1992) Homeobox genes and axial patterning. *Cell* **68**, 283–302.

Morgan, T.H., Bridges, C.B. and Sturtevant, A.H. (1988) *The Genetics of Drosophila*. New York: Garland.

Needham, J. (1933) On the dissociability of the fundamental processes in ontogenesis. *Biological Reviews (Cambridge)* **8**, 180–223.

Raff, R.A. (1992) Direct-developing sea urchins and the evolutionary reorganization of early development. *BioEssays* **14**, 211–217.

Simon, H.A. (1973) The organization of complex systems. In H.H. Pattee, ed., *Hierarchy Theory*. New York: Braziller, pp. 3–27.

Slack, J.M.W., Holland, P.W.H. and Graham, C.F. (1993) The zootype and the phylotypic stage. *Nature* **301**, 440–492.

Game Theory in Biology

G.W. Rowe

1 Introduction

Darwin's theory of evolution by natural selection is now almost universally accepted by the scientific community as the means by which life on Earth has attained its current myriad of forms, yet the theory does not have as firm a mathematical foundation as do most theories in the physical sciences. This is due mainly to the nature of biology: Living things are inherently very complicated compared to even the most intricate inanimate system, so it is difficult to formulate precise quantitative theories about them.

This chapter describes one of the theoretical approaches to providing a more quantitative theory of evolution. On the face of it, game theory (which is, as its name implies, a theory of how to play games) may seem to have little to do with biology, except on the human level. However, as we shall see, it can provide a useful framework in which to analyze the behavior of certain animals.

To understand how game theory may be applied to biology, it is first necessary to understand a few of the fundamentals of the theory of evolution by natural selection. Two main forces are at work in evolution: *mutation* and *selection*. To see what is meant by these terms, let us consider some hypothetical species of asexual organism called a *hyporg* (for *hyp*othetical *org*anism).[a] In an ideal world, asexual organisms reproduce by forming exact copies of themselves. Since their world is ideal, the hyporgs live a life of luxury (except for the lack of sex).

However, life is seldom ideal. If we move our population of hyporgs into a different environment by, say changing the weather conditions or introducing a predator, the hyporg population will likely go extinct if its members cannot change their way of life. We shall assume that both the physical form and the behavior patterns of hyporgs are determined genetically, and that they do not possess enough intelligence to cope with changes in their environment by any feats of engineering.

By assuming that the essence of a hyporg is stored in its genes, we are saying that in reproduction, hyporg offspring are given an exact copy of the DNA sequence of the parent. The only way, therefore, for the hyporg population to

[a]In science, as in life, sex serves to complicate things.

change is by altering its DNA sequence from one generation to the next. One way this can be done is by a mutation in the DNA.

There are several kinds of mutation, some of which produce major (and usually fatal) changes in the resulting organism. Others have little or no effect. Regardless of the biochemical details, however, the point is that a mutation in a DNA sequence has the potential to produce a change in the organism, either in physical form or behavior. If mutations are introduced into our hyporg population, the offspring will no longer be exact copies of their parents. Rather, each parent will give birth to several different offspring that will be similar to, but subtly different from it. Some of these offspring may be better able to cope with the changes in the environment. This is the essence of mutation in the theory of evolution; it is the generator of diversity in the population.

The second factor in evolution is natural selection. In one sense, natural selection is the inverse to mutation: Rather than increasing diversity, it removes those varieties of organism which cannot cope with the environment in which they find themselves.

Each of the mutated hyporgs that appear in a generation must face the environment and attempt to survive. Because they are genetically different, they will have different physical abilities and behaviors. Some of them may be able to cope better with colder temperature (because of a thicker fur coat, for example). Others may be able to run faster and therefore escape from predators more easily. Still others may possess a behavioral trait which causes predators to avoid or overlook them. Some, of course, will find themselves worse off than their parents and siblings, and will not survive.

All these factors play a role in natural selection. Those organisms that are best adapted to the environment will stand the best chance of survival, and those that survive to a reproductive age will be able to pass along their genes, possibly with further improvements through mutation, to the next generation.

Although natural selection may seem to be a destructive force because it kills off those organisms that are unfit, it is perhaps more useful from an evolutionary viewpoint to think of it as a creative force, since in the long run it improves the overall fitness of a population.

We have pointed out that both the physical form and the behavioral aspects of an organism in our hypothetical population are genetically determined, and that both can contribute to the organism's fitness, or ability to survive. It is with the behavioral aspect of fitness, and its investigation by means of game theory, that we will be concerned in this chapter. Game theory was founded in 1944 with the publication of a book by von Neumann and Morgenstern called *Theory of Games and Economic Behavior*. The original intention of game theory, as can be deduced from the title of the 1944 book, was as a means of analyzing the behavior of individuals and businesses in the world of commerce. However, any game which has sets of rules and possible strategies to pursue during play can, in theory, be analyzed as well.

The goal of game theory is to find the best strategy or combination of strategies that a player should use at each stage of the game. Anyone who watches sports on television will have seen commentators practising an amateur form of game theory when they analyze the play of some sporting contest. The sort of game theory which we will consider here, however, requires some mathematical analysis to find the best strategy. Since the understanding of physical processes rarely if ever uses models based on the notion of a strategic contest, game theory is a subject new to most physical scientists — albeit ultimately a fascinating one. Readers meeting game theory here for the first time will find the essentials set out in the Appendix. While brief, the summary is self-contained and provides the background needed for the material I cover here, and for wider reading on evolutionary games.

2 Dynamics and Evolution

Game theory, as it was initially formulated, will predict the optimal strategy or strategies for a particular game. If the players of the game are intelligent and know the optimal solution, they will use it from the very first time they play the game. However, the theory of evolution assumes that its two main driving forces, mutation and natural selection, are blind and purposeless, lacking direction and intelligence. As such, they cannot recognize an optimal solution to any problem; they must arrive at such a solution slowly, step by step. The vital missing element from the application of game theory to evolution is therefore this dynamic element. Given a set of strategies and their corresponding payoffs, we can use game theory to determine the optimal strategies that should be employed by both contestants, but in the absence of any intelligent driving force, how is the population to arrive at these solutions?

The answer to this question relies on the biological interpretation of the payoffs as evolutionary fitnesses, or measures of success in passing along one's genes to the next generation. Before we can give the equations for implementing this interpretation as a dynamics occurring over evolutionary time, we must clarify one point concerning the concept of mixed behavioral strategies.

You will recall that, for the basic Hawk-Dove game (Appendix), when the fitness value V of the contested resource and the fitness cost $-C$ of an injury from fighting one such that $V < C$, the optimal solution, or ESS, is a mixed strategy in which the hawk strategy should be used a fraction V/C of the time. In our discussion of mixed strategies in the Appendix, we stated that a mixed strategy is implemented in a game by a player choosing randomly which strategy to use in any particular instance of the game, with the probability of choosing each strategy given by the vector of probabilities defining the mixed strategy. We can use the same interpretation in the biological case by assuming that an animal's genetics are such that the frequencies with which various pure strategies contributing to the mixed strategy will be used are predetermined. In that way, one animal can truly be a mixed strategist.

An alternative interpretation is that, in a large population, all animals are pure strategists but that the fraction of animals using each of the pure strategies is equal to the corresponding probability of that pure strategy being used in the optimal mixed strategy. For example, if we have a population of hyporgs playing the Hawk-Dove game in a scenario where $V < C$, a fraction V/C of the hyporgs will be pure hawks, and the remaining $1 - V/C$ are pure doves. Since one of the requirements of an optimal mixed strategy is that all pure strategies will get the same average payoff when confronting a mixed strategist, and any one animal will confront an opponent playing any one strategy in exactly the frequency dictated by the mixed strategy, the payoffs to all animals in the population, on average, should be equal. It would seem, therefore, that the two ways of interpreting a mixed strategy are the same.

In a static population this would be true. However, we are considering organisms in which the behavioral strategy, pure or mixed, used by an animal is determined by their genetics, and that the frequencies with which strategies are used in the population must change slowly, by means of mutation and natural selection. If we assume that all animals are pure strategists, and therefore that a mixed strategy can arise only as a consequence of different animals in a population using different pure strategies, then we are assuming that any mutation can only change one pure strategy into another. In other words, in the gene (or genes) controlling the behavior of the animal, there are only as many alleles as there are pure strategies.

On the other hand, if we assume that a single animal can play a mixed strategy, there are in principle an infinite number of possible genotypes, one for each mixed strategy. For example, in the Hawk-Dove game, one animal may be playing the strategy $[p_1, p_2]$, where p_1 and p_2 are the frequencies with which the two strategies are played, and a mutation may cause its offspring to play $[p_1 + \epsilon, p_2 - \epsilon]$, where ϵ is some very small quantity.

With these cautions in mind, let us now consider the dynamics of biological games. In such modeling, it is usually assumed that the proportion of the population playing a particular pure strategy in the offspring of the current generation is proportional to the frequency and fitness of that strategy in the parent generation. If, as before, p_i is the fraction of the time that strategy

i is used in the parent generation and p'_i is the corresponding fraction in the offspring, then we assume that (Rowe, Harvey and Hubbard 1985)

$$p'_i \propto p_i W_i \tag{1}$$

where W_i is the fitness of strategy i, given in general by

$$W_i = W_0 + \sum_{j=1}^{n} E_{ij} p_j \tag{2}$$

in a game where there are n strategies.

Since we must have $\sum_i p'_i = 1$, we can replace the proportion by an equation

$$p'_i = \frac{p_i W_i}{\sum_j p_j W_j} \tag{3}$$

where, here and in all that follows, a summation without an upper limit implies summation over all pure strategies available. Using the definition of W_j we have

$$p'_i = \frac{p_i\left(\sum_k E_{ik} p_k + W_0\right)}{\sum_{k,m} E_{km} p_k p_m + W_0} \tag{4}$$

If we specify some initial conditions, equation (4) is a recurrence relation which allows us to determine the values of p_i for all future generations. We can use this relation to plot trajectories of strategy usage in the population, or we can attempt to find the equilibrium, or fixed, points, i.e. those points where the frequencies do not change from one generation to the next.

We can find the fixed points by requiring that $p'_i = p_i$ for all strategies i. Doing this in equation (4) gives the condition

$$p_i = \frac{p_i\left(\sum_k E_{ik} p_k + W_0\right)}{\sum_{k,m} E_{km} p_k p_m + W_0} \tag{5}$$

If all values of $p_i \neq 0$, we can cancel p_i and get the condition

$$\sum_{k,m} E_{ik} p_k p_m = \sum_k E_{ik} p_k \tag{6}$$

After some matrix algebra, we can solve this relation for p_i:

$$p_i = \frac{\sum_k (E^{-1})_{ik}}{\sum_{k,m} (E^{-1})_{km}} \tag{7}$$

where $\mathbf{E}^{-1}$ is the inverse matrix of $\mathbf{E}$.

If one or more of the p_i is zero at the equilibrium, we can proceed as follows. Suppose we have a four-pure-strategy game in which we seek an equilibrium with $p_2 = p_4 = 0$. Take the original payoff matrix and cross out those columns and rows corresponding to the absent strategies. We get the reduced 2×2 matrix $\mathbf{E}'$:

$$\mathbf{E}' = \begin{bmatrix} E_{11} & E_{13} \\ E_{31} & E_{33} \end{bmatrix} \tag{8}$$

We may now apply equation (7) to this reduced system to obtain values for p_1 and p_3.

A complete analysis of an n-strategy game requires determining all possible equilibria in which none, one or more of the pure strategies are absent. The number of subgames in which r of the original n strategies are present is given by the binomial coefficient $\binom{n}{r}$.

B.1 Stability of Equilibria

As with a fixed point in any dynamical system, an equilibrium point in a game dynamics may be asymptotically stable, neutrally stable, or unstable. The standard method of checking stability by linearizing the system near the fixed point may be applied to these equilibria with the following results (Rowe et al. 1985).

We will define an equilibrium point $\mathbf{s}$ as a vector whose components are determined according to the method in the previous subsection. That is, if we set $p_i = s_i$ for all i, the resulting pure (if all $s_i = 0$ except one) or mixed (if more than one $s_i \neq 0$) strategy is an equilibrium point. For each such equilibrium point, we must construct a matrix $\mathbf{B}$, with elements B_{ij}, as follows. If $s_i = 0$ then

$$B_{ij} = \frac{\delta_{ij}\left(W_0 + \sum_k E_{ik}s_k\right)}{W_0 + \sum_{k,m} E_{km}s_k s_m} \tag{9}$$

If $s_i \neq 0$ then

$$B_{ij} = \delta_{ij}\left[\frac{s_i\left(E_{ij} - \sum_k (E_{jk} + E_{kj})s_k\right)}{W_0 + \sum_{k,m} E_{km}s_k s_m}\right] \tag{10}$$

where δ_{ij} is the Kronecker delta: $\delta_{ij} = 1$ if $i = j$ and zero otherwise. Note that $\mathbf{B}$ is *always* a full $n \times n$ matrix, even if some of the pure strategies are absent at the corresponding equilibrium.

To use $\mathbf{B}$ to test the stability of a fixed point, its eigenvalues must be calculated. Using the shorthand notation $\sum_{k,m} E_{km}s_k s_m \equiv \mathbf{sEs}$, we have the following conditions for stability: One of the eigenvalues should always be equal to $W_0/(W_0 + \mathbf{sEs})$. If this eigenvalue occurs only once and if all *other* eigenvalues have modulus less than one, the equilibrium point is dynamically stable, i.e. a slight perturbation in any combination of strategies from this point will damp out after a time, and the population will return to that point. If this eigenvalue occurs more than once, then *all* eigenvalues, including this one, must have modulus less than one for the point to be stable. Since some payoffs may be negative (e.g. the hawk-hawk payoff in the Hawk-Dove game when $V < C$) it is possible for this central eigenvalue to be greater than one; however, if the payoffs are defined such that $\mathbf{sEs} > 0$ then a simple check that all eigenvalues have modulus less than one is all that is required for the point to be stable.

B.2 Dynamic Stability and the ESS Conditions

These stability conditions were defined without any reference to Maynard Smith's ESS conditions (Maynard Smith 1982), so it is reasonable to ask how the two conditions compare with each other. It turns out that the two sets of conditions *do* agree with each other in many cases, but there are some payoff matrices with ESS points that do not satisfy our stability conditions, and others with points that *are* dynamically stable but do not satisfy the ESS conditions. We have coined the term *evolutionarily stable point* or ESP (Rowe et al. 1985) to stand for a fixed point that satisfies our stability conditions, whether or not that point is an ESS. As we will see, the ESP is in many ways a simpler and more consistent motion of evolutionary stability than that provided by ESS.

As an example of a payoff matrix that gives different results depending on whether it is analyzed by the ESS or ESP methods, consider the matrix originally studied by Zeeman (1980):

$$E = \begin{bmatrix} 0 & 6 & -4 \\ -3 & 0 & 5 \\ -1 & 3 & 0 \end{bmatrix} \tag{11}$$

This matrix may be analyzed by the methods in this section and is found to have two ESPs: [1,0,0] and [1/3,1/3,1/3]. The first of these points is also an ESS, but the second is not. Thus we have a point which is dynamically stable but does not satisfy the ESS conditions. How can this be?

Examining the dynamics of the system near the second point shows that the population spirals elliptically into the stable point, rather than approaching the point directly (as is the case with the first point). Since the spiral is elliptical rather than circular, with the fixed point at the center of the ellipse, there will be points on the trajectory where the system actually moves further from the fixed point before approaching it again. If one attempts to use the ESS conditions to analyze where the system will go at each point in a trajectory, we can see that they require that *every step taken by the population must be towards a strategy that gives higher payoffs than the previous one.* Thus the ESS conditions will miss stable points such as the one here, where the payoffs may decrease slightly before increasing again.

It is also worth noting that the parameter W_0, although not affecting the location of the equilibria, can affect their stability. For example, in the payoff matrix

$$E = \begin{bmatrix} 0 & 2 & -1 \\ -1 & 0 & 2 \\ 2 & -1 & 0 \end{bmatrix} \tag{12}$$

a single fixed points occurs at [1/3,1/3,1/3]. This point is always an ESS. However, if $W_0 = 1$, the point is unstable, i.e. is not an ESP. However, if $W_0 > 2$, the point becomes stable. Thus we have an example of a point that, under certain conditions on the basic fitness parameter W_0, satisfies the ESS conditions but is not an ESP.

B.3 The Dynamics of Mixed Strategies

The analysis up to now has implicitly assumed that each animal in the population plays only a pure strategy. This assumption has crept in through the use of the dynamics (4), since the only strategies included in that equation are the n pure strategies. As we pointed out earlier, although a distinction between populations where animals play only pure strategies and those where they are allowed mixed strategies may not seem important for static populations, it can make quite a difference when dynamics are introduced. An example may help to clarify this point.

Consider the payoff matrix (Maynard Smith, personal communication)

$$E = \begin{bmatrix} 0 & 5 & -4 \\ -7 & 0 & 8 \\ -1 & 2 & 0 \end{bmatrix} \tag{13}$$

This matrix has two ESPs: [1/3,1/3,1/3] and [1,0,0]. The first ESP does not satisfy the ESS conditions, while the second does. Maynard Smith states that the first point is stable if only pure strategists are allowed in the population, becoming unstable if mixed strategists are introduced. Thus, if mixed strategies are allowed, the only stable point is [1,0,0] and after a while, the entire population should be using only the first strategy.

Let us examine the reasoning behind this analysis. Let $I =$ [1/3,1/3,1/3] and $M =$ [1,0,0]. Further, let the notation $E(I, M)$ be defined as the payoff to a strategy I user against a strategy M opponent. If we apply the ESS conditions to these two strategies, we find that $E(M, M) > E(M, I) > E(I, M) > E(I, I)$, which should imply that strategy M could always invade strategy I, contrary to what our ESP conditions say.

However, what we have done here is to replace the original three-strategy game by a two-strategy game, with strategies I and M. Players in this game are allowed to play either of these strategies, but no others. In other words, a mutation can only cause the offspring of an I strategist to be an M strategist, or vice versa. Animals playing pure strategies 2 or 3, or any mixed strategy other than I, are not allowed.

Let us broaden our horizons a bit and allow a few more types of mixed strategy in the game. In particular, let us start with a strategy $P_1 = [0.3, 0.3, 0.4]$. We now calculate another mixed strategy P_2 by applying the recursion relation (4) to P_1. If we allow this new strategy P_2 to exist, and use the ESS conditions to compare it with P_1, we find that P_2 can invade P_1. Iterating this procedure, we generate a series of strategies $P_3, P_4, \ldots$, each of which, according to the ESS conditions, can invade the previous one. Since we are using the dynamical equations developed in this section to calculate the series of strategies, we know they will converge on the ESP [1/3,1/3,1/3], as can be verified by numerical simulation.

Now we can begin to see the problems associated with the dynamics of mixed strategies. It is clearly not correct to simply identify the fixed points and then apply either the ESS or ESP conditions to each, if we are allowing individual animals to play mixed strategies.

Some approaches to a more correct analysis of mixed strategies have been made, but nothing truly satisfactory has emerged. Hines (1980) approached the problem from an analytic standpoint but had to make some compromising assumptions in order to produce tractable equations. Rowe (1985) took an approach suggested by the discussion in this section, in which a new payoff matrix is defined where many discrete mixed strategies are allowed. Neither of these approaches deals with what is ultimately an infinite-dimensional minimization problem.

Perhaps the best solution is simply to abandon analytical attempts to solve this problem in favor of computer simulation. The infinite number of dimensions in the full mixed strategy analysis only arises because we are assuming an infinite population size. Any real population is, of course, finite, and probably of a size that an average desktop workstation could realistically simulate it. By following the life histories of all animals in the population individually, we need never worry about how to analyze continuous distributions. A model in this spirit is presented below.

B.4 Sex Games

The analyses of the previous sections have all dealt with asexual organisms because we have assumed that the genotype of the offspring will differ from that of its single parent only through the effects of mutation. In sexual reproduction, mutation is still present, of course, but a much more potent generator of diversity is recombination. In every cell (with one important exception) of a sexual organism, each gene is represented twice: one copy inherited from the mother and one from the father. The important exception is the sex cell itself, where through the process of meiosis, only one copy of each gene is retained in each cell. However, in order to produce this single copy, the sex cell does not simply take either the mother's or father's genes from a diploid cell. Rather, it gets a collection of genes from both sources, resulting in a mixture of genes from both parents. Thus, when you pass along a copy of your genes to your children, you pass along only one copy of each gene present in each of your cells, but which copy is largely a matter of chance. This mixing and matching is known as *recombination*.

Since the units of recombination are entire genes rather than individual molecular components, as with mutation, it is not surprising that its effects can be much more profound. The uncertainty as to the origin of each gene also makes analysis much more complicated than in the asexual case. Most work on game theory in a sexual population has concentrated on applying the ESS concept to sexual evolution (for references see Rowe (1987)). An application of ESP theory to sexual reproduction has been carried out by Rowe (1987, 1988) but, due to space limitations, we will not pursue the topic further here.

3 Superparasitism in Parasitic Wasps: An Application of Game Theory

As an example of how game theory can be usefully applied to a real biological situation, we will examine an experiment in which the egg-laying behavior of a species of wasp has been successfully modeled using game theory (Hubbard, Marris, Reynolds and Rowe 1987).

The wasp *Nemeritis canescens* reproduces by laying its eggs in the larvae of the flour moth *Ephestia cautella*. Because the newly hatched wasps feed off the live larva, this wasp is known as a parasitic wasp. This species is particularly appropriate to use as the basis of a game theory model as its mode of reproduction is essentially asexual. Although these wasps are in fact diploid organisms, so that they do possess two sets of all their genes, and meiosis does occur, resulting in halving of the chromosome number in the sex cells, only a single sex cell is required to produce an offspring, as the chromosomes are duplicated without the contribution of a second sex cell. This means that, while the daughter wasp may not be genetically identical to its mother (if the two sets of chromosomes in the mother are not identical), the degree of relatedness between mother and daughter will be much higher than in sexually reproducing organisms. This high degree of relatedness will have consequences for our model, as we will see.

A single moth larva provides enough food to nourish one developing wasp larva, so it is to the benefit of a wasp mother to avoid laying an egg in a larva which already contains one. However, whenever two or more eggs are laid in a single host larva, an event known as *superparasitism*, since the unfortunate moth has been parasitized more than once, the first wasp to hatch will attack and destroy any other wasp eggs in the host. Should two eggs hatch at the same time, the two larvae will fight to the death. In practice, once a wasp larva is injured (the wasp larvae have quite effective biting mouthparts), the immune system of the host moth larva will enter the injured wasp larva and kill it, so the first wasp larva to successfully bite the other will win the fight.

A series of experiments was done to examine the behavior of several adult wasps laying eggs on a patch of moth larvae. It was discovered, not surprisingly, that a wasp prefers hosts that have no eggs laid in them. If there are no such hosts available, it was found that wasps will superparasitize a host, but that they tend to avoid hosts in which they *themselves* have already laid an egg. In other words, a wasp seems to have some way of detecting (1) whether or not an egg has been laid in a host and (2) if a host contains an egg, whether or not that egg is its own. It appears that a wasp secretes a substance into the host along with the egg which serves as a personal chemical marker, and which is detectable by other wasps.[b]

We may formulate a game theory model of egg-laying behavior by parasitic wasps as follows. We propose the three strategies:

1. Lay only in healthy hosts (i.e. ones that have no eggs in them).

2. Lay either in healthy hosts or any host that contains a single egg, whether or not the egg is your own.

3. Lay either in healthy hosts or in hosts that contain a single egg, provided

[b]Although it may seem unlikely that each wasp would have its own unique chemical marker, such diversity is not unknown. For example, in the human immune system there are several molecules forming the *major histocompatibility complex*, or MHC, that seem to be unique to each person. This feature allows the immune system to distinguish between cells that belong to its own body and foreign invaders.

that egg is not one of your own. If no such host is available, do not lay an egg.

We will not allow wasps to lay eggs in hosts that already contain more than one egg. This behavior occurs very rarely in nature anyway.

We now need to define a payoff matrix for this game. We define

- V = value of the resource (the moth larva). To normalize the other parameters we will take $V = 1$.
- W = cost of producing an egg. Since wasps can produce only a limited number of eggs, an egg is a valuable resource and should not be wasted. Therefore, whenever an egg is laid, it must be considered a cost. Since a wasp produces about 100 eggs in its life, we will take $W = 0.01$.
- C = cost of fighting between wasp larvae. The cost to the loser of the fight is, of course, absolute, since it dies. However, it would be expected that a winner of a fight would suffer some adverse effects compared to a larva that did not have to fight at all. This effect does not seem to be too serious, however, so we set $C = 0.1$.
- p = probability of the second-laid egg winning the fight. It is unlikely that the second egg will win, since the first laid egg will usually hatch sooner and kill the second larva. Occasionally, the second egg does win, though, so we take $p = 0.1$.

Since we will analyze this game by means of computer simulation, rather than by formal analysis of the payoff matrix, we will derive the average payoff to each strategy. We first classify the payoff to an egg which finds itself in one of three states:

1. A single egg. Since a single egg does not have to fight, the payoff is $V - W$, the value of the resource minus the cost of producing the egg.
2. First-laid egg with competition. Here a fight will result, with the probability of the first-laid egg winning being $1 - p$. The payoff is therefore $(1 - p)(V - C) - W$.
3. Second-laid egg with competition. Here the egg has probability p of winning; otherwise it is identical to the first-laid egg. The payoff is therefore $p(V - C) - W$.

We define N_{ij} to be the number of eggs, at the end of a generation of egg-laying, laid by a wasp using strategy i that fall into category j. We can now calculate the payoff (per egg laid) to each strategy in the game. For strategy 1, we have (we use lower case e for payoff here, as a further modification must be made to calculate the final payoff):

$$e_1 = \frac{N_{11}(V-W) + N_{12}[(1-p)(V-C)-W]}{N_{11}+N_{12}} \tag{14}$$

and for the other two strategies (i = 2,3):

$$e_i = \frac{N_{i1}(V-W) + N_{i2}[(1-p)(V-C)-W] + N_{i3}[p(V-C)-W]}{N_{i1}+N_{i2}+N_{i3}} \tag{15}$$

To calculate the final form of the payoffs, we take account of a phenomenon called *inclusive fitness*. Inclusive fitness was introduced into evolutionary theory by William Hamilton as an explanation for altruistic behavior in such species as ants and bees, where apparently sterile workers will serve the colony, sometimes with their lives, even though they have no possibility of ever having offspring and therefore would seem, at first sight, to have nothing to gain, genetically speaking, from serving other, fertile members of the colony. Hamilton therefore defined inclusive fitness to capture the notion of the total fitness accruing to an individual as a result of its own reproductive behavior and the reproductive proclivities of others, whom that individual might assist even at cost to itself, bearing copies of *its* genes (for additional background see e.g. Wilson (1975)).

The inclusive fitness concept can be applied to the case of individuals that are closely related, even if they are not genetically identical. For example, if two wasps share 75% of their genes, then anything helping the first wasp to have offspring will benefit three-quarters of the genes possessed by the second wasp. If the first wasp gains fitness x, therefore, the second wasp is deemed to gain fitness $0.75x$ merely because it shares 75% of its genes with the first. Therefore, in colonies or groups where the individuals are closely related, every individual gains some "vicarious" fitness from the successes of other members of the group.

Because of their method of reproduction, it is inevitable that parasitic wasps will be highly related. To simplify things for our model, we shall assume that all wasps share a fraction r (called the *coefficient of relatedness*) of their genes. For the wasps used in this experiment, it is likely that $r = 0.75$, approximately. The final fitness of each strategy will be, therefore:

$$E_i = e_i + r(e_j + e_k) \tag{16}$$

where i, j and k are all different.

Finally, we impose the same dynamics on our system as that used in the ESP model in the previous section. If there are p_i wasps of strategy i present in the current population, and p'_i in the next generation, then

$$p'_i = \frac{p_i E_i}{\sum_{j=1}^{3} p_j E_j} \; . \tag{17}$$

This model was simulated on a computer to see which of the three strategies would prove to be dominant, or if a mixed strategy would appear. It was found that after 10 to 20 generations, strategy 3 (that of avoiding your own eggs when superparasitizing) became the only strategy used. It would appear therefore that superparasitism, but avoiding of your own eggs, is evolutionarily stable. The rationale for this appears to be that, even though the probability of the second egg winning is low, this is offset by the fact that if the first egg wins, you are still getting three-quarters of the fitness gain through the inclusive fitness term. However, since 25% of the genes are still unique to you, you would prefer that your own egg won, if possible. It is therefore worth investing an egg (whose cost is relatively low) to see if you can win a fight with a first laid egg and thus gain the extra fitness increment of 0.25. However, there is no point in superparasitizing your own egg, since one of them must die and it is a waste of resources.

The model was examined for other parameter ranges just to see what it would predict. If the coefficient of relatedness r is reduced to 0.1, we find that strategy 1 becomes dominant. In this case, since very little fitness will be gained from someone else's egg winning a fight with yours, it is better to avoid fights altogether. Raising the probability of the second egg winning to 0.9 (other parameters at their original values) does not change the strategy — it is still strategy 3. Increasing the cost of an egg to 0.5 causes a shift to strategy 1. Now eggs cost so much you do not want to risk one on the slim chance of a second egg winning a fight.

Although wasps in these other parameter ranges are not known to exist, the predictions of the model in these areas seem realistic.

4 Summary and Conclusions

A great deal of mathematical analysis of the ESS concept has taken place over the years since it was introduced. Despite the undoubted value of evolutionary game theory as a means of understanding animal behavior, I believe that much of the ESS work is misguided, since the ESS is not the most natural measure of evolutionary stability. The ESP is based on the concept of dynamic stability, and also lends itself well to computer simulation of realistic biological models.

References

Hines, W.G.S. (1980) Strategy stability in complex populations. *Journal of Applied Probability* **17**, 600–610.

Hubbard, S.F., Marris, G., Reynolds, A. and Rowe, G.W. (1987) Adaptive patterns in the avoidance of superparasitism by solitary parasitic wasps. *Journal of Animal Ecology* **56**, 387– 401.

Jones, A.J. (1980) *Game Theory: Mathematical Models of Conflict.* Chichester, U.K.: Ellis Horwood.

Maynard Smith, J. (1982) *Evolution and the Theory of Games.* New York: Cambridge University Press.

Maynard Smith, J. and Price, G.R. (1973) The logic of animal conflict. *Nature* **246**, 15–18.

Rowe, G.W. (1985). Mutations, mixed strategies and game theory. *Journal of Theoretical Biology* **117**, 291–302.

Rowe, G.W. (1987) A dynamic game theory model of a diploid genetic system. *Journal of Theoretical Biology* **129**, 243–255.

Rowe, G.W. (1988) To each genotype a separate strategy — A dynamic game theory model of a general diploid system. *Journal of Theoretical Biology* **134**, 89–101.

Rowe, G.W., Harvey, I.F. and Hubbard, S.F. (1985) The essential properties of evolutionary stability. *Journal of Theoretical Biology* **115**, 269–285.

von Neumann, J. and Morgenstern, O. (1944) *Theory of Games and Economic Behavior.* Princeton, NJ: Princeton University Press.

Wilson, E.O. (1975) *Sociobiology: The New Synthesis.* Cambridge, MA: Belknap Press of Harvard University Press.

Zeeman, E.C. (1980) Population dynamics from game theory. In Z. Nitecki and C. Robinson, eds., *Global Theory of Dynamical Systems.* Proceedings, Northwestern University, 1979. New York: Springer-Verlag, pp. 471–497.

Appendix. Evolutionary Game Theory: A Physicist's Primer

A.1 Payoffs and Strategies

It is easiest to see how game theory works if we look at a simple example. Let us consider the game Red-Black (Jones 1980, p. 20).

Two players, which we shall call A and B, sit opposite each other at a table. Player A chooses one card from a standard deck of playing cards, looks at it and notes whether the card is red or black, but does not tell player B. If the card is red, A can either decide the winner of the game by tossing a coin, or can pass the move to player B. If the card is black, A must pass the move to B. If B receives the move, he must guess whether A's card is red or black.

We could analyze this game by stating under what conditions either A or B wins, but to apply game theory properly, we need to specify *how much* each player wins or loses, i.e. we need to put numerical values on each of the possible outcomes. These values are called the *payoffs* to each of the players.

To visualize the possibilities in a game, it is helpful to draw a *game tree*, which illustrates the possible paths that a typical game might take (see Figure 1). In the game tree we see that each node is labeled with one of the letters A, B or C. The letters A or B indicate that the corresponding player must make a choice at that point to decide which branch to follow. The letter C indicates that the outcome at that point is decided by a chance event (or at least an event outside either player's control). Thus the initial node (the selection of the playing card) is a chance event, since player A selects a card at random from the deck. The actual outcome, together with the probability of that outcome occurring, are used to label the branch emanating from the node. Thus, since there are equal numbers of red and black cards in the deck, the probabilities of both red and black are equal to 0.5.

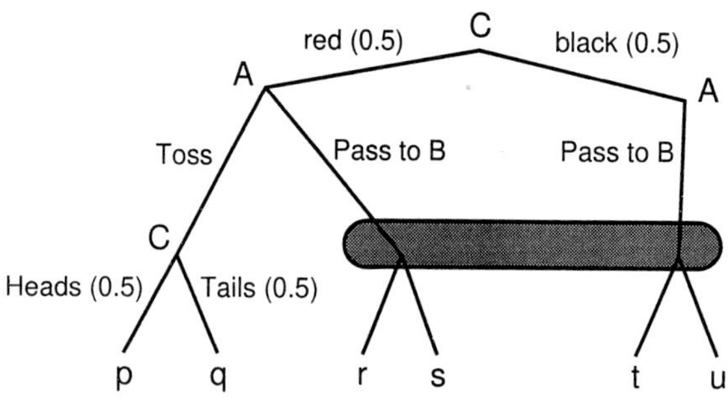

Figure 1: Game tree for the Red-Black game.

After A has made the initial selection of a card, he has control of the next stage in the game. If the card is red, he is allowed to choose whether to end the game by tossing a coin (another chance event) or to offer a turn to player B, while if the card is black, he must offer a turn to B. If B gets a turn, he now has control of the game and has two choices: guess red or guess black. However, B will not know exactly where in the tree he is when his turn comes, since he does not know the color of A's card. Since B could be at one of two nodes, we have enclosed both nodes in a shaded box in Figure 1 to emphasize B's uncertainty.

When either a coin is tossed or B has made his choice, the game is over. We see from the game tree that there are six possible outcomes for the game. We assign each of them a payoff which, for the moment, we will simply indicate by one of the letters from p to u, as shown. The convention in game theory is that the payoff is always the amount paid to player A. If A actually pays B something, the payoff to A will be negative. By following the rules of the game, it can be seen that only one of these six outcomes is possible in any one play of the game.

Now that we have seen what is possible in the game, both players would now like to know what their best course of action is at any stage in the game. For a game such as this, both players will have to make at most one decision, and possibly none (A's course of action is forced if he draws a black card, and if A draws a red card and decides to toss a coin, B has no say in the game). However, both players can decide before playing the game what they think is the optimal strategy in those cases where they must make a decision. A complete analysis of the possible strategies in this game gives us the following:

Strategies for A

A1 If red card, toss coin.

A2 If red card, pass to B.

Strategies for B

B1 If get a turn, guess red.

B2 If get a turn, guess black.

We can now work out the average payoff to each player for all possible combinations of strategies. This results in a *payoff matrix*, E. To calculate each of the elements in E we must trace the outcome of the game in the case where the corresponding strategies are used by each player, including any chance events as we go. For example, if player A chooses strategy A1, then player B will never get a turn if A draws a red card, so the payoff to A is independent of what B has decided to do in this case. The actual payoff to A for strategy A1 in any single play of the game depends on the outcome of the coin toss, but we are interested in the average payoff. Assuming a fair coin so that heads and tails are equally likely, the average payoff will be $(p+q)/2$.

However, we have overlooked the case where A draws a black card. Since A has no choice in this case, we have not included this case in the list of A's strategies, but of course we must include it in our analysis of the game. If A draws a black card and B is playing strategy B1, then the payoff is t, while if B is playing strategy B2, the payoff is u.

Let us assume that A is playing strategy A1 and B is playing B1. On average, A will draw a black card half the time and a red card the other half. If A draws a red card, we know the average payoff is $(p+q)/2$, and this payoff will be obtained half the time, so the average contribution from this section of the game tree will be $(p+q)/4$. Since B is playing strategy B1, the outcome of the game if A draws a black card will always be t, and since this occurs half the time, the average contribution of this part of the tree is $t/2$. Thus the average payoff in the case where A is playing A1 and B is playing B1 is $(p+q)/4+t/2$.

The other three entries in the matrix can be worked out in a similar way, yielding the overall payoff matrix:

	B1	B2
A1	$(p+q)/4+t/2$	$(p+q)/4+u/2$
A2	$(r+t)/2$	$(s+u)/2$

To make things definite, suppose that A pays B £1 if either a tail comes up on the coin or if B guesses correctly, and B pays A £1 otherwise. Then $p=s=t=1$ and $q=r=u=-1$. In this case, the payoff matrix becomes

	B1	B2
A1	1/2	-1/2
A2	0	0

Now we can plainly see what each player can expect if two strategies are employed against each other. If A plays A1 and B plays B1, then on average, A will win 50p on each round, while if A plays A1 and B plays B2, B will win 50p. If A plays A2, then no matter what B plays, both players will break even in the long run.

So what is the best strategy for each player? It may seem that A's best bet is to play A1, since that is the only strategy for which he can make a net gain. However, once B recognizes that A has decided to use A1, he will switch to using B2, since that will make him an overall winner. The best strategy would therefore seem to be for A to play A2, in which case no matter what B plays, they both break even.

There is a more systematic way to analyze payoff matrices. From A's point of view, he would like the highest return against whatever strategy B uses. B, of course, is interested in doing the exact opposite: Whatever strategy A uses, B will try to minimize the gains to A. A way to find the best solution for both parties is for each player to look at the *worst* he can do for each of his strategies, and then pick the best of the worst. Thus for each row of the payoff matrix, A will find the minimum element, and for each column, B will find the maximum. If we locate these elements for the matrix given above, we have:

	B1	B2	Row min
A1	1/2	−1/2	−1/2
A2	0	0	0
Col max	1/2	0	

Now, from A's point of view, the best of the worst outcome in each row is when he plays A2. From B's point of view the best of the worst is when he plays B2. The element of the matrix where the A2 row and the B2 column intersect happens to be the element which gives the best of the worst for both players, so this is the optimal combination of strategies. You might wonder why, if A is playing A2, B must play B2 and not B1, since the net payoffs are the same for both. If B played B1, however, A could improve his winnings by switching to A1, whereupon B would switch to B2 and A would counter by switching back to A2 again, thus returning the situation to the A2-B2 choice it started from. The point is that, if the players are playing A2-B2, neither player can improve his lot by switching strategies, so the agreement is stable.

A.2 Mixed Strategies

The optimal strategies for players A and B that we found for the special case of the red-black game are an example of a solution in *pure strategies*, meaning that each player uses one, and only one, of his strategies in every play of the game. A solution in pure strategies is only possible if the same element in the payoff matrix results in the best of the worst cases for both players, as it did in the previous example. Most payoff matrices, as you might expect, do not have such a simple solution. In such cases, it is possible to show, as von Neumann did as far back as 1928, that an optimal solution to the game can always be found if one or more *mixed strategies* are used.

A mixed strategy is one where the player varies the strategy used from one play of the game to the next. The strategy to be used in any given play is chosen at random, but the probability that a particular strategy, say strategy i, is chosen is given by a fixed probability a_i for player A or b_i for player B. For a game with n_A strategies for A and n_B for B, we can therefore characterize a mixed strategy by a set of probabilities for each player, where the probabilities satisfy the usual conditions: $\sum_{i=1}^{n_A} a_i = \sum_{j=1}^{n_B} b_j = 1$ and $0 \leq a_i, b_j \leq 1$ for $i = 1, \ldots, n_A$, $j = 1, \ldots, n_B$. A more convenient notation often used for mixed strategies is to represent the set of probabilities for each player as a vector. Thus we could write the probabilities for players A and B as $\mathbf{a} = [a_1, a_2, \ldots, a_{n_A}]$ and $\mathbf{b} = [b_1, b_2, \ldots, b_{n_B}]$.

To see how an optimal mixed strategy can be defined, suppose that player A has decided on some particular mixed strategy $\mathbf{a}_0$ (which need not be an optimum). Then player B would like to choose his mixed strategy $\mathbf{b}$ to minimize the gains to A. Let the element of the payoff matrix $\mathbf{E}$ which corresponds to player A using strategy i and player B using strategy j be E_{ij}. Then the average payoff to player A in the mixed strategy game considered here is

$$P(\mathbf{a}_0, \mathbf{b}) = \sum_{i=1}^{n_A} \sum_{j=1}^{n_B} (a_0)_i E_{ij} b_j \tag{A.1}$$

This follows from the fact that the average payoff to A if A is playing his pure strategy i and B is using mixed strategy $\mathbf{b}$ is $\sum_{j=1}^{n_B} E_{ij} b_j$. However, since A is playing a mixed strategy, the overall payoff is just the average of the payoffs to each pure strategy weighted by the probabilities with which each of A's strategies occurs.

Given that A's mixed strategy is fixed at $\mathbf{a}_0$, B will choose his mixed strategy $\mathbf{b}_0$ so that A's payoff is minimized, that is, so that

$$P(\mathbf{a}_0, \mathbf{b}_0) = \min P(\mathbf{a}_0, \mathbf{b}) \tag{A.2}$$

Player A will naturally try to choose a mixed strategy $\mathbf{a}_0$ so that this quantity is as large as possible. Thus the optimal mixed strategy for A is determined by the condition that

$$P(\mathbf{a}_0, \mathbf{b}) = \max_a \min_b P(\mathbf{a}, \mathbf{b}) \tag{A.3}$$

That is, a mixed strategy for A is chosen, whereupon B chooses his mixed strategy that minimizes the payoff to A. Then A chooses another mixed strategy, B minimizes the payoff to that new strategy and so on. When all of A's strategies have been tried, A will choose the one that gives the highest payoff subject to B's attempts to minimize the return.

Using a similar argument, B may choose a mixed strategy first and allow A to adjust his mixed strategy to maximize the returns. Then B will choose another strategy and A will counter by choosing another of his strategies that maximizes the returns and so on. This leads to the optimal strategy for B as the strategy $\mathbf{b}_0$ which satisfies the condition

$$P(\mathbf{a}, \mathbf{b}_0) = \text{min}_b \text{max}_a P(\mathbf{a}, \mathbf{b}) \tag{A.4}$$

Notice that the only difference between equations (A3) and (A4) is the order in which the min and max are taken. It may seem reasonable that the two operations give the same result and, in fact, a theorem known as the *minimax theorem* proves that this in indeed the case.

The fact that these two payoffs are equal means that an optimal solution for both players always exists and, therefore, that it is possible to "solve" any game in terms of mixed strategies even if there is no solution in pure strategies. The common value of these two payoffs is known technically as the *value* of the game. It is the expected average gain of the first player and loss of the second.

A game which has no pure strategy solution is the *Rock-Scissors-Paper* game. For those readers whose childhood did not include this game, we will include the rules here. Two players face each other and at the count of three, each person presents his right hand in one of three ways: (1) as a closed fist, representing a rock; (2) with the index and middle fingers extended, representing a pair of scissors; (3) with the palm open, representing a sheet of paper. If both players produce the same symbol, the game is a draw. If the players produce different symbols, however, the winner is determined by the following rules:

- Rock wins over scissors, since scissors cannot cut a rock.
- Scissors wins over paper, since scissors cut paper.
- Paper wins over rock, since paper can wrap up the rock.

If we represent the payoff for a win as +1, a loss as −1 and a draw as 0, the payoff matrix is[c]

	R	S	P
R	0	1	−1
S	−1	0	1
P	1	−1	0

[c]One of Gary Larson's classic cartoons in the *Far Side* series shows two cavemen playing this game in the days before paper or scissors were invented, with the result that both players always produce a rock. This reduced version of the game is somewhat easier to analyze!

By analyzing this matrix as in the Red-Black game above, we can see that all row minima are -1 and all column maxima are $+1$, so there is no solution in terms of pure strategies. The minimax theorem assures us that there is a solution in terms of mixed strategies however, although it does not tell us how to find it. In this game, however, the symmetry of the rules should tell us that all three strategies are equally likely to win, lose or draw, so a reasonable mixed strategy would appear to be one where the player uses all three strategies equally often. In vector form we may write this as $\mathbf{a} = \mathbf{b} = [1/3, 1/3, 1/3]$. The common payoff to both parties is zero, which is therefore the value of the game.

A.3 Other Types of Games

The games we have considered in this section are of a particularly simple type called *zero-sum games.* The term "zero-sum" means that the sum of the payoffs to the winner and loser is always zero: what one players loses, the other gains. As we will see in the next section, most biological games are not of this type, so the simple methods of this section will often not suffice to analyze them.

A.4 Evolutionary Games

A.4.1 The Hawk-Dove game

From what we have seen of game theory so far, it may not be obvious how it can applied to biology. The first such application was made by Maynard Smith and Price in 1973. A more comprehensive review of the applications of game theory to biology is given in Maynard Smith's 1982 book *Evolution and the Theory of Games.*

Recall from the first section of this chapter that we drew a distinction between two main properties, or *phenotypes* as they are more properly known, on which natural selection can act: *physical* properties such as body size, fur color, etc., and *behavioral* properties such as aggression, hiding under a rock to escape detection from a predator, etc. We assume that behaviors are genetically determined, rather than learned, so that we apply the theory of evolution by mutation and natural selection to them as well as to physical traits. To what extent this is true in the "real world" of live animals is still a matter of debate amongst biologists, but we shall take the easy way out and avoid this controversy. The discussion in this chapter, then, applies *if* behavioral traits are inherited.

The idea behind Maynard Smith and Price's application of game theory to evolution is that, when an animal is faced with some stressful situation, the genetics of its species may allow several possible behaviors to be used. A simple example, one which was used originally by Maynard Smith and Price (1973) and which has found its way into practically every discussion of game theory in biology ever since, is the *Hawk-Dove game*. Despite the name of the game, we are actually dealing with only a single species of animal, which need not be either a hawk or a dove (or even a bird). We will use our hypothetical organism, the hyporg, again to illustrate how the game works.

Every so often in the life of a hyporg it will need to gather food, stake out some living territory, or find a mate. If there are no other hyporgs around (except for the prospective mate in the latter case), there is no problem: The hyporg simply eats the food, marks out the territory or claims the mate. However if another hyporg appears with similar intentions to the first, some means of resolving the conflict is needed. The Hawk-Dove game provides two behaviors.

Hawk. A hyporg using the hawk behavior will immediately attack its opponent. The fight will continue until either the opponent flees or one or the other combatant is injured. In the latter case, the injured party loses and retreats.

Dove. Those hyporgs using the dove behavior will initially stand their ground and make threatening gestures towards their opponents (the hyporg equivalent of saber-rattling). However, if a dove comes up against a hawk opponent, the dove will flee at the first sign that the hawk really means to attack. If two doves confront each other, the threats will continue until one of the contestents gives up, which could, of course, take some time.

These two behavior patterns may be considered as strategies in a game. If we can define a payoff matrix for this game, we can use the results of the previous section to find the optimal strategies for each hyporg.

It may still not be clear how these ideas can be related to the theory of evolution. The key is in what we mean by a "payoff" in this context. Hyporgs (or even real animals) are obviously not interested in money. From an evolutionary point of view, what matters is the ability to pass on your genes to the next generation. Any activities during your lifetime which will enhance your chances of doing so will be favored by natural selection or, to use the technical term, will improve your *fitness*.

If a game-theoretic analysis of the Hawk-Dove game should show that one of the strategies is optimal, this means that any hyporg using that strategy will have a higher fitness than animals using the other strategy, which will, in turn, mean that that strategy allows its users to pass along more of their genes than their opponents. Natural selection should, therefore, tend to favor users of the optimal strategy and cause their numbers to increase over time.

Let us now examine how we can define payoffs in a biologically meaningful way. As usual, we consider two opponents, or "players" in the Hawk-Dove game, which we shall call players A and B. All payoffs will be to player A. Let us say that the value of the object over which the hyporgs are contesting is V. What this means is that, should a hyporg win the contest, the resource so obtained (food, territory or mate) would increase its chances of passing along its genes by an amount V. The precise meaning of this "value of the resource" term is necessarily vague in a general example like this. As a specific case illustration, suppose the hyporgs are fighting over some food which is scare at this time of year. Getting the food may mean that, due to improved health, the winner will be able to have five more offspring than the loser. In this case, $V = 5$. It is more usual, however, to simply set $V = 1$ and use this definition as a scale by which other parameters in the game are measured.

We have four possible contests:

Hawk vs. Hawk. By the rules of the game, two hawks will attack each other and continue fighting until one of them is injured. If we assume that all hyporgs are equally fit before the contest, any one hawk has an equal chance of winning or losing the contest. The winner will, of course, obtain a fitness gain of V. The loser will gain nothing, but must also sustain an injury which will impair its ability to survive and have offspring. Thus the loser will actually *lose* fitness rather than simply coming out empty-handed. If we say that an injury results in a fitness cost $-C$ ($C > 0$), then in a hawk-hawk contest, the average fitness gain to player A is $(V - C)/2$.

Hawk vs. Dove. As soon as the hawk attacks the dove, the dove will flee. Thus the hawk gains the resource at no cost, giving it a fitness gain of V.

Dove vs. Hawk. Since the dove flees immediately, it sustains no injury, but also gains nothing, so its payoff is 0.

Dove vs. Dove. Here no physical fighting occurs so no injury is sustained. If we assume that either dove is equally likely to give up, the average fitness gain is $V/2$.[d]

[d]It has been pointed out that the time taken to decide a dove-dove contest should incur a cost, especially if the season is short and the animals do not live very long. However, we will ignore this effect in our simple model.

The payoff matrix for the hawk-dove game is therefore

	Hawk	Dove
Hawk	$(V-C)/2$	V
Dove	0	$V/2$

The introduction of the cost term C means that the Hawk-Dove game is *not* a zero-sum game, as defined at the end of the last section. The fitness lost by the injured party in a hawk-hawk fight is not gained by its opponent—it is simply lost fitness. This means that we cannot use the simple methods discussed above to analyze the game and find the optimal strategies. We shall resort to a more biological interpretation of the payoffs to do this.

One feature of our earlier analysis must, however, still be true: In an optimal solution, the average payoffs to both contestants must be equal. If this were not the case, the player with the lower payoff could change its strategy in an attempt to equalize the payoffs. If the solution is in terms of pure strategies, this is easy to check since we can simply read the payoffs from the payoff matrix and check that they are equal. To allow for the possibility of a mixed-strategy solution, let us define p_1 to be the probability that the hawk strategy is used and $p_2 = 1 - p_1$ to be the corresponding probability for the dove strategy. In order for an optimum to occur, we require that the average payoff to both strategies be equal, whenever they are used. That is, if we define the fitness W_i of strategy i to be

$$W_i = W_0 + \sum_{j=1}^{2} p_j E_{ij} \tag{A.5}$$

where W_0 is the fitness of the animal before the contest, we require $W_1 = W_2$. The solution of this pair of linear equations gives us the condition

$$p_1 = \frac{E_{22} - E_{12}}{E_{11} + E_{22} - E_{12} - E_{21}} \tag{A.6}$$

which for the hawk-dove payoff matrix, gives us

$$p_1 = V/C \tag{A.7}$$

Obviously equation (A7) can only apply in the case where $0 \leq V \leq C$, since, as p_1 is a probability, it must lie in the unit interval. We therefore know that a mixed-strategy solution cannot exist if $V > C$. However, if $V < C$, then the optimal strategy is for both players to use the hawk strategy a fraction V/C of the time and the dove strategy a fraction $1 - V/C$. We can check from equation (A5) that the common payoff in this case is $V/2 - V^2/2C$.

What happens when $V > C$? Since a mixed strategy is not possible, the only solution must be one in pure strategies. Suppose there were only doves in the population. Then all hyporgs would get a payoff of $V/2$. If there were only hawks, the average payoff would be $(V - C)/2$. You might think that, in the case where $V > C$, to decide which of the two strategies is the optimal one, you would simply take that strategy that gave the larger payoff. However, that is misleading, because what you really need to consider is not only which strategy does better against itself, but also against all other strategies. Since a hawk always obtains a payoff of V against a dove, a population consisting entirely of doves would be defenceless against a hawk that arose, say, by mutation. A population of 100% hawks, on the other hand, although each individual may get a smaller payoff than a dove in a totally dove population, cannot be invaded by any number of doves. In other words, so long as the resource is worth more than the cost of injury, it is worth fighting for.

A.4.2 Evolutionarily Stable Strategies

It was the idea of defence against invasion that Maynard Smith and Price used as the foundation of their application of game theory to evolution. If it could be shown that a particular strategy (pure or mixed) did better against itself than any other strategy, then that strategy was to be considered optimal. More precisely, Maynard Smith and Price (1973) defined an *evolutionarily stable strategy* or *ESS* as a strategy i which satisfied the conditions:

- Either $E_{ii} > E_{ji}$ for all $j \neq i$ (i.e. strategy i does better against itself than any other strategy), or

- If $E_{ii} = E_{ji}$, then $E_{ij} > E_{jj}$ (if there is a strategy j which does as well against i as does i itself, then i does better against j than j does against itself).

We have already seen that the hawk strategy is an ESS when $V > C$, since it satisfies the first ESS condition. For the mixed strategy in the case where $V < C$, we can see that it does better against both hawks and doves than they do against themselves. Since $p_1 = V/C$, we have that the payoff to a mixed strategist against pure strategy j is $\sum_{i=1}^{2} p_i E_{ij}$. Inserting the quantities required, we find that the payoff to a mixed strategist against a pure hawk is $(V/C)(V - C)/2 > E_{11} = (V - C)/2$, since $V < C$. Against a pure dove, the payoff is $V^2/2C + V/2 > E_{22} = V/2$. Thus we know that the mixed strategy gives better payoffs against both hawks and doves than these pure strategies do against themselves. What we have *not* shown, however, is that the mixed strategy we are using does better than any other mixed strategy. This turns out to be a somewhat thornier problem.

Reflections

The Editors

The twentieth century has been a restless one for our inquisitive species, with social and political turmoil interlocking vast revolutions in technology and science. From the universalist confidence of modern times through to the trendy malaise of postmodernity, the spiral of hope has seemed to invert and press downward, the doomsday clock on the *Bulletin of the Atomic Scientists* edging ever closer to, ever back from, the stroke of midnight.[a] For now at least, humankind may have backed off that final attractor: At the millennium's close, one can sense renewed vigor and movement in the human spirit.

So, yes, our civilization has shown it can find out a lot about nature conceived as a great machine, but how much can we know about ourselves? Ballyhooed secrets of living matter notwithstanding, the ancient paradoxes of mind and spirit remain untouched. Can that be changed? Should it? In a highly unstable world of collapsing regimes and fragmentary returns to old traditions and the false comforts of ancient tribalism, the whole played out amidst ecological disasters and unresolved human conflict, where lies science?

We admit to the belief that unity, of a kind that springs from the scientist's pursuit of truth, can bend history toward futures worth living. In the great human adventure of discovery, creation, critique and interpretation, we would include the prospect of an ontic unity of a reality partly beyond and partly from the mind; an epistemic unity of knowledge shared by people seeking to understand and cooperate; even (could it be?) a deontic unity betokening the deep universals of all moral systems. What lies ahead for the disciplines of mind and spirit, any coming great synthesis that may join the human sciences and the great religions to the natural sciences, is a story to be written by all our children. That is not the stuff of history yet.

But it is now, in the vital present, that the new history of biology is being created and written down. The essays collected in this volume explore the realm where physics meets biology, not as a coincidental collision of scientific inquiries, but as a truly joint venture where questions and approaches kindled in physics illuminate biological problems and biology stimulates new physics. The barriers between physics and biology are struggled against because we

[a]As we write these words the clock stands at fourteen minutes to midnight. See http://neog.com/atomic/clock.html

believe that critical knowledge can be gained when the boundaries between subjects are overcome. This has been a lesson of scientific history: Mechanics on earth extends to planetary motion. Fluid dynamics can model traffic flow. Pattern formation in chemical solutions applies also to populations of animals. When the laws or models or metaphors appropriate to one regime or system are found to apply in a larger realm, new understanding and unity are attained.

In the twentieth century, and especially the last decades, boundaries between fragmented bodies of knowledge have been shifting. Our philosophy, our way of looking at the world, dividing it, and classifying its behaviors, is changing: Science is moving away from a prototype simple worldview characterized by stability and determinism to a view that emphasizes the ubiquity of complexity: unpredictable, nonlinear, bifurcating, chaotic, nondeterministic, undecidable, stochastic, nonequilibrium, self-organizing, emergent — these are the tropes of a new, post-industrial epistemology. And this complexity is recognized as such not just in the classic "complex" systems of biology and sociology, but also in the fundamental behavior of "simple" dynamical systems.

The worldview of simplicity that dominated for so long, though based on only a few "special cases" favored in classical physical theory, represented a dangerous prejudice. This prejudice, working as it did within the traditionally exclusive divisions of areas of study and concern — physics *versus* biology, economics *versus* the environment, etc. — limited not only our way of thinking, but also acting in the global environment (which respects no boundaries). Worldviews that portray nature as an infinite force, everlasting, ever-renewing, unperturbable, now seem radically insufficient. We are only now beginning to recognize how the complexities of our human activities can wrench the biosphere and challenge, rather than foster, the conditions for life.

What basic problems does this search for biological understanding face, to which theoretical physics might make substantial contributions? Happily, the best ones are still open. Some personal favorites:

- To what extent are living systems genuinely complex, having no descriptions more parsimonious than themselves, and to what extent have they been restricted to exploiting the advantages of apparent complexity, as realized in chaotic and fractal processes?

- Where and how is information stored in the developmental process? In tissue transplant and limb regeneration for example, how does the system "know" which parts are missing, which to regenerate or duplicate?

- What is the role of coherence in biological contexts? How does it relate, if at all, to the so-called Frölich condensation? How is the overall unity of an organism achieved from the activities and interactions of its many parts?

- What are the physical bases of biological signal transduction and the integration of signal transduction patterns in development, adaptation and behavior?

- Nature no longer appears, as it once did, infinitely resilient. It is often fragile, and surprisingly difficult to predict or control or heal. What is required to establish and maintain species diversity in ecosystems? How can perturbations be forecast and managed?

- How did life begin, and the genetic code organize itself in the form we know today?

- Does the cell use non-computational processes to duplicate itself, or the brain use them to create consciousness? Or can life and mind, as a computable dynamics, albeit a complex one, in Turing's sense? What is the role of quantum theory in answering these questions? Ultimately, is the human self (i.e., minimally, the self-organizing, self-interpreting narrative field instantiated by our human consciousness) then explicable in physical terms, or will its mysteries continue to elude conventional scientific thinking?

- Is life all mechanism? Or is there something more complex still little studied by physical theory?

- What types of cosmologies, and systems of physical laws, support the origin of life, the evolution of consciousness, and the appearance of social history?

Unanswered questions indeed, but the time of physics and biology, the sciences of atom and gene, is clearly at hand. To take but one example, molecular dynamics simulations of entire living cells are expected soon to become not only feasible, but routine tools of biophysical engineering. The implications for medicine, and the marketplace, require no emphasis. Similarly, as life's physiochemical map is completed our species becomes the first on this planet, perhaps the first in the universe, to willfully shape its own evolution. Where will that post-Darwinian journey carry us?

Index

C

D

H

I

O

P

Q

R

S

T

U

V

W

Y

Z